AMERICAN MATHEMATICAL SOCIETY TRANSLATIONS

Series 2

Volume 92

SEVENTEEN PAPERS ON TOPOLOGY AND DIFFERENTIAL GEOMETRY

by

A. M. Ančikov
A. V. Arhangel′skiĭ
S. M. Bahrah
H. Cišang [Heiner Zieschang]
L. N. Krivonosov
Ju. G. Lumiste
P. I. Petrov
K. Š. Ramazanova
E. G. Skljarenko
Su Bu-chin
V. I. Vedernikov
N. Veličko
Wu Wen-tsün

Published by the
AMERICAN MATHEMATICAL SOCIETY
Providence, Rhode Island
1970

Library of Congress Catalog Number A51-5559
International Book Number 0-8218-1792-2

Printed in the United States of America

TABLE OF CONTENTS

Amer. Math. Soc. Transl.
(2) Vol. 92, 1970

ON A CLASS OF SPACES CONTAINING ALL METRIC SPACES AND ALL LOCALLY BICOMPACT SPACES *

A. V. ARHANGEL'SKIĬ

Metric spaces on the one hand and locally bicompact spaces on the other are two classes of spaces which have long occupied an important place both in set-theoretic topology and in its applications. At the same time it is clear that in the whole family of spaces studied in topology these classes take antipodal places; their qualitative difference hits the eye even when comparing definitions. Obviously this is explained by the fact that until recently only one natural class of spaces was known encompassing all metric spaces and all locally bicompact spaces, namely, the class of completely regular spaces, introduced into topology by A. N. Tihonov. It is quite evident that the latter class is exceedingly broad; in it those specific features of locally bicompact spaces and metric spaces which make them interesting in themselves are lost. Locally bicompact spaces and metric spaces, in spite of all their dissimilarity, have quite a number of common properties; for example, closedness relative to the taking of a finite product, of a perfect image, of a closed subspace; the addition theorem on weight is true both for these and for other spaces. This series of examples can be augmented, but already we can note that these common properties seemingly stem from different causes and, at any rate, require different explanations. Therefore it is very tempting to try and find a class of spaces which would contain all metric spaces and all locally bicompact spaces and in which the facts stated would acquire a common meaning.

Such a class is proposed in the present paper.

In the body of the paper we give the exact definitions and theorem statements, together with necessary historical commentaries; here, however, we try to outline in general terms the basic concepts and results of the paper.

I. The concept of the feathering of a space in its compactification is defined: this is the family of coverings of this space by sets open in the compactification, satisfying a specified condition (Definition 1.1).

II. Pinnate spaces: these are the spaces which have a feathering in

* Translation of Mat. Sb. 67 (109) (1965), 55–88.

their own Wallman compactifications (Definition 1.4).

III. First result: the property that a space has a feathering is invariant relative to its bicompact Hausdorff compactifications (Theorem 1.7).

IV. Locally bicompact spaces, metric spaces and Čech-complete spaces are all pinnate spaces (Theorem 2.3).

V. Pinnateness is preserved under finite and countable multiplication (Theorem 2.4).

VI. The behavior of pinnateness under mappings is studied. In particular, it is proved that the closed image of a metric space is a pinnate space if and only if it is metrizable (Theorem 3.5).

VII. An addition formula is true for pinnate spaces: if such a space is a sum of τ copies of spaces of weight less than or equal to τ, then the weight of the space itself does not exceed τ (Theorem 4.4). (This solves the Aleksandrov-Uryson addition problem which led up to their memoir.)

VIII. Perfect preimages of metric spaces are, precisely, paracompact pinnate spaces (Theorem 5.1). (This yields the solution to P. S. Aleksandrov's problem, see [2].)

IX. Whereas a product of paracompacta is not necessarily a paracompactum, the product of paracompact pinnate spaces (even in a countable set) is a paracompact pinnate space (Theorem 5.4).

X. A number of characteristics of paracompact pinnate spaces is given, including an intrinsic one (Theorem 5.3). The case of finally compact spaces is considered (§6).

We make several remarks on the terminology and notation adopted in the article.

A. All spaces for which we do not specify precisely which axiom of separability is assumed to be fulfilled in them must be taken as being completely regular.

B. The mappings are all assumed in advance to be continuous.

C. The terminology relating to multivalued mappings is taken from the paper by V. I. Ponomarev [16].

D. By γA, where A is a set and γ is a covering of some enveloping space, we mean the set-theoretic sum of all elements of γ intersecting with A. Often γA is called the star of A relative γ.

E. The notation $G < \gamma$, where G is a set and γ is a covering, signifies that G is contained in some element of γ.

F. We shall write $\gamma_1 < \gamma_2$, where γ_1 and γ_2 are coverings of a space X, if $\gamma_1 x < \gamma_2 x$ for any point $x \in X$.

§ 1. Definitions. Invariance theorem

1.1. **Definition.** Let $X \subseteq Y$, where X and Y are topological spaces. The system $\phi = \{\gamma_n\}$ of coverings of X by sets open in Y is called a *feathering* of X in Y if for any point $x \in X$ we have $\bigcap_{n=1}^{\infty} \gamma_n x \subseteq X$.

1.2. **Remark.** Obviously, if $X = \bigcap_{i=1}^{\infty} G_i$, where G_i are sets open in Y, $i = 1, 2, \cdots, \infty$, then $\phi = \{\gamma_n\}$, $\gamma_n = (G_n)$, is a feathering of X in Y.

1.3. **Proposition.** *If $X \subseteq Y \subseteq Z$, where X, Y, Z are topological spaces and X possesses a feathering in Y and Y in Z, then X has a feathering in Z.*

Proof. Let $\phi = \{\gamma_n;\ n = 1, 2, \cdots, \infty\}$, where $\gamma_n = \{U_\alpha^n,\ \alpha \in M_n\}$, be a feathering of X in Y and let $\psi = \{\lambda_n;\ n = 1, 2, \cdots, \infty\}$, where $\lambda_n = \{V_\alpha^n,\ \alpha \in L_n\}$, be a feathering of Y in Z.

By virtue of Definition 1.1 every U_α^n is open in Y; by $\widetilde{U}_\alpha^n$ we denote some set open in Z such that $\widetilde{U}_\alpha^n \cap Y = U_\alpha^n$. Such a $\widetilde{U}_\alpha^n$ can be found, since Y is a subspace of Z. We set $\widetilde{\gamma}_n = \{\widetilde{U}_\alpha^n,\ \alpha \in M_n\}$ and $\widetilde{\phi} = \{\widetilde{\gamma}_n;\ n = 1, 2, \cdots, \infty\}$. The system $\eta = \widetilde{\phi} \cup \psi = \{\widetilde{\gamma}_n, \lambda_n;\ n = 1, 2, \cdots\}$ consisting of systems of sets occurring both in $\widetilde{\phi}$ and in ψ forms the feathering of X in Z. Indeed, η, as a sum of two countable sets, is countable, and therefore, by introducing new notation, the elements of system η can be numbered: $\{\mu_n;\ n = 1, 2, \cdots, \infty\}$. All the elements of η, i.e. all of the $\widetilde{\gamma}_n$ and λ_n, are coverings of space X, open in Z. It remains to verify that $\bigcap_{n=1}^{\infty} \mu_n x \subseteq X$, $x \in X$. But

$$\bigcap_{n=1}^{\infty} \mu_n x = \bigcap_{n=1}^{\infty} \widetilde{\gamma}_n x \cap \bigcap_{n=1}^{\infty} \lambda_n x \subseteq Y.$$

Therefore

$$\Big(\bigcap_{n=1}^{\infty} \widetilde{\gamma}_n x\Big) \cap \Big(\bigcap_{n=1}^{\infty} \lambda_n x\Big) = \Big(\bigcap_{n=1}^{\infty} \widetilde{\gamma}_n x\Big) \cap \Big(\bigcap_{n=1}^{\infty} \lambda_n x\Big) \cap Y$$

$$\subseteq \bigcap_{n=1}^{\infty} (\widetilde{\gamma}_n x \cap Y) = \bigcap_{n=1}^{\infty} \gamma_n x \subseteq X.$$

The relation $\bigcap_{n=1}^{\infty} \mu_n x \subseteq X$, $x \in X$, and by the same token Proposition 1.3, is proved.

1.4. **Definition.** A topological T_1-space X is called *pinnate* if X

possesses a feathering in ωX, i.e. in its own Wallman compactification (see [9][1]).

Before we go on to discuss an important point, namely, the width of the class defined, we should prove two propositions on featherings and state the invariance theorem based on them.

1.5. **Proposition.** *If the space X possesses a feathering in ωX, then X also has a feathering in any regular*[2] *bicompact extension of itself.*

Proof. Let $\phi = \{\gamma_n\}$, $\gamma_n = \{U^n_\alpha, \alpha \in M_n\}$, be a feathering of X in ωX, let mX be any regular bicompact extension of X, and let $f\colon \omega X \to mX$ be the natural mapping of ωX onto mX, i.e. the continuous closed mapping for which $f^{-1}x = x$ if $x \in X$.

We set $V^n_\alpha = mX\setminus f(\omega X\setminus U^n_\alpha)$ for all n and $\alpha \in M_n$. Obviously, for any n, $\lambda_n = \{V^n_\alpha, \alpha \in M_n\}$ is a covering of X by sets open in mX: indeed, if $x \in X$, then for every n we can find $U^n_\alpha \in \gamma_n$, $U^n_\alpha \ni x$. Then $f(\omega X\setminus U^n_\alpha) \not\ni x$, since $f^{-1}x = x \subseteq U^n_\alpha$. Therefore $V^n_\alpha = mX\setminus f(\omega X\setminus U^n_\alpha) \ni x$, and by the same token we have proven that λ_n covers X. The openness of V^n_α follows from the fact that the mapping f is closed.

We verify that $\psi = \{\lambda_n\}$ is a feathering of X in mX. Let us assume the contrary, namely, that we can find a point $x_0 \in X$ for which $\bigcap_{n=1}^\infty \lambda_n x_0 \ni x_1$, where x_1 is some point of $mX\setminus X$. This is equivalent to the fact that $x_0 \in \bigcap_{n=1}^\infty V^n_{\alpha_n}$ and $x_1 \in \bigcap_{n=1}^\infty V^n_{\alpha_n}$ hold for some sequence $\{V^n_{\alpha_n}\}$. Hence $U^n_{\alpha_n} \ni x_0$ and $U^n_{\alpha_n} \supseteq f^{-1}x_1$, $n = 1, 2, \cdots, \infty$; the latter is proved in the following manner: since $mX\setminus f(\omega X\setminus U^n_{\alpha_n}) \ni x_1$, we have $f(\omega X\setminus U^n_{\alpha_n}) \not\ni x_1$, whence $(\omega X\setminus U^n_{\alpha_n}) \cap f^{-1}x_1 = \Lambda$, i.e. $U^n_{\alpha_n} \supseteq f^{-1}x_1$, which is what we had to prove.

From the relations obtained and from the fact that ϕ is a feathering of X in ωX, we obtain $x_0 \in \bigcap_{n=1}^\infty U^n_{\alpha_n} \subseteq \bigcap_{n=1}^\infty \gamma_n x_0 \subseteq X$, and consequently $f^{-1}x_1 \subseteq \bigcap_{n=1}^\infty U^n_{\alpha_n} \subseteq X$, whence $x_1 = ff^{-1}x_1 \subseteq X$; namely, we have arrived at a contradiction. Proposition 1.5 is proved.

1.6. **Proposition.** *Let X and Y be topological spaces, let $X_0 \subseteq X$ and*

---

1) *Translator's note.* Reference [9] nowhere contains the words "pinnate" and "Wallman" or, indeed, even "feathering." The author defines a "p-space" in a manner equivalent to his definition of a pinnate space here, but using the Čech compactification.

2) mX is a regular extension of the T_1-space X if there exists a closed mapping $f\colon \omega X \to mX$ (ωX is the Wallman extension of X) such that $f^{-1}x = x$ for all $x \in X$. In particular, the Wallman extension and all Hausdorff bicompact extensions are regular extensions (see [14]).

$Y_0 \subseteq Y$ be subspaces of them and let $f\colon X \to Y$ be a continuous mapping such that $f^{-1}Y_0 = X_0$.

Then if Y_0 has a feathering in Y, X_0 possesses a feathering in X.

Proof. We set $V_\alpha^n = f^{-1}U_\alpha^n$, $\lambda_n = \{V_\alpha^n,\ \alpha \in M_n\}$ and $\psi = \{\lambda_n;\ n = 1, 2, \cdots, \infty\}$, where $\phi = \{\gamma_n\}$, $\gamma_n = \{U_\alpha^n,\ \alpha \in M_n\}$, is some feathering of the space Y_0 in Y.

It turns out then that ψ is a feathering of X_0 in X. Indeed, for any positive integer n, for an arbitrary point $x_0 \in X_0$ we have $\lambda_n x_0 \subseteq f^{-1}\gamma_n f x_0$. Hence

$$\bigcap_{n=1}^{\infty} \lambda_n x_0 \subseteq \bigcap_{n=1}^{\infty} f^{-1}\gamma_n f x_0 = f^{-1}\bigcap_{n=1}^{\infty}\gamma_n f x_0 \subseteq f^{-1}Y_0 = X_0, \text{ i.e.}$$

$$\bigcap_{n=1}^{\infty} \lambda_n x_0 \subseteq X_0.$$

Since by virtue of the continuity of f and the relation $f^{-1}Y_0 = X_0$, λ_n is a covering of X_0 by sets open in X, we have thus proven Proposition 1.6: ψ is the desired feathering of X_0 in X.

1.7. **Theorem** (on invariance). *In order for a topological space to possess a feathering in any regular bicompact extension of it, it is sufficient that it have a feathering in some such extension.*

Proof. Let b_1X and b_2X be some bicompact extensions of space X, and, moreover, suppose that we know that X has a feathering in b_1X. By f_1 and f_2 we denote the natural mappings of the Wallman compactification ωX of the space X onto b_1X and b_2X: $f_1\colon \omega X \to b_1X$ and $f_2\colon \omega X \to b_2X$. Since $f_1^{-1}X = X$, X possesses a feathering in ωX by virtue of Proposition 1.6. But then it follows from Proposition 1.5 that X has a feathering also in b_2X. Theorem 1.7 is proved.

1.8. **Proposition.** *If X possesses a feathering in Y, where $X \subseteq Y$ and Y is a completely regular pinnate space, then X is a pinnate space.*

Proof. By bY we denote some Hausdorff bicompactification of Y. From Theorem 1.7 it follows that Y has a feathering in bY. Then, by virtue of Proposition 1.3, X possesses a feathering in bY. Obviously this feathering cuts out on $bX = [X]_{bY}$ a feathering of X in bX. But bX is a Hausdorff (consequently, regular) bicompactification of space X. By the same token, in view of Theorem 1.7, the proposition is proved.

§ 2. Width of a class

Which of the known forms of spaces relate to pinnate spaces? Two classes occupy a predominant place in topology: the class of bicompacta and the class of metric spaces. We establish first of all what relation they bear to pinnate spaces.

a) *Bicompacta are pinnate spaces.*

This follows immediately from Definition 1.4.

2.1. **Proposition.** *Among completely regular spaces the class of pinnate spaces is hereditrary relative to subsets of type* G_δ.

Proposition 2.1 follows from Remark 1.2 and Proposition 1.8.

b) *Čech-complete spaces are pinnate spaces.*

This follows from Proposition 2.1 and assertion a).

2.2. **Proposition.** *A closed subspace of an arbitrary pinnate space is pinnate.*

Proof. If P is a closed set in X and $\phi = \{\gamma_n\}$, $\gamma_n = \{U^n_\alpha,\ \alpha \in M_n\}$, is a feathering of X in ωX (see Definition 1.4), then, obviously, $\psi = \{\lambda_n\}$, $\lambda_n = \{V^n_\alpha,\ \alpha \in M_n\}$, where $V^n_\alpha = U^n_\alpha \cap [P]_{\omega X}$, is a feathering of the space P in $[P]_{\omega X}$. But, as is well known, $[P]_{\omega X} = \omega P$. By the same token, in view of Definition 1.4, Proposition 2.2 is proved.

Proposition 2.2 furnishes us with the knowledge of one more interesting property of pinnate spaces, but, under application only to a) and b), it does not yield new information on the composition of the class in which we are interested. Proposition 2.2 will be made use of later.

c) *All metric spaces are pinnate.*

Indeed, let X be a metric space and let bX be some Hausdorff bicompactification of it. By μ_n $(n = 1, 2, \cdots, \infty)$ we denote the collection of all sets open in bX, the intersection of each of which with X has a diameter not greater than $1/n$. Then $\psi = \{\mu_n\}$ also is a feathering of X in bX.

Let us prove this. Suppose that the point $x \in X$ is arbitrary and that Ox is any neighborhood of it in bX. We see at once that $\mu_{n'} x \subseteq Ox$ for some number n'; obviously, this will signify that $\bigcap_{n=1}^{\infty} \mu_n x = x \subseteq X$, which for us is sufficient.

We select a neighborhood $O'x$ of the point x, for which $[O'x]_{bX} \subseteq Ox$. The set $P = X \setminus O'x$ is closed in X, $P \not\ni x$, and therefore $\rho(x, P) > 0$. We take the number n' such that $1/n' < \rho(x, P)$; then, if $V \in \mu_{n'}$, $X \ni x$, we have $(V \cap X) \cap P = \Lambda$, or $V \cap X \subseteq O'x$. Since bX is an extension of X, X

is everywhere dense in bX, and therefore $V \subseteq [V \cap X]_{bX}$. All the more, $V \subseteq [O'x] \subseteq Ox$. In view of the arbitrariness in the choice of $V \in \mu_{n'}$, what we have proved signifies that $\mu_{n'} x \subseteq Ox$. Hence it follows, as we have seen, that ψ is a feathering of X in bX. In particular, by assuming $bX = \beta X = \omega X$ (for metric spaces, as also for any normal spaces, $\omega X = \beta X$; see [11]) we arrive, according to Proposition 1.4, at the conclusion that X is a pinnate space.

We collect together all that we have proven above.

2.3. **Theorem.** *The class of pinnate spaces includes all*: a) *bicompacta*, b) *locally bicompact spaces*, c) *Čech-complete spaces*, d) *metric spaces*.

Obviously, certain items of the theorem reinforce the others. We note that although the completeness of a space is a test of its pinnateness, the presence of completeness is not at all necessary. We see this clearly from the example that not only complete metric spaces but also arbitrary metric spaces are pinnate.

The information on the composition of the class of pinnate spaces which we gave in Theorem 2.3 is not exhaustive. In this respect the following theorem permits us to extend our horizon significantly.

2.4. **Theorem.** *The Tihonov product of a countable set of pinnate spaces is a pinnate space.*

We limit our proof to the case of completely regular factors.

Proof. We assume that X_i, $i = 1, 2, \cdots, \infty$, are pinnate spaces, βX_i are their Čech compactifications, and $\phi_i = \{\gamma_i^j,\ j = 1, 2, \cdots, \infty\}$, $\gamma_i^j = \{U_{i,\alpha}^j,\ \alpha \in M_{i,j}\}$, are their featherings in βX_i, $i = 1, 2, \cdots, \infty$. Let us show that then the space $X = \Pi_{i=1}^{\infty} X_i$ possesses a feathering in $bX = \Pi_{i=1}^{\infty} \beta X_i$, where, as follows from Tihonov's theorem, bX is a bicompact extension of X.

Indeed, for arbitrarily fixed j and k we denote by $\lambda^{j,k}$ the collection of all sets represented as the product of some elements taken one from each γ_i^j for which $i \leq k$, by all the βX_i satisfying the condition $i > k$.

Obviously, each $\lambda^{j,k}$ is a covering, open in bX, of the set X. The collection ψ of these coverings corresponding to all possible different pairs of integers j and k is countable. Let us verify that ψ is also a feathering of X in bX.

Let the points $x \in X$ and $x' \in bX \backslash X$ be chosen arbitrarily. We need to select j and k so as to fulfill the relation $\lambda^{j,k} x \not\ni x'$.

But since $x' = \{x_i'\} \in \Pi_{i=1}^{\infty} \beta X_i \backslash \Pi_{i=1}^{\infty} X_i$, we shall have $x_{i'}' \in \beta X_{i'} \backslash X_{i'}$

for some value $i = i'$. Since $\phi_{i'} = \{\gamma^j_{i'},\ j = 1, 2, \cdots, \infty\}$ is a feathering of $X_{i'}$ in $\beta X_{i'}$, the relation $\gamma^{j'}_{i'} x_{i'} \not\ni x'_{i'}$, where $\{x_i\} = x \in X$, will be fulfilled for some $j = j'$.

But then, as follows from the definition of the system $\lambda^{j,k}$,

$$\lambda^{j,i'} x \subseteq \prod_{i \leq i'} \gamma^{j'}_i x \times \prod_{i > i'} \beta X_i \subseteq bX \backslash x',$$

which is what we had to show. By this we have proven that ψ is a feathering of X in bX. But then, according to Theorem 1.7, X is a pinnate space. Theorem 2.4 is proved.

Theorems 2.3 and 2.4 have the following corollary.

2.5. Corollary. *The product of a countable set of spaces each of which is Čech-complete or measurable is a pinnate space.*

In [10] we introduced the concept of a space of pointwise-countable type, namely, a space which can be covered by bicompacta possessing a countable basis in it. Therein it was proven that if a set lying in a bicompactum is a sum of sets of type G_δ in this bicompactum, then this set, in the topology induced by the bicompactum, is a space of pointwise-countable type (see [10]).

Now let X be a completely regular pinnate space, bX its bicompact extension, and $\phi = \{\gamma_i\}$, $\gamma_i = \{U^i_\alpha,\ \alpha \in M_i\}$, a feathering of X in bX. Then, obviously, $X = \bigcup_{x \in X} (\bigcap_{i=1}^{\infty} \gamma_i x)$, i.e. X is a sum of sets of type G_δ in bX. Thus the following theorem is true, by what we have said above.

2.6. Theorem. *Every completely regular pinnate space is a space of pointwise-countable type.*

This theorem, together with one of the theorems from [10], yields an important corollary.

2.7. Corollary. *Every completely regular pinnate space is a k_2-space.*

§ 3. Pinnate spaces and mappings

Assertion 2.7, together with one of the theorems from [10], leads us to the following conclusion.

3.1. Theorem. *Every completely regular pinnate space is a factor space of some locally bicompact space.*

The next theorem establishes a curious connection between the classes of metric and pinnate spaces.

3.2. Theorem. *Every completely regular pinnate space is an open Y-bicompact image of some metric space.*

This theorem follows from Theorem 2.6 and one of the theorems in [10].

The following theorem is based on Propositions 1.5 and 1.6.

3.3. **Theorem.** *The preimage of a pinnate space under a perfect mapping is a pinnate space.*

Proof. Let $f\colon X \to Y$ be a perfect mapping and let Y be a pinnate space. As is well known, the mapping f can be continued onto the Wallman compactification, i.e. we can select a mapping $\tilde{f}: \omega X \to \omega Y$ such that $fx = \tilde{f}x$ for all $x \in X$. It turns out here (see [26,16]) that $\tilde{f}^{-1}Y = X$, i.e. the hypotheses of Proposition 1.6 are fulfilled for the mapping $\tilde{f}$, whence it follows that X is a pinnate space. The theorem is proved.

In connection with Theorem 3.3 we remark that we are not interested in whether the perfect image of a completely regular pinnate space is a pinnate space.

Theorem 3.3 can be given the form

3.3′. **Theorem.** *The preimage of a completely regular pinnate space under an arbitrary k-mapping*[1] *is a pinnate space.*

This fact is based on the following proposition.

3.4. **Proposition.** *An arbitrary k-mapping onto a completely regular pinnate space is closed.*

The latter follows from Corollary 2.7 and Theorem 1.3.

The results presented in this section, together with certain theorems stated at the end of the paper, exhaust all that we know on the preservation of pinnateness under mappings.

Several simple examples and one theorem, of which we shall speak below, reveal that the property of a space to be pinnate, in spite of its wide prevalence, is very tenuous and fragile and is easily destroyed even under comparatively good mappings.

In the first place, pinnateness is not preserved when passing to the preimage even under open-closed mappings. This follows from the fact that the mapping of any space, including a nonpinnate one, onto a one-point space is open-closed, while the latter, of course, is pinnate. Nor is pinnateness necessarily preserved when passing to the image under open or closed mappings.

In the case of open mappings we have V. I. Ponomarev's theorem [18]: any T_1-space satisfying the first axiom of countability is the open image

1) See [10].

of a metric (i.e. pinnate) space. Obviously, among T_1-spaces with the first axiom of countability there are as many nonpinnate ones as desired; for example, such is the nonnegative reals with the topology generated by the left-closed right-open intervals (see [12], p. 631). We establish the latter later.

In the case of closed mappings we have the following theorem.

3.5. Theorem. *A completely regular space Y which is the closed image of some metric space X is metrizable if and only if it is a pinnate space.*

Theorem 3.5 is based on assertions 2.3 and 2.6 and on a theorem in [10].

By the example of Theorem 3.5 we see that in spite of the extensiveness (see Theorem 2.3, Corollary 2.5) of the class of pinnate spaces, the requirement that a space belong to this class can prove to be a decisive factor for obtaining significantly deeper information on its structure, no matter what information we gain in our case by knowing that Y is metrizable.

Theorem 3.5 allows us to point out a simple method for constructing nonpinnate spaces, the essence of which is apparent from the following illustration of it: we take the partition space Z of the ordinary plane E^2, a unique non-one-point element of which is some straight line l. The topology is the quotient topology. The corresponding factor mapping f is closed. Here, as is easy to see, the first axiom of countability is not fulfilled in the partition space Z at the point (l), i.e. Z, all the more, is not metrizable. Consequently, in view of the closedness of f, Z is also not a pinnate space by Theorem 3.5. Thus we see that the closed image of a pinnate space is not necessarily pinnate.

§ 4. Addition formula for the weight of pinnate spaces

The following problem goes bact to the memoir by P. S. Aleksandrov and P. S. Uryson on compact topological spaces. It is known that $X = \bigcup_{\alpha \in M} X_\alpha$, where the cardinality of M and the weight of each of the X_α do not exceed a certain cardinal number τ. Under what additional constraints on X will it follow from this that the weight of the whole space X does not exceed τ? At that time the answer to this question was as yet unknown even for the case of only two terms in all, each of which possesses a countable basis, under the assumption that X is a bicompactum, which was in fact the situation for which the general problem presented in the memoir was formulated. The latter problem was solved only after 30 years, in 1956, by Ju. M. Smirnov, who proved that a bicompactum which is a sum of a countable set of spaces with a countable basis itself possesses a countable basis [21]. Three years later the author succeeded in obtaining the following theorem by a completely different method: if a Čech-complete space is the sum of not more than τ copies of spaces of weight not exceeding τ, then the weight of the whole space also does not exceed τ.

In the achievement of this result a very helpful role way played by the concept of a net[1)] in a space, introduced in [5]. Let us recall it.

4.1. **Definition.** A system s of subsets of a space X is called a *net* in X if for any x and Ox, i.e. a point and its neighborhood in X, we can find $P \in s$ such that $x \in P \subseteq Ox$.

4.2. **Fundamental Theorem.** *If a completely regular pinnate space X possesses a net s of cardinality less than or equal to τ, then there is an outer basis of X in bX of cardinality less than or equal to τ, where bX is any Hausdorff bicompactification of X.*

Proof. Let $\phi = \{\gamma_n\}$, $\gamma_n = \{U^n_\alpha, \alpha \in M_n\}$, $n = 1, 2, \cdots, \infty$, be a feathering of X in bX (it exists by virtue of Theorem 1.7). Let us show that from each γ_n we can select a covering γ'_n of X whose cardinality does not exceed τ, $\gamma'_n = \{U^n_{\alpha'}, \alpha' \in M'_n\}$. Indeed, with every element of s for which this is possible we can associate some completely determined element of γ_n containing it. From the fact that s is a net in X it follows easily that the collection γ'_n of the elements of γ_n thus selected forms a covering of X. Obviously the cardinality of γ'_n does not exceed τ.

We set $\lambda_n = \{bX \setminus U^n_\alpha, U^n_\alpha \in \gamma'_n\}$ and $\lambda = \bigcup_{n=1}^\infty \lambda_n$. Obviously λ is a system of bicompact sets of cardinality less than or equal to τ. Let $\bar{s} = \{[P]_{bX}, P \in s\}$ and $\mu = \bar{s} \cup \lambda$. Finally, for each pair of nonintersecting elements of μ we fix some completely determined pair of their nonintersecting neighborhoods in bX. This is possible since all the sets occurring in μ are bicompacta and bX is a Hausdorff space. We denote the collection of the selected open sets of the space bX by g'. The cardinality of g' does not exceed τ, as is easy to see. Therefore the system g, consisting of the elements of g', of their complements with respect to their closures in bX and all possible finite intersections of the sets indicated, has a cardinality less than or equal to τ.

Let us show that g forms a point basis of X in bX. Let $x_0 \in X$ and let its neighborhood Ox_0 be chosen arbitrarily. We set $F = bX \setminus Ox_0$. For any point $x \in F$ we can find an element $V_x \in g$ such that $x \in V_x$ and

1) *Editor's note.* This usage should be distinguished from that in the theory of convergence ([13], Chapter 2).

$[V_x]_{bX} \not\ni x_0$. Indeed, if $x \in X$, then we can find neighborhoods $Ux \ni x$ and $Ux_0 \ni x_0$ such that $[Ux]_{bX} \cap [Ux_0]_{bX} = \Lambda$. In s we can choose elements s_x and s_{x_0} for which $x \in s_x \subseteq Ux$ and $x_0 \in s_{x_0} \subseteq Ux_0$. Then $\Phi_{x_0} = [s_{x_0}]_{bX}$ and $\Phi_x = [s_x]_{bX}$ are nonintersecting bicompacta occurring in $\bar{s}$, and hence also in μ.

But then by definition there exist in g' sets V_x and V_{x_0} for which $\Phi_x \subseteq V_x$ and $\Phi_{x_0} \subseteq V_{x_0}$. Then, obviously, $x \in V_x$, $[V_x]_{bX} \not\ni x_0$ and $V_x \in g$, i.e. V_x is the desired set. If $x \in bX\backslash X$, then since $\phi' = \{\gamma_n'\}$ is a feathering of X in bX, we shall have $\gamma'_{n_x} x_0 \not\ni x$ for some number n_x. Consequently in λ_{n_x} we can find a bicompactum F_x for which $x \in F_x$ but $x_0 \notin F_x$. Making use of the regularity of bX, we select a neighborhood $O'x_0$ such that $x_0 \in O'x_0 \subseteq [O'x_0]_{bX} \subseteq bX\backslash F_x$, and in s we find an element $s'_{x_0} \in s$ for which $x_0 \in s'_{x_0} \subseteq O'x_0$. We set $F_{x_0} = [s'_{x_0}]_{bX}$. Then, obviously, $x \in F_x$, $x_0 \in F_{x_0}$, $F_x \in \mu$, $F_{x_0} \in \mu$ and $F_x \cap F_{x_0} = \Lambda$. Consequently, in g' we can find sets $V_{x_0} \supseteq F_{x_0}$ and $V_x \supseteq F_x$, $V_{x_0} \cap V_x = \Lambda$. Then, obviously, $x \in V_x \subseteq [V_x]_{bX} \subseteq bX\backslash x_0$, $V_x \in g$, i.e. the set V_x is the desired one.

From the covering of F by the sets V_x, $x \in F$, in view of the bicompactness of F and the openness of V_x, we can select a finite covering $V_{x_1}, \cdots, V_{x_k}$, $\bigcup_{i=1}^k V_{x_i} \supseteq F$.

Here from the definition of the sets V_x it follows that $[V_{x_i}]_{bX} \not\ni x_0$, $i = 1, 2, \cdots, k$.

Consequently, since $[\bigcup_{i=1}^k V_{x_i}]_{bX} = \bigcup_{i=1}^k [V_{x_i}]_{bX}$, the set $V^0_{x_0} = bX\backslash[\bigcup_{i=1}^k V_{x_i}]_{bX}$ is a neighborhood of the point x_0. Here obviously the condition $x_0 \in V^0_{x_0} \subseteq Ox_0$ is fulfilled. But $(bX\backslash[V_{x_i}]_{bX}) \in g$, $i = 1, 2, \cdots, k$, by the definition of g and since all $V_{x_i} \in g'$. Since $V^0_{x_0} = bX\backslash[\bigcup_{i=1}^k V_{x_i}]_{bX} = \bigcap_{i=1}^k (bX\backslash[V_{x_i}]_{bX})$, also $V^0_{x_0} \in g$. Consequently $V^0_{x_0}$ also is the desired neighborhood of the point x_0. In view of the arbitrariness in the choice of the point $x_0 \in X$ and of its neighborhood Ox_0, we have proven that the system g forms an outer basis of the space X in bX of cardinality less than or equal to τ. Theorem 4.2 is proved.

4.3. Corollary. *If a completely regular pinnate space possesses a net of cardinality less than or equal to τ, then its weight does not exceed τ.*

This follows at once from Theorem 4.2, since the outer basis of X in bX (some bicompact extension of X, which exists by Tihonov's theorem) cuts out of X a basis of the space X in the usual sense, obviously of the required cardinality.

4.4. Addition theorem for pinnate spaces. *If $X = \bigcup_{\alpha \in M} X_\alpha$, where X is* *a completely regular pinnate space, the weights of the* X_α *are less than or equal to* τ, $\alpha \in M$, *and the cardinality of* M *is less than or equal to* τ, *then the weight of* X *also does not exceed* τ.

This theorem follows immediately from Corollary 4.3 and from the next trivial lemma (see [5]).

4.5. Lemma. *If in the assumption of Theorem 4.4, for each* $\alpha \in M$, B_α *is some basis of the space* X_α, *of cardinality not greater that* τ, *then the system* $B = \bigcup_{\alpha \in M} B_\alpha$ *forms a net in the space* X *of cardinality not greater than* τ.

From Theorem 4.2 together with Theorem 4.4, there results the following corollary.

4.6. Corollary. *The weight of a completely regular pinnate space does not exceed its cardinality.*

Indeed, the collection of all one-point sets of the space forms a net in it.

In particular, we obtain the next corollary.

4.7. Corollary. *A completely regular countable pinnate space possesses a net basis.*

As Uryson's well-known example [23] shows (also see [10]), a countable space, in general, cannot even be a Fréchet-Uryson space (see [10]).

Finally, the Fundamental Theorem 4.2 leads to the following important result.

4.8. Theorem on the behavior of the weight under mappings. *Under a continuous mapping* $f\colon X \to Y$ *of an arbitrary topological space* X *onto a completely regular pinnate space* Y *the weight of* Y *always does not exceed the weight of* X.

Indeed, the following proposition holds.

4.9. Proposition. *If a topological space* Y *is a continuous image of a space* X *of weight* τ, *then it possesses a net of cardinality not greater than* τ. *Conversely, if in* Y *there is a net of cardinality* τ, *then* Y *is a condensation (a one-to-one continuous image) of some space of weight not greater than* τ.

Indeed, if s is a net in Y of cardinality not greater than τ, then by taking the collection of all possible finite intersections of its elements as the basis in a new topology, we also obtain the desired space.

Conversely, from the continuity of f and from the definitions of basis and net it follows immediately that the image of the basis in X is a net in Y. The rest is obvious.

Theorem 4.8 obviously follows from Theorem 4.2 and Proposition 4.9.

4.10. Corollary. *A completely regular pinnate space which is a continuous image of a space with countable basis, itself possesses a countable basis.*

In conclusion we note that Theorems 4.2, and 4.4 and 4.9 cover, by virtue of Theorem 2.3, the results previously published by the author [5,7].

§ 5. Paracompact and pinnate spaces

1°. Formulation of the fundamental results. P. S. Aleksandrov posed the problem: what spaces may be perfectly mapped onto metric spaces? Obviously, paracompactness is a necessary property of such spaces. Aleksandrov's problem acquired a particular interest after it became known, as Frolík [24] proved, that a necessary and sufficient condition for a Čech-complete space to be perfectly mapped onto a metric space is its paracompactness. Here the image admits of the introduction of a complete metric.

On the other hand, simple examples (see [12]) show that paracompactness alone of a space is still too little for us to be able to map it perfectly onto a metric space.

The general solution of the problem, including, in particular, Frolík's theorem, allows us to make use of the concept of a pinnate space.

5.1. Theorem. *In order that a topological space can be perfectly mapped onto a metric space it is necessary and sufficient that it be paracompact and pinnate simultaneously.*

It is evident that this theorem should be considered as a fundamental result on paracompact pinnate spaces.

Together with the extrinsic characterization of the class of paracompact pinnate spaces which is given by Theorem 5.1, we have succeeded in giving an intrinsic characterization of the spaces of this class, which is considerably more succinct than afforded by the definition. We need the following definition.

5.2. Definition. A system $\mathfrak{D}$ of open sets of a topological space X is said to be *k-fine* if for any point $x \in X$ we can find a bicompactum F such that whatever be the neighborhood $U \supseteq F$ there are at most finitely many $V \in \mathfrak{D}$ such that $V \cap (X \setminus U) \neq \Lambda$ and $V \ni x$ simultaneously, and there is at least one $V_1 \in \mathfrak{D}$ for which $x \in V_1 \subseteq U$.

In this case we shall sometimes say that system $\mathfrak{D}$ is *regular on the pair* (x, F).

Definition 5.2 is a generalization of the notion of a uniform basis, introduced by P. S. Aleksandrov [3], which we obtain from Definition 5.2 by requiring that as the bicompactum F indicated therein, corresponding to the point $x \in X$, we can take this point itself.

5.3. Theorem. *In order that a paracompactum can be perfectly mapped onto a metric space it is necessary and sufficient that a k-fine system of sets exist in it.*

The following important result is based on Theorem 5.1.

5.4. Theorem. *The product of a countable set of paracompact pinnate spaces is a paracompact pinnate space.*

In connection with Theorem 5.4 we recall that even the product of only two paracompacta is not necessarily paracompact (see [12,24]).

Theorem 5.1 has the following consequence.

5.5. Theorem. *The preimage of a paracompact pinnate space under a perfect mapping is a paracompact pinnate space.*

Another group of results of this article is connected with the representation of paracompact pinnate spaces as good, although multivalued, images of metric spaces (see [10,15,16]). The most interesting of these are the following two results.

5.6. Theorem. *An open, continuous, two-sided bicompact*[1] *image, of a metric space is a pinnate space.*

5.7. Theorem. *The class of paracompact pinnate spaces coincides with the class of continuous, open, closed, two-sided bicompact images of metric spaces.*

§ 6 of the paper is devoted to the countable case, while in §7 we discuss in detail the behavior of pinnateness under mappings, give examples, and derive a number of positive results.

2°. Proofs. The proofs of the theorems presented in subsection 1° are put together from a number of propositions, which we shall label simply by the letters A, B, C, D, etc.

A. *The preimage of a metric space under a perfect (single-valued) mapping can be represented as an open, closed, two-sided bicompact (continuous and multivalued) image of a metric space.*

---

1) *Editor's note.* This means that the mapping in question preserves bicompactness in both directions.

This follows immediately from the definitions of openness, closedness and continuity of multivalued mappings (see [16]). In order to obtain the desired mapping it is sufficient to take the mapping inverse to the original one.

B. *A continuous, open, closed, two-sided bicompact (multivalued) image of a metric space is a paracompactum.*

This follows from the paracompactness of metric spaces and from the result of V. I. Ponomarev on the preservation of paracompactness under mappings of the type mentioned above (see [16]).

C. *A continuous, open, two-sided bicompact (multivalued) image of a metric space forms a k-fine system.*

(Note that the closedness of the mapping is not assumed.)

Proof of Proposition C. Let $f\colon X \to Y$ be a mapping such as in the statement of Proposition C. By B we denote any regular basis in X (recall that a basis B in X is said to be regular if for any point $x \in X$ and neighborhood Ox of it we can find a neighborhood $O_1x \ni x$ such that in B there is at most a finite set of elements intersecting O_1x and the complement with respect to Ox simultaneously; see [8]).

We set $\gamma = \{\Phi \subseteq Y \mid \Phi = fx,\ x \in X\}$.

In view of the assumptions γ is a covering of space Y by bicompacta. The collection of the images of the elements of B under the mapping f is denoted by $\mathfrak{D}$. We now prove that $\mathfrak{D}$ is regular on every pair (y, Φ), where $\Phi \in \gamma$ and $y \in \Phi$ (obviously, this is an even stronger property than k-fineness).

Indeed, let the neighborhood $V \supseteq \Phi$ be selected arbitrarily. The set $F = f^{-1}y$ is a bicompactum, and, by virtue of the continuity of f, for every point $x \in F$ we can find a neighborhood $Ox \ni x$ such that $fOx \subseteq V$. From the covering of F by such neighborhoods we choose the finite covering $Ox_1, \dots, Ox_k$ and we set $G = \bigcup_{i=1}^{k} O_{x_i}$. Then $fG \subseteq V$.

From the regularity of the basis B it follows that for every point $x \in F$ we can find a neighborhood $O'x$ for which the set of elements of B intersecting simultaneously with $O'x$ and $X \backslash G$ is empty or finite. By G' we denote the sum of some finite set of such neighborhoods, covering F. Then it is obvious that the set of elements of B intersecting G' and $X \backslash G$ simultaneously also is empty or finite. We denote the collection of them by B_y.

Let $U \in B \backslash B_y$ and assume that $fU \ni y$. Then $U \cap F \neq \Lambda$ and all the more $U \cap G' \neq \Lambda$. Hence from the definition of B_y it follows that $U \subseteq G$. Therefore $fU \subseteq fG \subseteq V$, i.e. $fU \cap (Y \backslash V) = \Lambda$. Thus we have proven that among the elements of $\mathfrak{D}$ only the images of the elements of B_y may (although not necessarily) intersect $Y \backslash V$ and simultaneously contain the point y. From the finiteness of B_y it follows that such elements in $\mathfrak{D}$ form a finite (or empty) set, which proves the regularity of system $\mathfrak{D}$ on the pair (y, Φ) and, together with this, Proposition C.

D. *A space possessing a k-fine system is pinnate.*

Lemma. *If $\mathfrak{D}$ is a k-fine system, then the collection $\widetilde{\mathfrak{D}}$ of all possible finite intersections of the elements of $\mathfrak{D}$ also forms a k-fine system.*

Proof of the lemma. We consider some pair (x, F) on which the system $\mathfrak{D}$ is regular and we show that $\widetilde{\mathfrak{D}}$ too is regular on (x, F). We take any open $U \supseteq F$. The regularity of $\mathfrak{D}$ on (x, F) signifies that the collection $\mathfrak{D}_{(U,x)}$ of elements of $\mathfrak{D}$ simultaneously intersecting $X \backslash U$ and x is not more than finite. But any element of $\widetilde{\mathfrak{D}}$ intersecting $X \backslash U$ and simultaneously containing x can be represented, by virtue of the definition of $\widetilde{\mathfrak{D}}$, as a finite intersection of elements of $\mathfrak{D}$, each of which, as is easy to see, necessarily belongs to $\mathfrak{D}_{(U,x)}$. In view of the finiteness of $\mathfrak{D}_{(U,x)}$ the collection of all possible finite intersections of the elements of $\mathfrak{D}_{(U,x)}$ also is finite. This proves the regularity of $\widetilde{\mathfrak{D}}$ on (x, F) and consequently also the lemma.

Proof of Proposition D. By virtue of the lemma we can assume that in the space X being considered we are given a k-fine system $\mathfrak{D}$ containing all possible finite intersections of its own elements.

By B we denote the collection of all locally bicompact elements of $\mathfrak{D}$.

We now determine by induction the sequences of set systems $\{C_k\}$ and $\{\mathfrak{D}_k\}$:

$$C_1 = \mathfrak{D},\ \mathfrak{D}_k = \phi C_k,\ C_{k+1} = \mathfrak{D} \backslash \bigcup_{i=1}^{k} \mathfrak{D}_i,\ k = 1, 2, \cdots, \infty.$$

(By $\phi\xi$, where ξ is a system of sets, we denote the collection of maximal elements of system ξ, i.e. those not contained in any other element of ξ.)

Finally, we set $\Pi_k = \mathfrak{D}_k \cup B$, $k = 1, 2, \cdots, \infty$, and we prove that Π_k is a covering of X for any k.

Let the point $x \in X$ be arbitrary and let F be some bicompactum such that $\mathfrak{D}$ is regular on (x, F). We shall construct a set sequence $\{V_n\}$ from $\mathfrak{D}$ in the following manner: as V_1 we fix an element of $\mathfrak{D}$ containing x and, assuming that the sets V_n have already been determined for $n \leq k$, $V_n \ni x$, we consider $U = \bigcap_{n=1}^{k} V_n$, $U \ni x$. By assumption, $U \in \mathfrak{D}$.

Two cases are possible here:

1) U is not locally bicompact; then $U \cap (X \setminus F) \neq \Lambda$. In view of the regularity of $\mathfrak{D}$ on (x, F) we can find $V \in \mathfrak{D}$, $V \ni x$, for which $U \cap (X \setminus V) \neq \Lambda$. We then set $V_{k+1} = V$.

2) U is locally bicompact; then $x \in U \in B \subseteq \Pi_i$ and consequently the point x is covered by the system Π_i for any i.

In summary we see that either I) the process described for the construction of the sequence $\{V_n\}$ can be prolonged to infinity, and, moreover, in such a way that not one of its elements will be locally bicompact, or II) it breaks off, but then the point x will be covered by some element of Π_i for each i, which is what we required.

Let us now consider case 1).

We set $V_n = \bigcap_{i=1}^n V_i$. From the definition of the elements of sequence $\{V_n\}$ it follows that the sequence $\{U_n\}$ is a strictly decreasing one; here all the U_n belong to $\mathfrak{D}$. Hence it obviously follows that $U_k \in C_k$ for any k. But $x \in U_k$ and $U_k \cap (X \setminus F) \neq \Lambda$. Therefore every U_k can be contained in a not more than finite set of elements of $\mathfrak{D}$ and, all the more, of C_k, since otherwise, by taking any neighborhood $G \supseteq F$ such that $U_k \cap (X \setminus G) \neq \Lambda$, we could obtain that an infinite set of elements of $\mathfrak{D}$ intersects with $X \setminus G$ and simultaneously contains the point x, in contradiction to the regularity of the system $\mathfrak{D}$ on the pair (x, F). Consequently for every U_k we can find, in correspondence with the system C_k, the maximal element containing it: $U_k^0 \supseteq U_k$. Then $x \in U_0^k \in \phi C_k = \mathfrak{D}_k \subseteq \Pi_k$. By the same token we have proven that Π_k is a covering of X for any k.

The covering Π_k possesses the following obvious, but important, property: $\Pi_i \cap \Pi_j \subseteq B$ when $i \neq j$. In other words, if a set from X belongs to two different coverings from the sequence $\{\Pi_k\}$, then it is locally bicompact without fail.

Let us turn to βX, the Čech compactification of space X. For every element $\Gamma_\alpha^k \in \Pi_k$, by $\widetilde{\Gamma}_\alpha^k$ we denote some set, open in βX, such that $\widetilde{\Gamma}_\alpha^k \cap X = \Gamma_\alpha^k$. Here we set $\widetilde{\Gamma}_\alpha^k = \Gamma_\alpha^k$ if $\Gamma_\alpha^k \in B$ (since the set Γ_α^k is locally bicompact in this case, it itself will be open in βX).

Let us show that $\bigcap_{k=1}^\infty \widetilde{\Pi}_k x \subseteq X$ $(\widetilde{\Pi}_k = \{\widetilde{\Gamma}_\alpha^k \mid \Gamma_\alpha^k \in \Pi_k\})$ for any point $x \in X$. This will also signify that X is a pinnate space.

We assume the contrary; suppose that the stated relation is not fulfilled for some point $x \in X$, and consider a bicompactum $F \ni x$ such that the system $\mathfrak{D}$ is regular on (x, F). We have $\bigcap_{k=1}^\infty \widetilde{\Pi}_k x \cap (\beta X \setminus X) \neq \Lambda$. This is possible only if for some sequence $\psi = \{\widetilde{\Gamma}^k_{\alpha_k}\}$, where $\widetilde{\Gamma}^k_{\alpha_k} \in \widetilde{\Pi}_k$, $\widetilde{\Gamma}^k_{\alpha_k} \ni x$, $k = 1, 2, \cdots, \infty$, we shall have $(\bigcap_{k=1}^{\infty} \widetilde{\Gamma}^k_{\alpha_k}) \cap (\beta X \setminus X) \neq \Lambda$. We consider the sets $\Gamma^k_{\alpha_k} = \widetilde{\Gamma}^k_{\alpha_k} \cap X$, $\Gamma^k_{\alpha_k} \in \Gamma_k$ corresponding to the elements of this sequence. For any k we have $\widetilde{\Gamma}^k_{\alpha_k} \neq \Gamma^k_{\alpha_k}$, since $\widetilde{\Gamma}^k_{\alpha_k} \cap (\beta X \setminus X) \neq \Lambda$ while $\Gamma^k_{\alpha_k} \subseteq X$. Consequently, by virtue of the definition of the sets $\widetilde{\Gamma}^k_{\alpha}$, not one of these $\Gamma^k_{\alpha_k}$ will be locally bicompact, while since $\Pi_i \cap \Pi_j \subseteq B$ when $i \neq j$, all the $\Gamma^k_{\alpha_k}$ are mutually distinct. We take some point $y \in (\bigcap_{k=1}^{\infty} \widetilde{\Gamma}^k_{\alpha_k}) \cap (\beta X \setminus X)$. From $\widetilde{\Gamma}^k_{\alpha_k} \subseteq [\Gamma^k_{\alpha_k}]_{\beta X}$, $k = 1, 2, \cdots, \infty$, it follows that $y \in \bigcap_{k=1}^{\infty} [\Gamma^k_{\alpha_k}]_{\beta X}$. But $y \notin F$, since $F \subseteq X$. Therefore we can find a set $\widetilde{U}$, open in βX, such that $F \subseteq \widetilde{U} \subseteq [\widetilde{U}]_{\beta X} \subseteq \beta X \setminus y$. Then the set $U = \widetilde{U} \cap X$ will be open in X and will satisfy the relations $F \subseteq U \subseteq [U]_{\beta X} \subseteq [\widetilde{U}]_{\beta X} \subseteq \beta X \setminus y$. Consequently for each of the $\Gamma^k_{\alpha_k}$ being considered we have $\Gamma^k_{\alpha_k} \cap (X \setminus U) \neq \Lambda$, since otherwise for some k_0 it could turn out that $y \notin [\Gamma^{k_0}_{\alpha_{k_0}}]_{\beta X}$, which, as we have seen, is false. Thus we see that each element from the infinite set of distinct elements $\Gamma^k_{\alpha_k} \in \mathfrak{D}$, $k = 1, 2, \cdots$, intersects with $X \setminus U$ and at the same time contains the point x. Therefore the inference we have made contradicts the regularity of system $\mathfrak{D}$ on the pair (x, F). Consequently $\bigcap_{k=1}^{\infty} \widetilde{\Pi}_k x \subseteq X$ for any point $x \in X$. By the same token we have proven Proposition D.

E. *Let the feathering $\phi = \{\gamma_n\}$ of the space X in the bicompact extension mX of X, satisfy the condition*

$$[\gamma_{n_1} x]_{mX} < \gamma_{n_2} \text{ for } n_1 > n_2 \tag{e}$$

for any point $x \in X$. Then X can be perfectly mapped onto some metric space.

Proof of Proposition E. We set $Ax = \bigcap_{n=1}^{\infty} \gamma_n x$, $x \in X$. Since ϕ is a feathering, $Ax \subseteq X$. In the set of points of space X we introduce the equivalence relation $x \sim y$ for $x, y \in X$ if $Ax \ni y$. We check the symmetry, reflexivity and transitivity of the relation introduced:

a) If $x \sim y$, i.e. $Ax \ni y$, then $\gamma_n x \ni y$ for every n, and then $\gamma_n y \ni x$ for any n, i.e. $x \in \bigcap_{n=1}^{\infty} \gamma_n y = Ay$, and consequently $y \sim x$.

b) $Ax \ni x$ for any $x \in X$, since all the γ_n are coverings of X. Consequently $x \sim x$.

c) Transitivity (it is the least obvious) is proved thus. Let $x \sim y$ and $y \sim z$, where $x, y, z \in X$. We need to show that $x \sim z$. From the definitions of the sets Ax, Ay and Az we have $y \in \bigcap_{n=1}^{\infty} \gamma_n x$ and $z \in \bigcap_{n=1}^{\infty} \gamma_n y$. By assumption, $\gamma_{n+1}^2 x \subseteq \gamma_n x$, $n = 1, 2, \cdots, \infty$.[1] Since $\gamma_{n+1} \ni x$, it follows that $\gamma_{n+1}x \cup \gamma_{n+1}y \subseteq \gamma_{n+1}^2 x$. We get: $z \in \gamma_{n+1}y \subseteq (\gamma_{n+1}x \cup \gamma_{n+1}y) \subseteq \gamma_{n+1}^2 x \subseteq \gamma_n x$, i.e. $z \in \gamma_n x$ for any n, which is equivalent to $z \in Ax$; hence $x \sim z$. Transitivity is proved.

From properties a), b) and c) it follows that the relation defined is an equivalence relation, i.e. $X = \bigcup_{x \in X} Ax$, and for any $x, y \in X$ either $Ax = Ay$ or $Ax \cap Ay = \Lambda$; therefore there occurs a decomposition of the space X into pairwise nonintersecting sets.

Let us study the properties of this decomposition.

I) For any point $x \in X$ the set Ax is bicompact; this ensues from condition (e) in the following manner: $Ax = \bigcap_{n=1}^{\infty} \gamma_n x = \bigcap_{n=1}^{\infty} [\gamma_n x]_{mX}$ is bicompact (and here Ax is a set closed in mX and consequently also in X).

II) For an arbitrary set $U \supseteq Ax$ open in mX we shall have $\gamma_n x \subseteq U$ for some n; this also follows from condition (e), because otherwise

$$\left(\bigcap_{n=1}^{\infty} \gamma_n x\right) \cap (mX \setminus U) = \bigcap_{n=1}^{\infty} [\gamma_n x]_{mX} \cap (mX \setminus U) \neq \Lambda$$

as the intersection of a decreasing system of closed sets in a bicompact space.

III) The decomposition is continuous. By AV, $V \subseteq X$, we denote the sum of the decomposition elements intersecting with V. Then, obviously, $A\gamma_n x \subseteq \gamma_n^2 x$, $n = 1, 2, \cdots, \infty$. Let U be any neighborhood of the decomposition element Ax; we need to find a neighborhood V of it such that $AV \subseteq U$. But, as we have proven in II), for some n_1 we have $\gamma_{n_1} x \subseteq U$. From condition (e) ensues $\gamma_{n_1+1}^2 x \subseteq \gamma_{n_1} x$. Setting $V = \gamma_{n_1+1} x$, we obtain $AV = A\gamma_{n_1+1} x \subseteq \gamma_{n_1+1}^2 x \subseteq \gamma_{n_1} x \subseteq U$; consequently V is the desired neighborhood.

IV) The decomposition space Z is metrizable. By (V), $V \subseteq X$, we denote the collection of decomposition elements contained in V. We set $\lambda_n = \{(\gamma_n x), x \in X\}$, $n = 1, 2, \cdots, \infty$. Let a_x denote the decomposition element Ax as a point in the decomposition space and let Oa_x be some neighborhood of a_x. Then, by the definition of the topology in the decomposition space, $Oa_x = (V_{Ax})$, where V_{Ax} is a distinguished neighborhood (see [4]) of the set Ax in the space X (i.e. decomposing into a sum of decomposition elements). By virtue of II) we have $\gamma_{n_1} x \subseteq V_{Ax}$ for some n_1. From condition

1) γ_{n+1}^2 is the star of the point x relative to the covering of space X by the stars of all the points of X relative to the covering γ_{n+1}.

(e) it follows that then $\gamma^3_{n_1+3}x \subseteq V_{Ax}$.[1] Further,

$$\lambda^2_{n_1+3}x \subseteq (A\gamma^3_{n_1+3}x) \subseteq (AV_{Ax}) = (V_{Ax}) = Oa_x.$$

This proves that the system $\{\lambda_n;\ n = 1, 2, \cdots, \infty\}$ forms a fundamental set of coverings of space Z (see [6]).

The natural mapping f of the space X onto the decomposition space Z is continuous (by the definition of the topology in Z (see [13])), is closed in view of property II), and is bicompact by virtue of property I). Here the T_1-axiom of separability is fulfilled by virtue of the closedness of the decomposition elements Ax, proved in I). Consequently Z is metrizable as a T_1-space possessing a fundamental set of coverings, and $f\colon X \to Z$ is the desired perfect mapping.

F. *Let $\phi = \{\gamma_n\}$ be a sequence of open coverings $\gamma_n = \{U^n_\alpha,\ \alpha \in M_n\}$ of X such that $[\gamma_{i+1}U^{i+1}_\alpha] < \gamma_i$ for any $U^{i+1}_\alpha \in \gamma_{i+1}$. We consider the Wallman compactification ωX of X, and set*

$$\widetilde{U}^n_\alpha = \omega X \setminus [X \setminus U^n_\alpha]_{\omega X},\ \widetilde{\gamma}_n = \{\widetilde{U}^n_\alpha,\ \alpha \in M_n\},\ n = 1, 2, \cdots, \infty.$$

The sequence $\widetilde{\phi} = \{\widetilde{\gamma}_n\}$ thus defined from ϕ will be called the extension of ϕ to ωX in what follows.

First of all, $\widetilde{U}^n_\alpha \supseteq U^n_\alpha$, and therefore every $\widetilde{\gamma}_n$ is an open (in ωX) covering of X. Moreover,

$$[\widetilde{\gamma}_{n_1}\widetilde{U}^{n_1}_\alpha]_{\omega X} < \widetilde{\gamma}_{n_2} \quad \textit{for any } \alpha \in M_{n_1},\ n_1 > n_2. \tag{f}$$

Proof. From the definition of the sets $\widetilde{U}^n_\alpha$ it follows that $\widetilde{U}^n_\alpha \subseteq [U^n_\alpha]_{\omega X}$. Therefore

$$\widetilde{\gamma}_{n_1}\widetilde{U}^{n_1}_{\alpha_1} \subseteq \bigcup_{U^{n_1}_\alpha \cap U^{n_1}_{\alpha_1} \neq \Lambda} [U^{n_1}_\alpha]_{\omega X} \subseteq [\gamma_{n_1}U^{n_1}_{\alpha_1}]_{\omega X}.$$

But $[\gamma_{n_1}U^{n_1}_{\alpha_1}]_X < \gamma_{n_2}$ by hypothesis, i.e. we can find an element $U^{n_2}_{\alpha_2} \in \gamma_{n_2}$ such that $[\gamma_{n_1}U^{n_1}_{\alpha_1}]_X \subseteq U^{n_2}_{\alpha_2}$. Then

$$\Lambda = [[\gamma_{n_1}U^{n_1}_{\alpha_1}]_X]_{\omega X} \cap [X \setminus U^{n_2}_{\alpha_2}]_{\omega X} = [\gamma_{n_1}U^{n_1}_{\alpha_1}]_{\omega X} \cap [X \setminus U^{n_2}_{\alpha_2}]_{\omega X}.$$

---

1) $\gamma^3_{n_1+3}x$ is the star of the point x relative to the covering of space X by the sets $\gamma^2_{n_1+3}y$ when y ranges over the whole of X.

But $\omega X \setminus [X \setminus U^{n_2}_{\alpha_2}]_{\omega X} = \widetilde{U}^{n_2}_{\alpha_2}$. Consequently $[\widetilde{\gamma}_{n_1} \widetilde{U}^{n_1}_{\alpha_1}]_{\omega X} \subseteq \widetilde{U}^{n_2}_{\alpha_2}$, which is what we had to prove.

G. *Let the space* X *possess a sequence of open coverings* $\phi = \{\gamma_n\}$, $\gamma_n = \{U^n_\alpha,\ \alpha \in M_n\}$, *satisfying the following conditions*:

1) *For every point* $x \in X$ *we can find a bicompactum* $F \ni x$ *such that the system of sets* $\{\gamma_n F;\ n = 1, 2, \cdots, \infty\}$ *forms a basis of* F *in* X.

2) γ_i *is a star-refinement of* γ_j *when* $i > j$.

Then X *possesses a feathering in* ωX *satisfying condition* (e).

Proof of Proposition G. Under the hypotheses of Proposition G, for any integer $i > 0$ the relation $[\gamma_{i+4} G] < \gamma_i$ holds for any $G < \gamma_{i+4}$. This follows from trivial lemmas.

Lemma 1. *If* $G < \gamma_{i+2}$, *then* $[G] < \gamma_i$.

Lemma 2. *If* $G < \gamma_{i+2}$, *then* $\gamma_{i+2} G < \gamma_i$.

Therefore, from the very start we can take it that in Proposition G the sequence $\{\gamma_n\}$ satisfies the following additional condition:

3) $[\gamma_{i+4} G] < \gamma_i$ *if* $G < \gamma_{i+4}$, *for any integer* i.

We can always arrive at such a sequence, having kept terms with numbers which are multiples of four; here it is obvious that as before condition 1) will be fulfilled for the subsequence obtained.

By virtue of Proposition F, the extension $\widetilde{\phi}$ of ϕ in ωX satisfies condition (e) (see Propositions E and F).

Let us show that $\widetilde{\phi}$ is a feathering of X in ωX. Let the point $x \in X$ be arbitrary and let $F \ni x$; $F \subseteq X$ is some bicompactum for which condition 1) is fulfilled. We consider any point $y \in \omega X \setminus X$. Since $y \notin F = [F]$, we can find a closed (in X) set $P \subseteq X$ for which $y \in [P]_{\omega X}$ and $P \cap F = \Lambda$. From the bicompactness of F and the regularity of X it follows that we can find a neighborhood U of bicompactum F in X for which $[U]_X \cap P = \Lambda$. Then also $[U]_{\omega X} \cap [P]_{\omega X} = \Lambda$, whence $y \notin [U]_{\omega X}$. By virtue of condition 1), for some n_1 we have $\gamma_{n_1} F \subseteq U$. However, it is clear that if $U^{n_1}_\alpha \subseteq U$, then $\widetilde{U}^{n_1}_\alpha \subseteq [U]_{\omega X}$. Therefore $\widetilde{\gamma}_{n_1} x \subseteq [U]_{\omega X} \subseteq \omega X \setminus y$. Hence, in view of the arbitrariness in the choice of the point $y \in \omega X$, it follows that $\bigcap_{n=1}^{\infty} \widetilde{\gamma}_n x \subseteq X$, i.e. $\widetilde{\phi} = \{\widetilde{\gamma}_n\}$ is a feathering of X in ωX. Proposition G is proved.

H. *Let* X *be a paracompactum*, bX *its Hausdorff bicompactification, and let* $X \subseteq Y \subseteq bX$, *where* Y *is a pinnate space with feathering* (*in* bX)

$$\phi = \{\gamma_n\},\ \gamma_n = \{U^n_\alpha,\ \alpha \in s_n\}.$$

Then we can find a space X', $X \subseteq X' \subseteq Y \subseteq bX$, *with a feathering* (*in* bX) $\psi = \{\lambda_n\}$, *satisfying condition* (e).

Proof. For every set $M \subseteq X$ we put $\widetilde{M} = bX \setminus [X \setminus M]_{bX}$. Then:

1°. If $M_1 \subseteq M_2$, then $\widetilde{M}_1 \subseteq \widetilde{M}_2$ because $X \setminus M_1 \supseteq X \setminus M_2$, whence $[X \setminus M_1]_{bX} \supseteq [X \setminus M_2]_{bX}$ and $bX \setminus [X \setminus M_1]_{bX} \subseteq bX \setminus [X \setminus M_2]_{bX}$.

2°. For any $M \subseteq X$ the set $\widetilde{M}$ is open and $\widetilde{M} \cap X \subseteq M$.

3°. $\widetilde{M} \subseteq [M]_{bX}$, since if $x \notin [M]_{bX}$, then $x \in [X \setminus M]_{bX}$ because $x \in [X]_{bX} = [(X \setminus M) \cup M]_{bX} = [X \setminus M]_{bX} \cup [M]_{bX}$. Hence $x \notin bX \setminus [X \setminus M]_{bX} = \widetilde{M}$.

We define inductively a certain special sequence of coverings of X by the sets $\lambda_0, \lambda_1, \cdots, \lambda_n, \cdots$, open in bX. As λ_0 we take any covering of space X, open in bX. Let us assume that we have already defined what we shall mean by λ_k. By μ_k we denote the system of all sets open in X with closure (taken in bX) contained both in the elements of λ_k as well as simultaneously in the elements of γ_k (with $k > 0$). Let $\mu_{k+1} = \{V_\alpha^{k+1}, \alpha \in L_{k+1}\}$ be some covering of X such that $\mu_{k+1} V_\alpha^{k+1} < \mu_k$ for any $\alpha \in L_{k+1}$. We set $\lambda_{k+1} = \widetilde{\mu}_{k+1} = \{\widetilde{V}_\alpha^{k+1}, \alpha \in L_{k+1}\}$.

Consider any $\widetilde{V}_\alpha^{k+1} \in \lambda_{k+1}$. By assumption we can find $G \in \mu_k$ for which $\mu_{k+1} V_\alpha^{k+1} \subseteq G$, i.e., whatever be $V_{\alpha'}^{k+1} \in \mu_{k+1}$, $V_{\alpha'}^{k+1} \cap V_\alpha^{k+1} \neq \Lambda$, we necessarily have $V_{\alpha'}^{k+1} \subseteq G$. But, in view of 3°, if $V_{\alpha'}^{k+1} \cap V_\alpha^{k+1} \neq \Lambda$, then also $\widetilde{V}_{\alpha'}^{k+1} \cap \widetilde{V}_\alpha^{k+1} \neq \Lambda$. From 1° and from the inclusion $V_{\alpha'}^{k+1} \subseteq G$ it then follows that $\widetilde{V}_{\alpha'}^{k+1} \subseteq \widetilde{G}$. But, by virtue of 3°, $\widetilde{G} \subseteq [G]_{bX}$ and $[G]_{bX} < \lambda_k$. We have $\lambda_{k+1} \widetilde{V}_\alpha^{k+1} \subseteq \widetilde{G}$, whence $[\lambda_{k+1} \widetilde{V}_\alpha^{k+1}]_{bX} \subseteq [\widetilde{G}]_{bX} \subseteq [G]_{bX} < \lambda_k$. Thus a condition stronger than (e) is fulfilled for the sequence $\{\lambda_n\}$.

We set $X' = \bigcup_{x \in X} (\bigcap_{n=1}^{\infty} \lambda_n x)$ and we verify that $\psi = \{\lambda_n\}$ is a feathering of X' in bX.

Let x be any point of X'. By the definition of X' we can find a point $y \in X$ for which $x \in \bigcap_{n=1}^{\infty} \lambda_n y$. Then $\lambda_{n+1} x \subseteq \lambda_{n+1}(\lambda_{n+1} y)$, $n = 1, 2, \cdots, \infty$. But $\lambda_{n+1}(\lambda_{n+1} y) \subseteq \lambda_n y$. Therefore $\bigcap_{n=1}^{\infty} \lambda_n x \subseteq \bigcap_{n=2}^{\infty} \lambda_n x \subseteq \bigcap_{n=1}^{\infty} \lambda_n y \subseteq X'$ (the last inclusion follows from the definition of X'). Thus $\psi = \{\lambda_n\}$ is a feathering of X' in bX, satisfying condition (e), as we see from what we have proved.

But $\bigcap_{n=1}^{\infty} \lambda_n x \subseteq \bigcap_{n=1}^{\infty} \gamma_n x \subseteq Y$ when $x \in X$, since $X \subseteq Y$, λ_n is a refinement of γ_n, and $\{\gamma_n\}$ is a feathering of Y in bX. Therefore $X' = \bigcup_{x \in X} (\bigcap_{n=1}^{\infty} \lambda_n x) \subseteq Y$. The inclusion $X' \supseteq X$ is obvious; in fact, all the λ_n are coverings of X. Proposition H is completely proved.

With the aid of Propositions A through H we now prove the theorems stated in subsection 1° (and some others).

Proof of Theorem 5.1. The sufficiency ensues from Propositions H and E. Let us verify the necessity. Let X be a perfect preimage of Y. From Proposition A it follows that X can be treated as a continuous, open, closed, (multivalued), two-sided bicompact image of a metric space. Then, by virtue of Proposition B, X is a paracompactum, while from Proposition C it follows that X possesses a k-fine system. By now applying Proposition D we get that X is a pinnate space. This proves the necessity of the conditions in Theorem 5.1.

Remark 1. We could have obtained the necessity of the conditions in Theorem 5.1 in another way be referring to the theorem on the fact that the preimage of a pinnate space under a perfect mapping is a pinnate space (see Theorem 3.3), while the preimage of a paracompactum is a paracompactum (see [26]).

Theorem 5.1 can be given a formulation in which the requirements of paracompactness and of the existence of the feathering are replaced by constraints imposed on the structure of a certain family of coverings of the space. Thus the question is of the most general intrinsic criterion for the existence of a perfect mapping of a space onto a metrizable space.

5.1'. Theorem.[1] *In order for a regular topological space to admit of a perfect mapping onto a metrizable space it is necessary and sufficient that there exist a countable family* $\phi = \{\gamma_i\}$ *of open coverings of* X, *satisfying the followings conditions*:

1) *For every point* $x \in X$ *we can find a bicompactum* $\Phi \ni x$ *such that the sets* $\gamma_n \Phi$, $n = 1, 2, \cdots, \infty$, *form, in the aggregate, a basis of* Φ *in* X.

2) *For any integers* i *and* j, $i > j$, γ_i *is a star-refinement of* γ_j.

Proof of Theorem 5.1'. Necessity. Let $f: X \to Y$ be a perfect mapping of the topological space X onto a metric space Y. In Y, as in any metric space, there exists a sequence of coverings satisfying conditions 1) and 2) of Theorem 5.1'. For example, such is $\phi = \{\gamma_n;\ n = 1, 2, \cdots\}$, where γ_n is the collection of open sets from Y, the diameter of each of which is not greater that $1/2^n$. Here, as the bicompactum Φ corresponding to the point x in the sense of condition 1), we can take this point itself.

We now turn to the space X. We set $\lambda_n = \{f^{-1}U | U \subseteq X,\ U \in \gamma_n\}$. Let

---

1) Independently of the author, V. I. Ponomarev has obtained a similar characterization for perfect preimages of metric spaces.

the point $x \in X$ be arbitrary, $y = fx$, $\Phi = f^{-1}fx$, and let $U \supseteq \Phi$ be any neighborhood of the set Φ in X. From the closedness of the mapping f it follows that $f^{-1}Oy \subseteq U$ for some neighborhood Oy of y. But $\gamma_{n_0}y \subseteq Oy$ holds for some $n = n_0$. Then also $\lambda_{n_0}\Phi = f^{-1}\gamma_{n_0}y \subseteq f^{-1}Oy \subseteq U$. This verifies that condition 1) of Theorem 5.1′ is fulfilled for the sequence $\{\lambda_n\}$.

Obviously, $\lambda_{n+1}x \subseteq f^{-1}\gamma_{n+1}fx$, but in γ_n there exists G such that $\gamma_{n+1}fx \subseteq G$. Then $f^{-1}G \in \lambda_n$, and we get $\lambda_{n+1}x \subseteq f^{-1}\gamma_{n+1}fx \subseteq f^{-1}G \in \lambda_n$. By the same token we have verified condition 2) for $\{\lambda_n\}$, i.e. $\{\lambda_n\}$ also is the desired sequence of coverings of space X.

The necessity of the conditions of Theorem 5.1′ has been proved.

The sufficiency ensues from Proposition E and G.

Theorem 5.1′ is proved.

Proof of Theorem 5.3. The necessity ensues from Propositions A and C. The sufficiency follows from Propositions D and E.

Proof of Theorem 5.4. Let X_i, $i = 1, 2, \cdots, \infty$, be paracompact pinnate spaces. By M_i we denote metric spaces onto which the X_i, $i = 1, 2, \cdots, \infty$, are perfectly mapped. It is very well known (for example, see [24]) that then the Tihonov product $X = \Pi_{i=1}^{\infty} X_i$ will be perfectly mapped onto $M = \Pi_{i=1}^{\infty} M_i$, where M, as the product of a countable set of metric spaces, is itself metrizable. Consequently, by virtue of Theorem 5.1, X is a paracompact pinnate space. Theorem 5.4 is proved.

It is now easy to point out an example (of which we spoke in §1) of a paracompactum which cannot be perfectly mapped onto any metric space whatsoever. Indeed, as follows from Theorem 5.4, such a one will be any paracompactum whose topological square is not paracompact, for example, the nonnegative reals with the topology generated by the left-closed right-open intervals (see [12], p. 631).

Proof of Theorem 5.5. Theorem 5.5 follows from Theorem 5.1.

Proof of Theorem 5.6. Theorem 5.6 follows from Propositions C and D taken together.

Proof of Theorem 5.7. On the one hand, Theorem 5.7 ensues from Propositions E and A, while on the other, from Propositions B, C and D as an aggregate.

In conclusion we cite a theorem which has proven useful to us in another paper; here it can be of interest to us as an illustration of the interrelation between the classes of paracompact and of paracompact pinnate spaces.

5.8. **Approximation theorem.** *Let X be a paracompact space and let $Y \supseteq X$ be a completely regular pinnate space. Then there exists a paracompact pinnate space X' such that $X \subseteq X' \subseteq Y$.*

Proof. Let bY be some Hausdorff bicompactification of Y. By applying Proposition H to the triple $bY \supseteq Y \supseteq X$, we get that there exists a space X', $X \subseteq X' \subseteq Y \subseteq bY$, possessing a feathering in bY for which condition (e) is fulfilled. But then, by virtue of Propositions E and A, X' also is the desired paracompact pinnate space. Theorem 5.8 is proved.

§ 6. The countable case

After Theorem 5.1 there arises the question of what spaces admit of perfect mappings onto a space with countable basis.

6.1. **Theorem.** *In order that a topological space can be perfectly mapped onto a space with countable basis it is necessary and sufficient that it be a finally compact space.*

Proof of Theorem 6.1. The necessity and the sufficiency of the condition in Theorem 6.1 ensue from Theorem 5.1, from the known situation that the preimage (just as the image) of a finally compact space under a perfect mapping is a finally compact space (see [26,25]), and from the fact that the final compactness of a metric space is equivalent to the presence in it of a countable basis.

Theorem 6.1, of course, answers the question posed at the beginning of the section; however, it is a legitimate wish to try to pick out the desired criterion from Theorem 5.1 with the aid only of some constraints on the nature of the feathering of the space. The thought suggests itself whether it would not be possible to require that the feathering, i.e. each countable set of coverings occurring in it, be countable. But already the example of a complete paracompact space (in which every element of the feathering consists of only one set, and consequently in our case the feathering is countable) shows that our idea does not achieve the goal; in fact, a complete paracompact, even metrizable, space may not be a finally compact space.

Nevertheless, the following theorem holds.

6.2. **Theorem.** *In order that a paracompactum Y can be perfectly mapped onto a space X with countable basis it is necessary and sufficient that Y possess a countable k-fine system.*

Proof of Theorem 6.2. The necessity is proved in just the same way as

Proposition C; as the B of space X is chosen any regular countable basis; then the system $\mathfrak{D}$, as we have proven in Proposition C, will be k-fine in Y and, obviously, countable.

Sufficiency. We first prove the following lemma.

6.3. Lemma. *A paracompactum X possessing a countable k-fine system $\mathfrak{D}$ is finally compact.*

Proof of the lemma. It is obvious that the goal will be reached if we prove that any locally finite covering γ of X is countable. By $\mathfrak{D}_\gamma$ we denote the collection of elements of $\mathfrak{D}$ each of which intersects with a not more than finite set of elements of γ. From the countability of $\mathfrak{D}$ and from the inclusion $\mathfrak{D}_\gamma \subseteq \mathfrak{D}$ it follows that the system $\mathfrak{D}_\gamma$ is countable. If we now prove that $\mathfrak{D}_\gamma$ covers X, then from this will follow also the countability of the covering γ (since then every element of covering γ intersects with some element of $\mathfrak{D}_\gamma$).

Consider any point $x \in X$ and some bicompact F such that $\mathfrak{D}$ is regular on (x, F). By virtue of the local finiteness of γ there exists a covering of F by open sets each of which intersects with a not more than finite collection of elements of γ, and hence from the bicompactness of F it follows that we can find a neighborhood $U \supseteq F$ also intersecting with only a finite set of elements of γ. By virtue of the regularity of $\mathfrak{D}$ on (x, F) there exists $V \in \mathfrak{D}$ such that $x \in V \subseteq U$. Then V too intersects only with a finite set of elements of γ, whence it follows that $V \in \mathfrak{D}_\gamma$ by the definition of $\mathfrak{D}_\gamma$. Hence $\mathfrak{D}_\gamma$ covers the point x, and consequently, because x was chosen arbitrarily, it also covers the whole space X. By the same token, in view of the remarks made above, the lemma is proved.

The sufficiency of the condition in Theorem 6.2 follows from the lemma just proved and from Theorem 6.1.

The condition for the existence of a perfect mapping of a space onto a space with countable basis can be formulated without including an a priori assumption on its final compactness (similar conditions are given in Theorem 5.1). Such a condition is the presence in the space being mapped of a sequence of countable coverings for which condition (e) or (f) (see Propositions E and F) is fulfilled.

The proof completely repeats the arguments from E, except that in part IV it is advantageous to mean by λ_n the collection of all those open sets of the decomposition space whose bodies form maximal distinguished sets contained in the elements of γ_n. Then each λ_n is countable and in aggregate they also form a fundamental set of coverings in the decomposition space.

§ 7. Pinnate topological groups

In this section we consider topological groups whose spaces are pinnate; for brevity we shall call them pinnate groups.

The principal goal of this section is to prove the following theorem.

7.1. **Theorem.** *A finite-dimensional locally connected pinnate group is metrizable.*

7.2. **Fundamental Lemma.** *Let G be a locally connected topological group, $\Phi \ni e$ a bicompactum,*[1] *and let $\phi = \{U_n\}$ be a sequence of open sets of G satisfying the conditions*: 1) *for any open $U \supseteq \Phi$ we can find $U_{n'} \in \phi$ for which $\Phi \subseteq U_{n'} \subseteq U$*; 2) *$U_{n_1} \subseteq U_{n_2}$ when $n_1 > n_2$.*

Then we can find a closed connected subgroup P of G, $P \subseteq \Phi$, whose point set is of type G_δ in Φ.

Proof. In view of the local connectivity of G and the definition of a topological group, there exists a sequence $\{V_n\}$ of neighborhoods of unity of group G for which the following conditions are fulfilled: a) $[V_n]$ is connected; b) $[V_n] \subseteq U_n$; c) $[V_{n_1}] \subseteq V_{n_2}$ when $n_1 > n_2$; d) $V_{n+1} \cdot V_{n+1}^{-1} \subseteq V_n$. We set $P = \bigcap_{n=1}^{\infty} V_n$. From condition c) it follows that $P = \bigcap_{n=1}^{\infty} [V_n]$, i.e. P is closed in G. The relation $P \subseteq \Phi$ ensues from condition b) and from the fact that $\Phi = \bigcap_{n=1}^{\infty} U_n$. Obviously P is a set of type G_δ in Φ.

Let us prove that P is a subgroup of G. It suffices to verify (see [19], Russian p. 17) that $ab^{-1} \in P$ follows from $a \in P$ and $b \in P$.

Let the number n be arbitrary. From the conditions $a \in P$ and $b \in P$ it follows that $a \in V_{n+1}$ and $b \in V_{n+1}$. Then $ab^{-1} \in V_{n+1} \cdot V_{n+1}^{-1} \subseteq V_n$; the latter follows from condition d). This proves that $ab^{-1} \in V_n$ for any integer $n > 0$. Consequently $ab^{-1} \in P$. We have proven that P is a subgroup of group G.

Let us show that the set P is connected. For the sake of brevity we introduce the notation $[V_n] = Q_n$, $n = 1, 2, \cdots, \infty$. We assume that $P = P_1 \cup P_2$, where $P_1 \neq \Lambda$ and $P_2 \neq \Lambda$ are sets closed in G, $P_1 \cap P_2 = \Lambda$. From $P_1 \subseteq \Phi$ and from the regularity of space G it follows that there exist sets Γ_1 and Γ_2, open in G, such that $\Gamma_1 \supseteq P_1, \Gamma_2 \supseteq P_2$ and $\Gamma_1 \cap \Gamma_2 = \Lambda$. We set $\Gamma = \Gamma_1 \cup \Gamma_2$, $F = (G \setminus \Gamma) \cap \Phi$ and $W_n = [V_n] \cap F$.

Obviously $\{W_n\}$ is a decreasing sequence of closed sets of the

---

1) Here e denotes the unit of the group G.

bicompactum F. But $\bigcap_{n=1}^{\infty} W_n \subseteq P \cap F \subseteq \Gamma \cap F = \Lambda$. Consequently, for some n_0,

$$W_{n_0} = \Lambda, \text{ i.e. } [V_{n_0}] \subseteq \Gamma \cup (G \setminus \Phi). \tag{*}$$

But $V_{n_0} \supseteq P_1 \cup P_2$, and therefore $[V_{n_0}] \cap \Gamma_1 \neq \Lambda$ and $[V_{n_0}] \cap \Gamma_2 \neq \Lambda$. Hence from condition a) and from the relation $\Gamma_1 \cap \Gamma_2 = \Lambda$ it follows that $Q_{n_0} \setminus \Gamma \neq \Lambda$. The set $\Pi = Q_{n_0} \setminus \Gamma$ is closed in G and, as follows from relation (*), $\Pi \cap \Phi = \Lambda$. Consequently $U = G \setminus \Pi$ is a neighborhood of Φ in G. In accordance with the lemma's hypothesis we select a number n_1, such that we would have $\Phi \subseteq U_{n_1} \subseteq U$ and $n_1 > n_0$. Then $Q_{n_1} \subseteq Q_{n_0} \subseteq \Gamma \cup (G \setminus \Phi)$; $Q_{n_1} \setminus \Gamma \subseteq Q_{n_0} \setminus \Gamma \subseteq \Pi$. But $Q_{n_1} \subseteq U_{n_1} \subseteq G \setminus \Pi$, i.e. $Q_{n_1} \cap \Pi = \Lambda$. Consequently $Q_{n_1} = (\Gamma_1 \cap Q_{n_1}) \cup (\Gamma_2 \cap Q_{n_1})$, where $\Gamma_1 \cap Q_{n_1} \supseteq P_1 \neq \Lambda$ and $\Gamma_2 \cap Q_{n_1} \supseteq P_2 \neq \Lambda$. As is easy to see, this contradicts the connectedness of the set $Q_{n_1} = [V_{n_1}]$. We have proven that P is a connected set. Consequently P is the desired group.

Proof of the theorem. Since G is a pinnate group, its space is a space of pointwise countable type (see Theorem 2.6). Consequently there exists a bicompactum Φ of countable nature in G containing the unit of G. We find ourselves in the situation of the Fundamental Lemma. Hence there exists a connected bicompact subgroup P of G, contained in Φ and being a set of type G_δ in Φ. But then (see [10]) P is of countable nature in Φ. Since Φ is a finite-dimensional bicompactum, P is finite dimensional. Consequently (see [20], Russian p. 84) P is a space with countable base. The unity of group G is of countable nature in P, P is a bicompactum of countable nature in Φ, and Φ is a bicompactum of countable nature in G. By applying Theorem 6.2 twice we get that the first axiom of countability is fulfilled in G, whence it follows that the space of G is metrizable. The theorem is proved.

7.3. Corollary. *If the space of a locally connected finite-dimensional group admits of a perfect mapping onto a space of pointwise countable type, for example, onto a space with the first axiom of countability, then both spaces are metrizable.*

7.4. Corollary. *A Čech-complete locally connected finite-dimensional group is metrizable by a complete metric.*

Proof of the corollaries. Corollary 7.3 ensues from the fact that the preimage of a space of pointwise countable type under a perfect mapping is a space of pointwise countable type, from the fundamental lemma and from Stone's theorem on the preservation of metrizability under perfect mappings.

Corollary 7.4 ensues from Theorem 7.1 and from Čech's theorem on the existence of a complete metric on a Čech-complete metrizable space.

In conclusion we remark that not one of assertions 7.1–7.4 remains valid if we do not assume that G is the space of a group.

§ 8. The noncondensability theorem

The following theorem will be proven.

8.1. **Theorem.** *A weakly paracompact subspace of a pinnate space, condensing on a space with uniform basis, itself possesses a uniform basis.*

In view of P. S. Aleksandrov's theorem on the metrizability of a paracompactum with uniform basis and on the existence of a uniform basis in every metrizable space, from Theorem 8.1 ensues an essential corollary.

8.2. **Corollary.** *A paracompact subspace of a pinnate space, condensing onto a metric space, is metrizable.*

In particular, we get that every paracompact pinnate space which can be continuously mapped one-to-one onto a metrizable space is metrizable.

It is clear that if the word "pinnate" is omitted in the formulations of assertions 8.1 and 8.2 they cease to be valid (Uryson's example).

8.3. **Fundamental Lemma.** *Let X be a topological space, $\{\lambda_n\}$ a system of open coverings of X for which* a) $\bigcap_{n=1}^{\infty}\lambda_n x = x$, $x \in X$, *and* b) *for any $x \in X$ and for the number n_1 we can find a number n_2 such that $[\lambda_{n_2}x] \subseteq \lambda_{n_1}x$, and $\lambda_{n_2}x \subseteq \lambda_{n_1}x$ for any $n_2 > n_1$.*

Further, let bX be a Hausdorff bicompactification of X, X_1 a subspace of X, and $\{\eta_n\}$ a sequence of coverings of X by sets open in bX such that c) $\bigcap_{n=1}^{\infty}\eta_n x \subseteq X$ *and* d) $[\eta_{n_1}x]_{bX} \subseteq \eta_{n_2}x$ *when $n_1 > n_2$, $x \in X_1$.*

Then the sequence $\{\xi_n\}$, where $\xi_n = \eta_n \cap \lambda_n$, $n = 1, 2, \cdots, \infty$, can be refined at the points of X_1 (and consequently $\{\xi'_n\}$, where $\xi'_n = \xi_n \cap X_1$, is a refined sequence of coverings of X_1).

Proof. Let the point $x \in X_1$ and its neighborhood Ox (in the space bX) be chosen arbitrarily. We set $\Phi = \bigcap_{n=1}^{\infty}\eta_x x$. From conditions c) and d) it follows that Φ is a bicompactum, $\Phi \subseteq X$, and $\{\eta_n x\}$ is a basis of Φ in bX.

By $O_1 x$ we denote some neighborhood of x in bX with closure lying in Ox. From b) it follows that the sequence $\{Q_n\}$, where we have set $Q_n = \Phi \cap \lambda_n x$, forms a basis of the point x in Φ. Consequently we can find numbers n_0 and n'_0 for which $Q_{n'_0} \subseteq O_1 x$ and $[\lambda_{n_0}x]_X \subseteq \lambda_{n'_0}x$. We set $[\lambda_{n_0}x \setminus Ox]_{bX} = P$. From the definition of the number n_0 (and from the closedness of Φ in X) it follows that $\Phi \cap [\lambda_{n_0}x]_X \subseteq Ox \cap \Phi$, i.e.

$([\lambda_{n_0}x]_X \backslash Ox) \cap \Phi = \Lambda$, while since Φ is bicompact and $\Phi \subseteq X$, we get

$$P \cap \Phi = [\lambda_{n_0}x \backslash Ox]_{bX} \cap \Phi \subseteq [[\lambda_{n_0}x]_X \backslash Ox]_{bX} \cap \Phi = \Lambda.$$

Consequently $U = bX \backslash P$ is a neighborhood of Φ in bX. Since $\{\eta_n x\}$ is a basis of Φ in bX, we can find a number n_1 for which $\eta_{n_1}x \subseteq U$. We set $n_2 = n_0 + n_1$. Then, by virtue of conditions b) and d), $\lambda_{n_2}x \subseteq \lambda_{n_0}x$ and $\eta_{n_2}x \subseteq \eta_{n_1}x$. Hence from the definition of ξ_n we get

$$\xi_{n_2}x \backslash Ox \subseteq (\lambda_{n_2}x \backslash Ox) \cap (\eta_{n_2}x \backslash Ox) \subseteq P \cap \eta_{n_2}x \subseteq (X \backslash U) \cap \eta_{n_2}x = \Lambda.$$

The Fundamental Lemma is proved.

We pass on to its applications.

8.4. **Proposition.** *Let a completely regular pinnate space X be condensed onto some regular space Y with a refined sequence of coverings, and let X_1 be a weakly paracompact subspace of X.*

Then X_1 also possesses a refined sequence of coverings.

Proof. There exists a sequence $\{\lambda_n\}$ of open coverings of X satisfying conditions a) and b) of the Fundamental Lemma. Let bX be an arbitrary Hausdorff bicompactification of X and let $\phi = \{\gamma_n\}$, $\gamma_n = \{U^n_\alpha,\ \alpha \in M_n\}$, $n = 1, 2, \cdots, \infty$, be some feathering of X in bX.

We set $\eta_1 = \gamma_1$. We assume that the covering η_n of the space X_1 by sets open in bX has already been determined for every positive integer $n \leq k$. By μ_{k+1} we denote a covering of X_1 by all sets open in bX contained with their closures in the elements of η_k and lying simultaneously in the elements of γ_k. Since X_1 is weakly paracompact, we can refine μ_{k+1} by a certain pointwise finite covering η'_{k+1} of X_1 by sets open in X_1. We set $\eta_{k+1} = \{V \subseteq bX | (V \cap X) \in \eta'_{k+1},\ V = bX \backslash [X \backslash (V \cap X)]_{bX}\}$.

Continuing, we obtain a sequence $\{\eta_n\}$ of coverings of X_1 by sets open in bX, for which the conditions in the Fundamental Lemma are fulfilled.

Indeed, $\bigcap_{n=1}^\infty \eta_n x \subseteq \bigcap_{n=1}^\infty \gamma_n x \subseteq X$, since η_n is a refinement of γ_n and $\phi = \{\gamma_n\}$ is a feathering of X in bX. Condition c) is fulfilled. From the construction of η_{k+1} we see that the star $\eta_{k+1}x$ of an arbitrary point $x \in X_1$ decomposes into a finite number of elements of the covering η_{k+1}, each of which is contained with its closure (in bX) in some element of the covering η_k. Since the sum of the closures of a finite number of sets equals the closure of their sum, $[\eta_{k+1}x]_{bX} \subseteq \eta_k x$. By the same token we have verified condition d).

Now we have the right to apply the Fundamental Lemma. Proposition 8.4 is proved.

Our two fundamental theorems are easily derived from Proposition 8.4.

Proof of Theorem 8.1. P. S. Aleksandrov has proven [3] that in every space with uniform basis there exists a refined sequence of coverings. Consequently, by virtue of Proposition 8.4, there is a refined sequence of coverings in X_1. But another theorem of P. S. Aleksandrov in the same paper asserts that a weakly paracompact space with a refined sequence of coverings possesses a uniform basis. Theorem 8.1, and together with it also Corollary 8.2, is proven.

Corollary 8.2 admits of an obvious generalization.

8.2'. **Theorem.** *A paracompact subspace of a pinnate space, condensed onto a space with uniform basis, is metrizable.*

It is curious that here the space with uniform basis being talked about itself may not be metrizable.

Supplement. Featherings and mappings

Let f be a mapping of a space X onto a space Y. Under what assumptions concerning X and f will the space Y be a pinnate space?

In §3 (Theorem 3.3) we saw that if Y is a pinnate space and $f: X \longrightarrow Y$ is a perfect mapping, then X will be a pinnate space. On the other hand, by the example of a mapping at one point we ascertained that in this aspect pinnateness is not preserved both under open as well as under closed mappings.

When passing to the image pinnateness behaves itself worse yet. Here we do not even know whether the perfect image of a pinnate space will be pinnate; we do not know whether this is so even for paracompact pinnate spaces. Incidently, the author thinks it is most probable that the answers to both questions are negative.

Examples of the fact that both the closed as well as the open image of a paracompact pinnate space may not be pinnate were presented in §3. Nevertheless, we have succeeded in obtaining certain positive results which are the best possible to the extent outlined by the examples mentioned.

S.1. **Lemma.** *In any paracompact pinnate space X there exists a feathering $\phi = \{\gamma_n;\ n = 1, 2, \cdots, \infty\}$ in βX, satisfying (simultaneously) the conditions* 1) $[\gamma_{n+1}\Phi]_{\beta X} \subseteq \gamma_n\Phi$, $n = 1, 2, \cdots, \infty$, *and* 2) $\bigcap_{n=1}^{\infty}\gamma_n\Phi \subseteq X$, *whatever the bicompactum $\Phi \subseteq X$ may be.*

Proof. By using the paracompactness of X it is easy to construct a sequence $\{\gamma_n\}$ of open (in βX) coverings of X, locally finite on X and such that the closure in βX of each element of γ_{n+1} is contained in some element of the system γ_n, $n = 1, 2, \cdots, \infty$. If $\{\lambda_n\}$ is some feathering of X in βX satisfying condition 2), then it presents no difficulty to achieve that γ_n will be a refinement of λ_n, $n = 1, 2, \cdots, \infty$. Then the system $\{\gamma_n\}$ will be a feathering of X in βX. This feathering is the one desired.

Indeed, let us consider an arbitrary bicompactum $\Phi \subseteq X$. In view of the local finiteness of γ_{n+1}, Φ intersects only with a finite set of elements of γ_{n+1}, each of which is contained with its closure (in βX) in some element of γ_n, by the same token intersecting with Φ. Then the closure of their sum, i.e. $[\gamma_{n+1}\Phi]_{\beta X}$, is contained in the sum of their closures and, all the more, is contained in $\gamma_n\Phi$. Condition 1) is fulfilled.

We have $\gamma_n\Phi \subseteq \lambda_n\Phi$, whence $\bigcap_{n=1}^{\infty}\gamma_n\Phi \subseteq \bigcap_{n=1}^{\infty}\lambda_n\Phi$, and since $\bigcap_{n=1}^{\infty}\lambda_n\Phi \subseteq X$ by assumption, also $\bigcap_{n=1}^{\infty}\gamma_n\Phi \subseteq X$ for any bicompactum $\Phi \subseteq X$.

Thus the lemma will be proven if we construct a feathering of X satisfying condition 2). To do this we consider some perfect mapping $f\colon X \to Y$ of X onto some metric space Y and its continuation onto the Čech compactifications: $\tilde{f}\colon \beta X \to \beta Y$.

Let μ_n be a collection of all sets open in βY cutting out on the space Y sets of diameter less than $1/n$, and let $\psi = \{\mu_n | n = 1, 2, \cdots, \infty\}$.

We set $\lambda_n = \{\tilde{f}^{-1}G | G \in \mu_n\}$, $\phi = \{\lambda_n | n = 1, 2, \cdots, \infty\}$. It turns out that ϕ is a feathering of X in βX satisfying requirement 2). From the continuity of $\tilde{f}$ it follows that all the λ_n, $n = 1, 2, \cdots, \infty$, are open (in βX) coverings of X.

Let Φ be any bicompactum from X. We set $F = f\Phi$ and $P = f^{-1}F$. Since f is a perfect mapping, $\tilde{f}^{-1}Y = X$, and therefore $P = \tilde{f}^{-1}F$.

It suffices for us to prove that $\bigcap_{n=1}^{\infty}\mu_n F \subseteq Y$, since

$$\lambda_n\Phi \subseteq \lambda_n P = \tilde{f}^{-1}\mu_n F, \quad \bigcap_{n=1}^{\infty}\tilde{f}^{-1}\mu_n F = \tilde{f}^{-1}\bigcap_{n=1}^{\infty}\mu_n F \text{ and } \tilde{f}^{-1}Y = X.$$

Let y_0 be any point of $\beta Y \setminus Y$. Then for some set Q closed in Y hold the relations $y \in [Q]_{\beta Y}$ and $Q \cap F = \Lambda$. By virtue of the bicompactness of F, $\rho(Q, F) > 0$. We select an integer n_0 such that $1/n_0 < \rho(Q, F)$. Obviously then $\mu_{n_0}F \subseteq [O_{1/n_0}F]_{\beta Y}$, where $O_{1/n_0}F$ is a point set of Y lying at a distance less that $1/n_0$ from F. But $[O_{1/n_0}F]_Y \cap Q = \Lambda$; then, by a well-known property of the Čech compactification, $[[O_{1/n_0}F]_Y]_{\beta Y} \cap [Q]_{\beta Y} = \Lambda$, whence $[O_{1/n_0}F]_{\beta Y} \subseteq \beta Y \setminus y_0$. We get $\mu_{n_0}F \subseteq [O_{1/n_0}F]_{\beta Y} \subseteq \beta Y \setminus y_0$. By the same token, in view of the arbitrariness in the choice of the point

$y_0 \in \beta Y \setminus Y$, we have proven that $\bigcap_{n=1}^{\infty} \mu_n F \subseteq Y$. The lemma is proved.

S.2. Theorem. *An open bicompact image of a paracompact pinnate space is a pinnate space.*

Proof. Let $f\colon X \to Y$ be an open bicompact mapping of a paracompact pinnate space X, $\tilde{f}\colon \beta X \to \beta Y$ its continuation onto the Čech compactifications of spaces X and Y, and let $\phi = \{\gamma_n;\ n = 1, 2, \cdots, \infty\}$, $\gamma_n = \{U_\alpha^n,\ \alpha \in M_n\}$, be a feathering of X in βX satisfying the conditions of Lemma S.1: $[\gamma_{n+1}\Phi]_{\beta X} \subseteq \gamma_n \Phi$, and $\bigcap_{n=1}^{\infty} \gamma_n \Phi \subseteq X$ for any bicompactum $\Phi \subseteq X$, $n = 1, 2, \cdots, \infty$.

By V_α^n, $n = 1, 2, \cdots, \infty$, $\alpha \in M_n$, we denote some set, open in βY, for which $V_\alpha^n \cap Y = f(U_\alpha^n \cap X)$; such a set exists since the set $f(U_\alpha^n \cap X)$, as the image of an open set under an open mapping, is open in Y.

We set $\lambda_n = \{V_\alpha^n,\ \alpha \in M_n\}$, $\psi = \{\lambda_n,\ n = 1, 2, \cdots, \infty\}$. Let us prove that ψ is a feathering of Y in βY.

We assume the contrary; for some point $y \in Y$ let the set $\bigcap_{n=1}^{\infty} \lambda_n y$ contain some point y_0 of $\beta Y \setminus Y$. We set $\Phi = f^{-1} y$ and we consider the set $P = \bigcap_{n=1}^{\infty} \gamma_n \Phi$. From conditions 1) and 2) it follows that P is a bicompactum, $P \subseteq X$. Consequently $F = fP$ also is a bicompactum, $F \subseteq Y$. In βY we can find a neighborhood $G \supset F$ such that $[G]_{\beta X} \subseteq \beta Y \setminus y_0$. In view of the continuity of the mapping $\tilde{f}$, we shall have $\tilde{f} H \subseteq G$ for some neighborhood H of the set P (in βX). We can find a number n_0 for which $\gamma_{n_0} \Phi \subseteq H$, because otherwise we would get that $P \cap (\beta X \setminus H) = \bigcap_{n=1}^{\infty} (\gamma_n \Phi \cap (\beta X \setminus H)) \neq \Lambda$, in contradiction with $H \supseteq P$. We have $\tilde{f} \gamma_{n_0} \Phi \subseteq G$. But it is easy to see that $\gamma_{n_0} y \subseteq [\tilde{f} \gamma_{n_0} \Phi]_{\beta Y}$. From these two relations it follows that $\lambda_{n_0} y \subseteq [\tilde{f} \gamma_{n_0} \Phi]_{\beta Y} \subseteq [G]_{\beta Y} \subseteq \beta Y \setminus y_0$.

Consequently $\bigcap_{n=1}^{\infty} \lambda_n y \subseteq \beta Y \setminus y_0$, whence, in view of the arbitrariness in the choice of the point y_0, it follows that $\bigcap_{n=1}^{\infty} \lambda_n y \subseteq Y$. The latter signifies, by definition, that ψ is a feathering of Y in βY. The theorem is proved.

The author chose to prove Theorem S.2 in this way mainly because he wished to set forth Lemma S.1, which, it seems to him, has some independent interest. If we had so desired we could have restricted ourselves to pointing out a suitable combination of Propositions A–F and a theorem of V. I. Ponomarev.

Indeed, let Y be a paracompact pinnate space. Then, in view of Theorem 5.1, $Y = fX$, where X is some metric space and f is a multivalued, continuous, open, closed, two-sided bicompact mapping. Further, let $Z = gY$, where g is a single-valued, continuous, open, bicompact mapping. We set $\phi = gf$, $\phi\colon X \to Z$. Being the superposition of continuous open mappings, the mapping ϕ is continuous and open. Let us show that it is bicompact on both sides.

1) Let $x \in X$ be any point. The set $\Phi = fx$ is a bicompactum. Then, in view of the continuity and single-valuedness of mapping g on Φ, $g\Phi = gfx = \phi x$ also is a bicompactum.

2) Consider any $z \in Z$ and $F = g^{-1}z$, $F \subseteq Y$. By virtue of the assumptions on g, F is a bicompactum. The mapping $f\colon X \to Y$ satisfies the hypotheses of a well-known theorem of V. I. Ponomarev [15]; by virtue of this theorem, $\Phi = f^{-1}F$ is a bicompactum. But $\phi^{-1}z = f^{-1}(g^{-1}z) = f^{-1}F = \Phi$.

The bicompactness of the mapping ϕ and of the mapping inverse to it have been proved.

We see that $\phi\colon X \to Z$ satisfies all the hypotheses of Proposition C. By applying Propositions C and D successively, we get that Z is a pinnate space.

Theorem S.2 permits us to extend our knowledge of the composition of the class of pinnate spaces.

S.3. Theorem. *A space with uniform basis is a pinnate space.*

The definition of a uniform basis was given some years ago by P. S. Aleksandrov (see [3]).

Proof. The theorem follows immediately from Theorem S.2 and from the characteristics of spaces with uniform basis as open, bicompact (continuous and single-valued) images of metric spaces [28].

It is known that among spaces with uniform basis there are no paracompact ones (every paracompact space with uniform basis is metrizable; this is a theorem of P. S. Aleksandrov). Consequently we can accompany Theorem S.2 with an essential remark: an open bicompact image of a paracompact pinnate space may not be a paracompact pinnate space. This remark indicates the definitiveness, in the sense defined, of the following result.

S.4. Theorem. *An open k-image of a paracompact pinnate space is a paracompact pinnate space.*

Proof. Let the mapping $f\colon X \to Y$ satisfy the theorem's hypotheses. The space X (as a pinnate space) is a k-space (see Corollary 2.7). An open image of a k-space is a k-space (see [10]). Finally, a k-mapping onto a k-space is always closed (see [10]). From these three remarks it follows that also the mapping $f\colon X \to Y$ will be closed. Then Y (as a closed image of a paracompactum) itself is a paracompactum. By virtue of Theorem S.2, Y is a pinnate space. The theorem is proved.

S.5. **Theorem.** *A space without isolated points which is an open-closed image of a paracompact pinnate space, itself is a paracompact pinnate space.*

Proof. The mapped space, as a Dieudonné-complete paracompactum (i.e. complete relative to the maximal structure compatible with its topology) and as a pinnate space by Corollary 2.7, is a k-space. But under an open-closed mapping of a Dieudonné-complete k-space the image of every point not isolated in the image is a bicompactum. Hence it follows that our mapping is perfect, and consequently (see [10]) is a k-mapping. It remains to refer to Theorem S.4. The theorem is proved.

The matter is by far simpler and better in the class of pinnate spaces which are spaces of topological groups.

B. A. Pasynkov has remarked that if the space of a group contains a bicompactum of countable nature, then this space is paracompact and pinnate. This assertion can be obtained easily with the aid of Theorem 5.1$'$.

But since any pinnate space is a space of pointwise countable type (see Theorem 2.6), there ensues from this the following theorem.

S.6. **Theorem.** *A pinnate group is paracompact.*

As is well known (see [19]), the factor space of a topological group is the open image of the space of this group. Under an open mapping a space of pointwise countable type goes over into a space of pointwise countable type; this is trivial.

By applying Theorem 2.6 and the results of B. A. Pasynkov mentioned, we discover the validity of the following theorem.

S.7. **Theorem.** *The factor group of a pinnate group is a pinnate group.*

In conclusion we present two unsolved questions, the first of which is another variant of the problem of which we spoke at the beginning of this Supplement.

1) Let the absolute[1)] of a space be perfectly mapped onto some metric space. Does it follow from this that this space itself can be perfectly mapped onto a suitable metrizable space? (Obviously, the converse is true by virtue of Theorem 5.1.)

2) In the formulation of Theorem S.5 could we drop the assumption that

---

1) See [17] for the concept of the absolute of a space.

the image does not contain isolated points?

BIBLIOGRAPHY

[1] P. S. Aleksandrov and P. S. Uryson, *Mémoire sur les espaces topologiques compacts*, Verh. Kon. Akad. Wetensch. Amsterdam. Afd. Natuurk., Sectie 1, 14 (1929), no. 1; Russian rev. transl., *On compact topological spaces*, Trudy Mat. Inst. Steklov. 31 (1950); also in P. S. Uryson, *Works in topology and other fields of mathematics*. Vol. 2, GITTL, Moscow, 1951, pp. 854–963. (Russian) MR 13, 264; 14, 122.

[2] P. S. Aleksandrov, *On some results concerning topological spaces and their continuous mappings*, Proc. Sympos. General Topology and its Relation to Modern Analysis and Algebra (Prague, 1961), Academic Press, New York and Publ. House Czech. Acad. Sci., Prague, 1962, pp. 41–54. MR 26 #3003.

[3] ———, *On the metrization of topological spaces*, Bull. Acad. Polon. Sci. Sér. Sci. Math. Astronom. Phys. **8** (1960), 135–140. MR 22 #5024.

[4] ———, *On bicompact extensions of topological spaces*, Mat. Sb. 5 (47) (1939), 403–423. (Russian) MR 1, 318.

[5] A. V. Arhangel'skiĭ, *An addition theorem for the weight of sets lying in bicompacts*, Dokl. Akad. Nauk SSSR 126 (1959), 239–241. (Russian) MR 21 #5176.

[6] ———, *New criteria for the paracompactness and the metrizability of an arbitrary* T_1*-space*, Dokl. Akad. Nauk SSSR 141 (1961), 13–15 = Soviet Math. Dokl. **2** (1961), 1367–1369. MR **24** #A1113.

[7] ———, *External bases of sets lying in bicompacta*, Dokl. Akad. Nauk Nauk SSSR 132 (1960), 495–496 = Soviet Math. Dokl. **1** (1960), 573–574. MR 22 #9954.

[8] ———, *On the metrization of topological spaces*, Bull. Acad. Polon. Sci. Sér. Sci. Math. Astronom. Phys. **8** (1960), 589–595. MR 23 #A2861.

[9] ———, *On a class of spaces containing all metric spaces and all locally bicompact spaces*, Dokl. Akad. Nauk SSSR 151 (1963), 751–754 = Soviet Math. Dokl. 4 (1963), 1051–1055. MR 27 #2959.

[10] ———, *Bicompact sets and the topology of spaces*, Trudy Moskov. Mat. Obšč. 13 (1965), 3–55 = Trans. Moscow Math. Soc. 1965, 1–62. MR 33 #3251.

[11] C. H. Dowker, *Topology of metric complexes*, Amer. J. Math. 74 (1952), 555–577. MR 13, 965.

[12] R. H. Sorgenfrey, *On the topological product of paracompact spaces*, Bull. Amer. Math. Soc. 53 (1947), 631–632. MR 8, 594.

[13] J. L. Kelley, *General topology*, Van Nostrand, Princeton, N. J., 1955. MR 16, 1136.

[14] P. K. Osmatesku, *$\omega\alpha$-compactifications*, Vestnik Moskov. Univ. Ser. I Mat. Meh. 1963, no. 6, 45–54. (Russian) MR 28 #579.

[15] V. I. Ponomarev, *A new space of closed sets and multivalued continuous mappings of bicompacta*, Mat. Sb. 48 (90) (1959), 191–212; English transl., Amer. Math. Soc. Transl. (2) 38 (1964), 95–118. MR 22 #8473.

[16] ———, *Properties of topological spaces preserved under multivalued continuous mappings*, Mat. Sb. 51 (93) (1960), 515–536; English transl., Amer. Math. Soc. Transl. (2) 38 (1964), 119–140. MR 22 #12512.

[17] ———, *On the absolute of a topological space*, Dokl. Akad. Nauk SSSR 149 (1963), 26–29 = Soviet Math. Dokl. 4 (1963), 299–302. MR 28 #589b.

[18] ———, *Axioms of countability and continuous mappings*, Bull. Acad. Polon. Sci. Sér. Sci. Math. Astronom. Phys. 8 (1960), 127–134. MR 22 #7109.

[19] L. S. Pontrjagin, *Continuous groups*, 2nd ed., GITTL, Moscow, 1954; English transl. Gordon and Breach, New York, 1966. MR 17, 171; 34 #1439.

[20] E. G. Skljarenko, *On the topological structure of locally bicompact groups and their factor spaces*, Mat. Sb. 60 (102) (1963), 63–88; English transl., Amer. Math. Soc. Transl. (2) 39 (1964), 57–82. MR 26 #5090.

[21] Ju. M. Smirnov, *On the metrizability of bicompacts decomposable into a sum of sets with countable basis*, Fund. Math. 43 (1956), 387–393. MR 18, 813.

[22] A. Tychonoff (Tihonov), *Über die topologische Erweiterung von Räumen*, Math. Ann. 102 (1929), 544–561.

[23] P. S. Uryson, *Über die Mächtigkeit der zusammenhängenden Mengen*, Math. Ann. 94 (1925), 262–295; Russian transl., Works on topology and other fields of mathematics. Vol. 1, GITTL, ,Moscow, 1951, pp. 177–218. (Russian) MR **14**, 122.

[24] Z. Frolík, *On the topological product of paracompact spaces*, Bull. Acad. Polon. Sci. Sér. Sci. Math. Astronom. Phys. 8 (1960), 747–750. MR 23 #A2859.

[25] F. Hausdorff, *Mengenlehre*, 3rd ed., de Gruyter, Berlin, 1937; English transl., Chelsea, New York, 1957; Russian transl., ONTI, Moscow, 1937. MR 7, 419; **19**, 111.

[26] M. Henriksen and J. R. Isbell, *Some properties of compactifications*, Duke Math. J. 25 (1957), 83–105. MR 20 #2689.

[27] E. Čech, *On bicompact spaces*, Ann. of Math. (2) 38 (1937), 823–844.

[28] A. V. Arhangel'skiĭ, *On mappings of metric spaces*, Dokl. Akad. Nauk SSSR 145 (1962), 245–247 = Soviet Math. Dokl. 3 (1962), 953–956. MR **25** #2579.

Translated by:
N. H. Choksy

Amer. Math. Soc. Transl.
(2) Vol. 92, 1970

ON EXTENSION OF MAPPINGS OF TOPOLOGICAL SPACES*

UDC 513.83

N. VELIČKO

§1

The purpose of this article is to generalize the well-known theorem of A. D. Taĭmanov [1] on the extension of continuous functions. In particular, it is proved that the theorem of Taĭmanov holds when Y (the image space) is completely separable, and H is closed and the mapping f is θ-continuous[1)] (in the sense of S. V. Fomin [2]). For this purpose we make a slight change in the well-known necessary and sufficient conditions for H-closedness of Hausdorff spaces.

Definition 1. The point x is called a point of θ-contact of the set S in the space X if every closed neighborhood of the point intersects S. The set $[S]_\theta$ of all points of θ-contact of S is called the θ-closure of S. The set S will be said to be θ-closed if $[S]_\theta = S$.

It is clear that $[S]_\theta \supseteq [S]$, where $[S]$ is the closure of the set S in X. If S is open, then $[S]_\theta = [S]$. In a regular space the operations of θ-closure and of closure are identical.

Definition 2. The point x is called a point of θ-contact of the filter[2)] $\mathfrak{F}$ in space X, if it is a point of θ-contact of every set of the filter. The filter $\mathfrak{F}$ is θ-convergent to the point x if every closed neighborhood of the point x belongs to $\mathfrak{F}$.

Lemma 1. *Point x is a point of θ-contact of the filter $\mathfrak{F}$ if and only if there exists a filter $\mathfrak{F}'$ which majorizes $\mathfrak{F}$ and is θ-convergent to x.*

---

*Translation of Sibirsk. Mat. Ž. 6 (1965), 64–69.

1) By a space we always mean, in the present article, a topological space. By a neighborhood of a point x in the space X we mean a set V which contains an open set W with $x \in W$ (compare [3]). The space X is said to be completely separable if any two points in it have nonintersecting closed neighborhoods. Concerning H-closed and completely separable spaces see [3], Russian 130. The mapping f of the space X into the space Y is said to be θ-continuous if for an arbitrary point $x \in X$ and an arbitrary neighborhood Oy of the point $y = f(x) \in Y$ there exists a neighborhood Ox of the point x such that $f(Ox) \subseteq [Oy]$.

2) For the definition of a filter and of related concepts see [3].

Proof. Let x be a point of θ-contact of F. The filter $\mathfrak{F}'$ generated by $\mathfrak{F}$ and by the set of closed neighborhoods of the point x majorizes $\mathfrak{F}$ and is θ-convergent to x. The converse statement is obvious.

Corollary. *For the ultrafilter $\mathfrak{U}$ to be θ-convergent to the point x it is necessary and sufficient that x be a point of θ-contact of the filter.*

Theorem 1. *The following statements are equivalent:*

(A) *The space X is H-closed.*

(B) *Every filter in X has a point of θ-contact.*

(C) *Every family of sets such that the interiors of their closures cover the space X contains a finite subfamily such that the closures of its sets cover X.*

(D) *Every ultrafilter is θ-convergent.*

(E) *Every family of sets in X such that the θ-closures of its members have an empty intersection contains a finite subfamily with empty intersection.*

Proof. (A) $\rightarrow$ (D). Let X be H-closed but let there exist a ultrafilter $\mathfrak{U}$ not containing any point of θ-contact (i.e. not θ-convergent). For every point $x \in X$ we chose an open neighborhood Ox such that $[Ox]$ does not intersect any set $\Phi_x \in \mathfrak{U}$. From the covering of X by open sets Ox we select a finite system $\{Ox_i\}_1^n$ such that $\bigcup_{i=1}^n [Ox_i] = X$. Then $\bigcap_{i=1}^n \Phi_{x_i} = \emptyset$ (where $\emptyset$ is the empty set), since $X \setminus \Phi_{x_i} \supseteq [Ox_i]$ for every $i = 1, \cdots, n$, in contradiction to the definition of a filter.

(D) $\rightarrow$ (B). Every filter is contained in some ultrafilter. It only remains to apply Lemma 1.

(B) $\rightarrow$ (E). The negation of (E) implies that the given family generates a filter not containing any point of θ-contact, i.e. implies the negation of (B).

(E) $\rightarrow$ (C). Let $\{\Gamma_\alpha\}_{\alpha \in I}$ be a system of sets such that $\bigcup_{\alpha \in I} \langle[\Gamma_\alpha]\rangle = X$ ($\langle P \rangle$ is the interior of the set P in the space X). The θ-closures of the members of the system $\{X \setminus [\Gamma_\alpha]\}_{\alpha \in I}$ have an empty intersection (since $[X \setminus [\Gamma_\alpha]]_\theta = [X \setminus [\Gamma_\alpha]] = X \setminus \langle[\Gamma_\alpha]\rangle$), and therefore this system contains a finite subsystem $\{X \setminus [\Gamma_{\alpha_i}]\}_1^n$ with empty intersection. Then

$$X \setminus \left(\bigcap_{i=1}^n X \setminus [\Gamma_{\alpha_i}] \right) = X = \bigcup_{i=1}^n X \setminus (X \setminus [\Gamma_{\alpha_i}]) = \bigcup_{i=1}^n [\Gamma_{\alpha_i}].$$

(C) $\rightarrow$ (A). Every open covering of the space X satisfies the condition: the nuclei of the closures of the sets cover X. The theorem is proved.

Corollary. *Every θ-closed set P in an H-closed space X is an H-subspace.*

By an H-subspace L of a space X we here mean a set $L \subseteq X$ with the following property: if the system $\{T_\alpha\}_{\alpha \in I}$ of open sets in X is such that $L \subseteq \bigcup_{\alpha \in I} T_\alpha$, then it is possible to choose a finite subsystem $\{T_{\alpha_i}\}_1^n$ such that $L \subset \bigcup_{i=1}^n [T_{\alpha_i}]$.

Theorem 2. *In an H-closed, completely separable space X the class of H-subspaces coincides with the class of θ-closed sets.*

Proof. Every θ-closed set in X is an H-subspace (Corollary of Theorem 1).

Now let P be an H-subspace in X and let $x \in X \setminus P$. For every point $y \in P$ we select an open neighborhood Oy such that $[Oy]$ does not intersect the closure of some neighborhood Oy_y of the point x. From the covering of P by the neighborhoods Oy we select a finite system $\{Oy\}_1^n$ such that $\bigcup_{i=1}^n [Oy_i] \supseteq P$. We set $Ox = \bigcap_{i=1}^n Ox_{y_i}$. Then $[Ox] = [\bigcap_{i=1}^n Ox_{y_i}] \subseteq \bigcap_{i=1}^n [Ox_{y_i}]$ and $[Ox] \cap P = \emptyset$, since $(\bigcup_{i=1}^n [Oy_i]) \cap ([\bigcap_{i=1}^n [Ox_{y_i}]) = \emptyset$. The theorem is proved.

Lemma 2. *For the filter $\mathfrak{F}$ to be θ-convergent in an H-closed, completely separable space X it is necessary and sufficient that this filter have a unique point of θ-contact.*

Proof. Let the filter $\mathfrak{F}$ be θ-convergent in X to the point x. For every $y \neq x$ there exists a neighborhood Oy such that $[Oy]$ does not intersect some closed neighborhood of the point x, and therefore y cannot be a point of the filter $\mathfrak{F}$.

Conversely, let x be the unique point of θ-contact of the filter $\mathfrak{F}$, and let Ox be an arbitrary neighborhood of x. If $[Ox]$ did not completely contain any set of the filter, then the trace $\mathfrak{F}_B$ of the filter $\mathfrak{F}$ on the set $B - [X \setminus [O_x]]$ would be a filter in B which would have a point of θ-contact in B (since the set B is H-closed), and this point would obviously be a point of θ-contact of the filter $\mathfrak{F}$ distinct from x. The lemma is proved.

§2

Theorem 3. *For the θ-continuous mapping f of an everywhere dense set S of the space X into an H-closed, completely separable space Y to be extendable to the whole of X with preservation of its θ-continuity, it is necessary and sufficient that for every two nonintersecting θ-closed sets B_1 and B_2 in Y the preimages $f^{-1}(B_1)$ and $f^{-1}(B_2)$ have nonintersecting closures in X.*

Proof. The necessity is clear from the following remark: if $x \in [B]$, $B \subseteq X$, and the mapping f of X into Y is θ-continuous at the point x, then $f(x) \in [f(B)]_\theta$.

Sufficiency. Let $x \in X$, and let $\mathfrak{F}$ be the trace on S of the filter of neighborhoods $\mathfrak{U}(x)$ of the point x. Then $f(\mathfrak{F})$, since it is the basis of a filter in Y, has points of θ-contact. Let us show that $\bigcap [f(V_\alpha)]_\theta$ consists of the one point y (where $V_\alpha = S \cap W_\alpha$, and W_α runs through $\mathfrak{U}(x)$). Let $y_1 \neq y_2$; $y_1, y_2 \in \bigcap [f(V_\alpha)]_\theta$. We chose open neighborhoods Oy_1 and Oy_2 of the points y_1 and y_2 such that $[Oy_1] \cap [Oy_2] = \emptyset$ (as is possible, since Y is completely separable). Since $[Oy_1]$ and $[Oy_2]$ are θ-closed (Theorem 2), therefore $[f^{-1}([Oy_1])] \cap [f^{-1}([Oy_2])] = \emptyset$; but since every $f(V_\alpha)$ intersects both $[Oy_1]$ and $[Oy_2]$, therefore every neighborhood of x intersects both $f^{-1}([Oy_1])$ and $f^{-1}([Oy_2])$, which means that $x \in [f^{-1}([Oy_1])] \cap [f^{-1}([Oy_2])]$. The contradiction thus obtained shows that $\bigcap [f(V_\alpha)]_\theta$ consists of the one point y. By Lemma 2 we see that $f(\mathfrak{F})$ is θ-convergent at the point y. Let us set $\tilde{f}(x) = y$. It is clear that $\tilde{f}(x) = f(x)$ on S. We now show that $\tilde{f}$ is a θ-continuous extension of the mapping f onto X. For this purpose we choose an arbitrary open neighborhood Oy of the point $y = \tilde{f}(x)$.

There exists an open neighborhood W_α of the point x such that $f(V_\alpha) \subseteq [Oy]$, where $V_\alpha = W_\alpha \cap S$. For an arbitrary point $z \in W_\alpha$ we have $\tilde{f}(z) \in [f(V_\alpha)]_\theta \subseteq [[Oy]]_\theta = [Oy]$ (since $z \in [V_\alpha]$), or in other words $\tilde{f}(W_\alpha) \subseteq [Oy]$. The theorem is proved.

Theorem 1 of A. D. Taĭmanov follows at once from what we have just proved, if we take into account that a θ-continuous mapping of the space X into a regular space Y is continuous and the classes of closed sets and θ-closed set coincide in a regular space.

Theorem 4. *For the θ-continuous mapping f on an everywhere dense set S of the space X into a θ-closed completely separable space Y to be extendable to the whole of X with preservation of the θ-continuity, it is necessary and sufficient that the image of the trace on S of the filter of neighborhoods of every point $x \in X$ have in Y a unique point of θ-contact.*

Proof. Necessity. From the definition of θ-continuity of the mapping F at the point $x \in X$ it follows that $f(\mathfrak{B}(x))$ is θ-convergent at the point $y = f(x)$ (where $\mathfrak{B}(x)$ is the filter of neighborhoods of the point x); consequently, y is the unique point of θ-contact of $f(\mathfrak{B}(x))$ (Lemma 2) and $f(\mathfrak{F})$, where $\mathfrak{F}$ is the trace of $\mathfrak{B}(x)$ on S.

Sufficiency. Again $\mathfrak{F}$ is the trace on S of the filter of neighborhoods $\mathfrak{B}(x)$ of the point $x \in X$; by Lemma 2 and the condition of the theorem, $f(\mathfrak{F})$

is θ-convergent to a certain point $y \in Y$. Setting $\tilde{f}(x) = y$ and arguing as in the proof of Theorem 3, we see that $\tilde{f}$ is a θ-continuous extension of f to the whole of X.

Corollary. *A continuous mapping f of an everywhere dense set S of a space Y into a bicompact has a continuous extension to X if and only if the image of the trace on S of the filter of neighborhoods of an arbitrary point $x \in X$ has an unique point of contact in Y.*

Theorem 5. *In order that a continuous mapping f of an everywhere dense set S in a space X into a regular space Y be extendable to X with preservation of continuity, it is necessary and sufficient that for every system of closed sets in Y with empty intersection the intersection of the closures in X of the preimages of the sets of the systems be empty.*

Proof. Necessity. If the system $\{\Phi_\alpha\}_{\alpha\in I}$ of closed sets in Y is such that $\bigcap_{\alpha\in I}\Phi_\alpha = \emptyset$, but $\bigcap_{\alpha\in I}[f^{-1}(\Phi_\alpha)] \neq \emptyset$, then f cannot be extended to $\bigcup_{\alpha\in I}[f^{-1}(\Phi_\alpha)]$.

Sufficiency. Let $x \in X$, and let $\mathfrak{F}$ be the image on S of the filter of neighborhoods $\mathfrak{B}(x)$ of the point x. The system of closed sets $\{[f(V_\alpha)]\}$ (where $V_\alpha = W_\alpha \cap S$, and W_α runs through $\mathfrak{B}(x)$) has a nonempty intersection in Y, since otherwise the system $\{[f^{-1}([f(V_\alpha)])]\}$ would have an empty intersection in X, in contradiction to the fact that x is a point of contact of $\mathfrak{F}$. We show that $\bigcap_{V_\alpha\in F}[f(V_\alpha)]$ consists of a unique point y. Let $y_1 \neq y_2$, with $y_1, y_2 \in \bigcap[f(V_\alpha)]$. Making use of the regularity of Y, we chose open neighborhoods Oy_1 and Oy_2 of the points y_1 and y_2 such that $[Oy_1] \cap [Oy_2] = \emptyset$. As in the proof of Theorem 3, it turns out that $x \in [f^{-1}([Oy_1])] \cap [f^{-1}([Oy_2])]$, in contradiction to the condition of the theorem. We show that $f(\mathfrak{F})$ converges to the point y. Let W be a arbitrary open neighborhood of the point y. If no set of $f(\mathfrak{F})$ lies entirely in W, then the system $\{Y \setminus (W \cap f(V_\alpha))\}_{V_\alpha\in\mathfrak{F}}$ consists of nonempty sets and has an empty intersection (in view of the fact that $f(\mathfrak{F})$ has only one point of contact). Setting

$$V_\alpha' = f^{-1}(Y \setminus (W \cap [f(V_\alpha)])), \tag{1}$$

we have $\bigcap[V_\alpha'] = \emptyset$. This means that there exists a neighborhood W_{α_1} of the point x and a set V_{α_0} such that $W_{\alpha_1} \cap V'_{\alpha_0} = \emptyset$; but then we would have $f(V_{\alpha_0} \cap V_{\alpha_1}) \subseteq W$ (where $V_{\alpha_1} = S \cap W_{\alpha_1}$, V_{α_0} is one of the sets corresponding to V'_{α_0} in the formula (1) and $V_{\alpha_0} \cap V_{\alpha_1} \in \mathfrak{F}$), in contradiction to our assumption. We can then complete the proof by making use of the corollary of Theorem 1 from [3], Russian p. 71.

Theorem 6. *A continuous mapping f of an everywhere dense set S of a space X into a final-compact (in the sense of coverings) regular space Y has a continuous extension to X if and only if, for every countable system of closed sets in Y with empty intersection, the intersection of the closures in X of the preimages of the sets of the system is empty.*

The proof of the necessity is the same as in the case of Theorem 5.

Sufficiency. Let us show that the conditions of Theorem 5 are satisfied. Let $\{\Phi_\alpha\}_{\alpha\in I}$ be an arbitrary system of closed sets Y with empty intersection. Then, since Y is final-compact, we may select from it a countable subset $\{\Phi_{\alpha_n}\}$ with empty intersection. Since $\bigcap_{n=1}^{\infty}[f^{-1}(\Phi_{\alpha_n})] = \emptyset$, therefore $\bigcap_{\alpha\in I}[f^{-1}(\Phi_\alpha)]^{\Gamma} = \emptyset$, so that everything is proved.

Theorem 2 of [1] can also be generalized.

We note that if a mapping f of a space X into a space Y is θ-continuous, than the preimage of every θ-closed set in Y is closed in X.

Definition 3. Let X and Y be two spaces. Two nonintersecting sets B_1 and B_2 contained in X are said to be Y_θ-separable if there exists a θ-continuous mapping from X into Y such that B_1 and B_2 are preimages of θ-closed sets in Y.

Theorem 7. *In order that every θ-continuous mapping of an everywhere dense set S of the space X into an H-closed, completely separable space Y may admit a θ-continuous extension to X, it is necessary and sufficient that every pair of Y_θ-separable sets S have nonintersecting closure in X.*

Proof. Necessity. If some pair of Y_θ-separable sets in S had intersecting closures in X, then a function F for which they were preimages of θ-closed sets in Y could not have a θ-continuous extension to X.

Sufficiency. Let f be an arbitrary θ-continuous mapping of S into Y. The images of two arbitrary nonintersecting θ-closed sets in Y are Y_θ-separable sets in S and have nonintersecting closures in X. The theorem is proved.

Corollary. *Every θ-continuous (or continuous, respectively) mapping of a Hausdorff space X into an H-closed, completely separable space Y (or into a bicompact) can be extended with preservation of the θ-continuity (or of the continuity) to every combinatorial, and in particular to every hypercombinatorial,*[1)] *H-closed extension of the space X.*

---

1) An extension cX of a space X is said to be combinatorial if the fact that a family of closed sets $\{F_i\}$, $i = 1, \cdots, n$, in X has as empty intersection implies that the intersection of the closures of the sets of the family is empty in cX, and it is

BIBLIOGRAPHY

[1] A. D. Taĭmanov, *On extension of continuous mappings of topological spaces*, Mat. Sb. **31** (73) (1952), 459–463. (Russian) MR 14, 395.

[2] S. Fomin, *Extensions of topological spaces*, Ann. of Math. (2) 44 (1943), 471–480. MR 5, 45.

[3] N. Bourbaki, *Topologie générale*. Chaps. I–III, 2ième ed., Actualités Sci. Indust., nos. 1142, 1143, Hermann, Paris, 1951; Russian transl., Fizmatgiz, Moscow, 1958. MR 3, 55; 5, 102.

Translated by:
S. H. Gould

---

said to be hypercombinatorial if the fact that the intersection of the sets of such a system $\{F_i\}$, $i = 1, \cdots, n$ is nowhere dense in X implies that $\bigcap_{i=1}^n F_i = \bigcap_{i=1}^n [F_i]_{cX}$ (where $[P]_{cX}$ is the closure of the set $P \subseteq X$ in the space cX).

Amer. Math. Soc. Transl.
(2) Vol. 92, 1970

ON PONTRJAGIN CLASSES. I*†

WU WEN-TSÜN

A sphere bundle $\mathfrak{B}$ (or a vector bundle in the sense of [1]) with orthogonal group or general linear group as structural group has as invariants a system of integral cohomology classes in the base space, besides the Stiefel-Whitney classes which are essentially mod. 2, a fact which was first discovered by Pontrjagin in 1942 [2]. Among these integral classes some of dimensions multiples of 4 are of particular importance which we have called Pontrjagin classes, denoted by $P^{4k}(\mathfrak{B})$, of the given bundle. We know little about these classes, and the fundamental problem about the topological invariance of these classes in the case of a differentiable manifold is not yet settled in spite of the fact that the same problem about the Stiefel-Whitney classes has been solved completely in quite a neat manner [3, 4].

According to the original definition of Pontrjagin these classes are defined only through the intervention of grassmannian manifolds and are therefore not intrinsically defined in the bundle. The present paper furnishes two different intrinsic definitions of these classes which will serve as the starting point of our investigations about their topological invariance in the case of a differentiable manifold.

We will adopt the following notation. For any integral homology or cohomology class A, let A_0 and A_p (p an integer > 1) denote respectively the real class and the mod p class obtained from A by reductions; the reductions themselves will be denoted by $\rho_0: A \to A_0$ and $\rho_p: A \to A_p$ respectively. We denote also by I, I_0, I_p the (additive) groups of integers, real numbers, and integers mod p respectively and $\rho_0: I \to I_0$, $\rho_p: I \to I_p$ the corresponding reductions.

I. The (real) grassmannian manifold $R_{n,m}$

Let $R_{n,m}$ be the grassmannian manifold of all m-elements in R^{n+m}. Let $\Omega_{n,m}$ be the set of all Schubert symbols $\mathrm{a} = (a_1, \cdots, a_m)$, where $0 \le a_1 \le \cdots \le a_m \le n$. For such an a we will put $\mathrm{a}_{(i)} = (a_1, \cdots, a_{i-1}, a_i - 1, a_{i+1}, \cdots, a_m)$ if $a_i > a_{i-1}$ and $\mathrm{a}^{(i)} = (a_1, \cdots, a_{i-1}, a_i + 1, a_{i+1}, \cdots, a_m)$

* First published in Chinese in Acta Mathematica Sinica, Vol. 3 (1953), 291–315.
† A facsimile of the translation in Scientia Sinica 3 (1954), 353–367.

if $a_i < a_{i+1}$; otherwise $\mathrm{a}_{(i)}$, $\mathrm{a}^{(i)}$ will be undefined. We call $D(\mathrm{a}) = \Sigma_{i=1}^m a_i$ the dimension of a. Suppose a system of coordinates in R^{n+m} is given, then a certain canonical subdivision of $R_{n,m}$ may be so defined that to each $\mathrm{a} \in \Omega_{n,m}$ there corresponds in the subdivision an open cell [a] of dimension $D(\mathrm{a})$. With these cells oriented in a definite manner we have the following boundary formulas:

$$\partial[\mathrm{a}] = \Sigma \epsilon_i (1 + (-1)^{a_i + i + m}) [\mathrm{a}_{(i)}], \ \epsilon_i = (-1)^{a_1 + \cdots + a_i}, \tag{1}$$

the summation being over all i for which $\mathrm{a}_{(i)}$ is defined. Let {a} be the cochain taking value 1 on [a] but 0 on all other cells with same dimension, then dually we have

$$\delta\{\mathrm{a}\} = \Sigma \eta_i (1 + (-1)^{a_i + i + m + 1}) \{\mathrm{a}^{(i)}\}, \ \eta_i = (-1)^{a_1 + \cdots + a_i - 1}, \tag{1$'$}$$

the summation being over all i for which $\mathrm{a}^{(i)}$ is defined.

Let

$$\begin{aligned} \mathfrak{p}_{4k} &= (\underbrace{0, \cdots, 0}_{m-2k}, \underbrace{2, \cdots, 2}_{2k}), \quad 0 \leq 2k \leq m, \\ \overline{\mathfrak{p}}_{4k} &= (\underbrace{0, \cdots, 0}_{m-2}, 2k, 2k), \quad 0 \leq 2k \leq n. \end{aligned} \tag{2}$$

Then $\{\mathfrak{p}_{4k}\}$ and $\{\overline{\mathfrak{p}}_{4k}\}$ are integral cocycles of dimension $4k$ whose classes will be denoted by P^{4k} and $\overline{P}^{4k}$ respectively. These classes are independent of coordinates chosen in R^{n+m} and will be called P-classes and $\overline{P}$-classes in $R_{n,m}$ respectively.

Chains of the form [a], $\mathrm{a} \in \Omega_{n,m}$ will be called elementary chains. The set of all elementary chains of same dimension will be arranged in an order $<$ such that

$$[a_1, \cdots, a_m] < [b_1, \cdots, b_m]$$

if and only if for some index i we have $a_m = b_m$, $a_{m-1} = b_{m-1}, \cdots, a_{i+1} = b_{i+1}$, $a_i < b_i$. Any chain $\mathfrak{x}$ may then be expressed in the "normal" form

$$\mathfrak{x} = \sum_{i=1}^k \lambda_i \mathfrak{x}_i, \ \lambda_i \neq 0,$$

in which all $\mathfrak{x}_i$ are elementary chains and $\mathfrak{x}_1 < \cdots < \mathfrak{x}_k$; then $\mathfrak{x}_1$ will be called the leading term of $\mathfrak{x}$. The set of all r-dimensional chains will be arranged in an order $<$ such that for any two such chains $\mathfrak{x} = \Sigma_{i=1}^k \lambda_i \mathfrak{x}_i$, $\mathfrak{x}' = \Sigma_{i=1}^{k'} \lambda'_i \mathfrak{x}'_i$, in normal forms we have $\mathfrak{x} < \mathfrak{x}'$ if and only if for some index i we have $\mathfrak{x}_1 = \mathfrak{x}'_1, \cdots, \mathfrak{x}_{i-1} = \mathfrak{x}'_{i-1}$ and either $\mathfrak{x}_i < \mathfrak{x}'_i$ or $i - 1 = k < k'$.

Let us express an elementary chain $\mathfrak{x}$ in the form

$$[0, \cdots, 0, \underbrace{b_s, \cdots, b_s}_{c_s}, \cdots, \underbrace{b_2, \cdots, b_2}_{c_2}, \underbrace{b_1, \cdots, b_1}_{c_1}], \tag{2}$$

where $0 < b_s < \cdots < b_2 < b_1$, and divide the set Ω^r of all r-dimensional elementary chains into three subsets Ω^r_i, $i = 0, 1, 2$ as follows: We put $\mathfrak{x}$ in Ω^r_0 if all b, c are even, while $\mathfrak{x} \in \Omega^r_1$ or Ω^r_2 according as the first odd integer in the sequence $(b_1, c_1, b_2, c_2, \cdots, b_s, c_s)$ happens among b's or c's. We say that a chain adheres to Ω^r_i if in the normal form all its terms are elementary chains of $\Omega^r_i (i = 0, 1, 2)$. It is easy to see that

(i) all chains adhering to Ω^r_0 are integral cycles.

(ii) no chain adhering to Ω^r_2 is an integral cycle, and the leading term of its boundary belongs to Ω^{r-1}_1.

Theorem 1. *If $r < n$ or $r = n =$ even, $C_r(R_{n,m})$ has a basis consisting of chains of the forms $\mathfrak{x}'$, $\frac{1}{2}\partial\mathfrak{x}''$, $\mathfrak{x}'''$ where $\mathfrak{x}'$, $\mathfrak{x}''$, $\mathfrak{x}'''$ are elementary chains of Ω^r_0, Ω^{r+1}_2 and Ω^r_2 respectively. If $r = n =$ odd, besides chains of the forms $\mathfrak{x}'$, $\frac{1}{2}\partial\mathfrak{x}''$, $\mathfrak{x}'''$ we should add the integral cycle $\mathfrak{x}_0 = [0, \cdots, 0, n]$ to form a basis of $C_r(R_{n,m})$.*

Proof. Let $\mathfrak{x} = \sum_{i=1}^k \lambda_i \mathfrak{x}_i \in C_r(R_{n,m})$, $r \leq n$, be in the normal form. If $r = n =$ odd and $\mathfrak{x}_1 = [0, \cdots, 0, n] = \mathfrak{x}_0$, then $\mathfrak{x} = \mathfrak{x}_0 \in \Omega^r_1$. Suppose that this is not so. If $\mathfrak{x}_1 \in \Omega^r_0$ or Ω^r_2, then $\mathfrak{x} - \mathfrak{x}_1$ is either 0 or $> \mathfrak{x}$. If $\mathfrak{x}_1 \in \Omega^r_1$, and $\mathfrak{x}_1$ is of the form (2) with b_j as the first odd integer in the sequence $(b_1, c_1, \cdots, b_s, c_s)$, we will take

$$\mathfrak{x}''_1 = [0, \cdots, 0, \underbrace{b_s, \cdots, b_s}_{c_s}, \cdots, \underbrace{b_j, \cdots, b_j}_{c_j - 1}, b_j + 1,$$

$$\underbrace{b_{j-1}, \cdots, b_{j-1}}_{c_{j-1}}, \cdots, \underbrace{b_1, \cdots, b_1}_{c_1}] \in \Omega^{r+1}_2.$$

We have then in normal form $\partial\mathfrak{x}''_1 = 2\sum_{i=1}^l \mu_i \mathfrak{y}_i$, with $\mu_i = \pm 1$ and $\mathfrak{y}_1 = \mathfrak{x}_1$. Hence $\mathfrak{x} - \mu_1 \cdot \frac{1}{2}\partial\mathfrak{x}''_1$ is either 0 or $> \mathfrak{x}$. It follows that by subtracting successively from $\mathfrak{x}$ chains of the form $\mathfrak{x}'$, $\frac{1}{2}\partial\mathfrak{x}''$, $\mathfrak{x}'''$ as stated in the theorem we get finally either 0 or $\mathfrak{x}_0$ in the case $r = n =$ odd. This proves our theorem.

From Theorem 1 and B we get easily the following

Corollary 1. *If $r < n$, then $Z_r(R_{n,m})$ has a basis consisting of cycles of the form $\mathfrak{x}'$, $\frac{1}{2}\partial\mathfrak{x}''$, where $\mathfrak{x}' \in \Omega_0^r$, $\mathfrak{x}'' \in \Omega_2^{r+1}$, for which $\mathfrak{x}'$ are of infinite order while $\frac{1}{2}\partial\mathfrak{x}''$ are of order* 2.

Corollary 2. *If $r < n$ then $H_r(R_{n,m}, I_p)$, p = odd or 0, has a basis with the elementary chains of Ω_0^r reduced* mod p *as the representative cycles. In particular, $H_r(R_{n,m}, I_p) = 0$ for $r < n$ and $r \not\equiv 0$* mod 4.

II. A certain canonical mapping $l: R_{n,m} \to C_{n,m}$

Let U^{n+m} be a unitary space of complex dimension $n+m$. The space $C_{n,m}$ of all complex m-elements in U^{n+m} is the (complex) grassmannian manifold defined in U^{n+m}. It is known that [5] the ring $H^*(C_{n,m})$ is generated by a system of classes $C^{2i} \in H^{2i}(C_{n,m})$, $0 \le i \le m$, which we will call C-classes of $C_{n,m}$. The C-classes $\overline{C}^{2i}$, $0 \le i \le n$, are then defined by the equations $\sum_{i+j=k} C^{2i} \cup \overline{C}^{2j} = 0$ or unit class 1, according as $k > 0$ or $k = 0$.

In the unitary space U^{n+m} and the euclidean space R^{n+m}, let us choose systems of complex and real coordinates $(z_1, \cdots, z_{n+m})$ and $(x_1, \cdots, x_{n+m})$, respectively. The equations $z_i = x_i$, $i = 1, \cdots, n+m$, then define a linear mapping

$$l: \begin{cases} R^{n+m} \to U^{n+m}, \\ (x_1, \cdots, x_{n+m}) \to (z_1, \cdots, z_{n+m}), \end{cases}$$

which induce naturally a canonical mapping

$$l: R_{n,m} \to C_{n,m}.$$

Though the definition of l depends on the choice of coordinates in R^{n+m} and U^{n+m}, it is easy to see that the homotopy class of l is independent of such choices.

Theorem 2. *Let $r = 0$ or an odd prime p_ν. Then*

$$l^* C_r^{2k} = \begin{cases} 0, & k \text{ odd}, \\ (-1)^{k/2} P_r^{2k}, & k \text{ even}, \end{cases} \quad (0 \le k \le m); \tag{1}$$

$$l^* \overline{C}_r^{2k} = \begin{cases} 0, & k \text{ odd}, \\ \overline{P}_r^{2k}, & k \text{ even}, \end{cases} \quad (0 \le k \le n). \tag{1$'$}$$

Proof. The case k odd being evident, we will suppose in what follows that k is even. We suppose also that $n > 2k$.

Let $\mathfrak{Z}_k^* \subset C_{n,m}$ be defined by the set of complex m-elements $\{Z\}$ satisfying

$$\text{Comp. dim}\,(Z \cap U^{n+k-1}) \geq k, \tag{2}$$

where U^{n+k-1} is an arbitrarily chosen $(n+k-1)$-element in U^{n+m}. Let $\mathfrak{Z}_k$ be the cycle represented by the naturally oriented $\mathfrak{Z}_k^*$ and $\mathfrak{X}_k$ the cycle $[\underbrace{0, \cdots, 0}_{m-k}, \underbrace{2, \cdots, 2}_{k}]$ defined in the canonical subdivision with respect to the given coordinate system $(x_1, \cdots, x_{n+m})$ of R^{n+m}. By Corollary 2 of §I, the second equation of (1) is equivalent to the following intersection formulas (I_r denotes the real or mod p index of intersection with respect to the natural orientation of $C_{n,m}$):

$$I_r(\mathfrak{Z}_k, l\mathfrak{X}) = 0, \quad \mathfrak{X} \in \Omega_0^{2k}, \ \mathfrak{X} \neq \mathfrak{X}_k; \tag{3}$$

$$I_r(\mathfrak{Z}_k, l\mathfrak{X}_k) = (-1)^{k/2}. \tag{3$'$}$$

For the proof of (3), let us consider a cycle $\mathfrak{X} = [a_1, \cdots, a_m] \in \Omega_0^{2k}$, $\mathfrak{X} \neq \mathfrak{X}_k$. Since $\sum_{i=1}^m a_i = 2k$ and each a_i is even, we must have $a_1 = \cdots = a_{m-k+1} = 0$. We know that $\mathfrak{X}$ may be geometrically represented by the pseudomanifold $\mathfrak{X}^*$ of which any m-element X should satisfy the condition

$$X \supset R^{m-k+1}, \tag{4}$$

where R^s is the element $x_{s+1} = \cdots = x_{n+m} = 0$. Since $\overline{l}(R^{m-k+1})$ is a complex $(m-k+1)$-element of U^{n+m}, we may choose in U^{n+m} a complex $(n+k-1)$-element U^{n+k-1} which intersects $l(\overline{R}^{m-k+1})$ in the origin of U^{n+m} only and determines $\mathfrak{Z}_k^*$ by (2). Comparing (2) and (4), we see easily that $\mathfrak{Z}_k^*$ and $l(\mathfrak{X}^*)$ can have no common points and (3) is thus proved.

To prove (3)$'$ let us remark that the cycle $\mathfrak{X}_k$ is geometrically represented by the pseudomanifold $\mathfrak{X}_k^*$ which is defined by

$$R^{m-k} \subset X \subset R^{m+2}.$$

Any such X has equations of the form

$$\left.\begin{aligned} &x_{m+3} = \cdots = x_{n+m} = 0, \\ &\sum_{i=m-k+1}^{m+2} a_{ji} x_i = 0, \quad j = 1, 2, \end{aligned}\right\} \tag{5}$$

in which a_{ji} are real numbers and the last two equations are linearly independent.

Define now $\mathfrak{z}_k^*$ by (2) with U^{n+k-1} as follows:

$$z_1 = \cdots = z_{m-k} = 0, \quad z_{m-k+1} = iz_{m-k+2}. \tag{6}$$

It is easy to see that there is only one element $X_0 \in \mathfrak{x}_k^*$ for which $l(x_0) = Z_0$ is common to $\mathfrak{z}_k^*$ and $l(\mathfrak{x}_k^*)$, namely the element given by the following equations:

$$x_{m-k+1} = x_{m-k+2} = 0, \quad x_{m+3} = \cdots = x_{n+m} = 0. \tag{7}$$

To determine the index of intersection of $l\,\mathfrak{x}_k^*$ and $\mathfrak{z}_k^*$ at Z_0 we take in U^{n+m} a complex coordinate system $(z'_1, \cdots, z'_{n+m})$ related to the previous system $(z_1, \cdots, z_{n+m})$ in the following manner:

$$\left.\begin{aligned} z'_i &= z_{m+i+2}, \quad i = 1, \cdots, n-2; \\ z'_{n-1} &= z_{m-k+2}; \\ z'_i &= z_{-n+m+i-k+3}, \quad i = n, \cdots, n+k-1; \\ z'_i &= z_{i-n-k}, \quad i = n+k+1, \cdots, n+m; \\ z'_{n+k} &= z_{m-k+1} - iz_{m-k+2}. \end{aligned}\right\} \tag{8}$$

By (6) and (7), U^{n+k-1} and $l(X_0) = Z_0$ are then given respectively by the following systems of equations:

$$U^{n+k-1}\colon z'_{n+k} = \cdots = z'_{n+m} = 0, \tag{9}$$

$$l(X_0) = Z_0\colon z'_1 = \cdots = z'_{n-1} = 0, \ z'_{n+k} = 0. \tag{10}$$

In $C_{n,m}$ any complex m-element neighboring Z_0 may be represented by the system of equations:

$$z'_j = \sum_{i=n}^{n+k-1} \zeta_{ji} z'_i + \sum_{i=n+k+1}^{n+m} \zeta_{ji} z'_i, \ j = 1, \cdots, n-1, n+k,$$

in which $\zeta_{ji} = \zeta'_{ji} + i\,\zeta''_{ji}$, with ζ', ζ'' real numbers. Such elements form an open cell L of dimension $2nm$, of which ζ_{ji} may be regarded as a system of complex coordinates. The cell L becomes then a unitary space with a natural orientation concordant with that of $C_{n,m}$. By (2), (9), $L \cap \mathfrak{z}_k^* = L_1$ is a complex element of L with equations

$$\zeta_{n+k,n} = \cdots = \zeta_{n+k,n+k-1} = 0. \tag{11}$$

On the other hand, by (5) and (7), the m-elements $X \in \mathfrak{x}_k^*$ in the neighborhood of X_0 may be represented by the following system of equations:

$$\left.\begin{aligned} x_j &= \sum_{i=m-k+3}^{m+2} \xi_{ji} x_i, \quad j = m-k+1,\ m-k+2, \\ x_j &= 0, \quad j = m+3, \cdots, n+m, \end{aligned}\right\} \tag{12}$$

where ξ_{ji} are real numbers. By (8), the equations of $l(X)$ for such X are

$$\left.\begin{aligned} & z_j' = 0,\ j = 1, \cdots, n-2; \\ & z_{n-1}' = \xi_{m-k+2,m-k+3} z_n' + \cdots + \xi_{m-k+2,m+2} z_{n+k-1}'; \\ & z_{n+k}' + iz_{n-1}' = \xi_{m-k+1,m-k+3} z_n' + \cdots + \xi_{m-k+1,m+2} z_{n+k-1}'. \end{aligned}\right\} \tag{13}$$

Such $l(X)$ form a complex element $L_2 = l(\mathfrak{x}_k^*) \cap L$ in the unitary space L. The elements L_1 and L_2 meet only in the origin $l(X_0) = Z_0$ of L, at which the index of intersection is easily seen to be $(-1)^{k/2}$. This proves (3)′ in the case $n > 2k$. As the restriction $n > 2k$ is easily removed, (1) is completely proved.

(1)′ may be proved similarly or deduced from (1) by considering the canonical mappings $R_{n,m} \to R_{m,n}$ and $C_{n,m} \to C_{m,n}$.

Theorem 3. *Let W^i ($\overline{W}^i$) be the* (mod 2) *W-classes ($\overline{W}$-classes) of $R_{n,m}$ and C_2^{2i} ($\overline{C}_2^{2i}$) the* mod 2 *C-classes ($\overline{C}$-classes) of $C_{n,m}$. If l: $R_{n,m} \to C_{n,m}$ is the canonical mapping mentioned before, then*

$$l^* C_2^{2i} = (W^i)^2, \quad l^* \overline{C}_2^{2i} = (\overline{W}^i)^2. \tag{14}$$

Proof. Using the same notation as before, let R^{2n+2m} be the euclidean space determined by U^{n+m}, with a (real) coordinate system $(x_1', \cdots, x_{n+m}', x_1'', \cdots, x_{n+m}'')$ such that $z_i = x_i' + ix_i''$, $i = 1, \cdots, n+m$. Consider R^{2n+2m} as a direct sum $R_1^{n+m} \oplus R_2^{n+m}$, where R_k^{n+m} is determined by $(x_1^{(k)}, \cdots, x_{n+m}^{(k)})$. Let $R_{n,m}^{(k)}$ and $R_{2n,2m}$ be the grassmannian manifolds defined in R_k^{n+m} and R^{2n+2m} respectively. Between the various grassmannian manifolds we have the following canonical mappings.

$$\begin{aligned} l&: R_{n,m} \to C_{n,m}, \\ k&: C_{n,m} \to R_{2n,2m}, \\ h&: R_{n,m}' \times R_{n,m}'' \to R_{2n,2m}, \end{aligned}$$

defined respectively as follows: for $Z \in C_{n,m}$, $X^{(k)} \in R_{n,m}^{(k)}$, $k(Z)$ is the $2m$-element *in* R^{2n+2m} determined by Z, $h(X', X'') = X' \oplus X''$ in R^{2n+2m}, and l is as before. Besides, the equations $x_i^{(k)} = x_i$, $i = 1, \cdots, n+m$; $k = 1, 2$ determine a linear mapping

$$j_k = \begin{cases} R^{n+m} \to R_k^{n+m}, \\ (x_1, \cdots, x_{n+m}) \to (x_1^{(k)}, \cdots, x_{n+m}^{(k)}), \end{cases}$$

which induces a mapping

$$j_k\colon R_{n,m} \to R_{n,m}^{(k)} \quad (k = 1, 2).$$

For $X \in R_{n,m}$ let $j(X) = j_1(X) \times j_2(X)$; then we have a mapping

$$j\colon R_{n,m} \to R'_{n,m} \times R''_{n,m}.$$

Let the W-classes (mod 2) in $R_{n,m}^{(k)}$, $R_{2n,2m}$ be respectively W_k^i $(k = 1, 2)$, W'^i; then

$$j^*(W_1^{i_1} \times W_2^{i_2}) = j_1^* W_1^{i_1} \cup j_2^* W_2^{i_2} = W^{i_1} W^{i_2}. \tag{15}$$

Since we have known [6] that

$$\begin{aligned} hj &= kl, \\ k^* W'^{2i} &= C_2^{2i}, \\ h^* W'^{2i} &= \sum_{i_1 + i_2 = 2i} W_1^{i_1} \otimes W_2^{i_2}, \end{aligned}$$

it follows easily from $l^* C_2^{2i} = l^* k^* W^{2i} = j^* h^* W^{2i}$ that

$$l^* C_2^{2i} = (W^i)^2, \tag{16}$$

which is the first formula of (14). Similarly we may prove the second formula of (14). This concludes Theorem 3.

Theorem 4.

$$l^* C^{2k} = \begin{cases} \frac{1}{2}\, \delta(S_q^{k-1} W^k) & k \text{ odd}, \\ (-1)^{k/2} P^{2k} + \frac{1}{2}\, \delta(W^{k-1} W^k), & k \text{ even}. \end{cases} \quad (k \le m) \tag{17}$$

$$l^* \overline{C}^{2k} = \begin{cases} \frac{1}{2}\, \delta(S_q^{k-1} \overline{W}^k), & k \text{ odd}, \\ \overline{P}^{2k} + \frac{1}{2}\, \delta(\overline{W}^{k-1} \overline{W}^k), & k \text{ even}. \end{cases} \quad (k \le n) \tag{17$'$}$$

Proof. Let $\mathbf{a}_i = (0, \cdots, 0, i) \in \Omega_{n,m}$ and $\mathbf{a}_{j,i} = (0, \cdots, 0, j, i) \in \Omega_{n,m}$, where $j \le i$. By (1)$'$ of §1, we have

$$\begin{aligned} &\delta\{\mathbf{a}_{2j,4i-2j-1}\} = \pm 2\{\mathbf{a}_{2j,4i-2j}\} \pm 2\{\mathbf{a}_{2j+1,4i-2j-1}\}. \\ &\delta\{\mathbf{a}_{2j-1,4i-2j}\} = 0. \end{aligned}$$

On the other hand, we know that [5]

$$\{a_{2i-1}\} \cup \{a_{2i}\} \sim \sum_{j=0}^{2i-1} \{a_{j,4i-j-1}\} \bmod 2,$$

$$\{a_{2i}\} \cup \{a_{2i}\} \sim \sum_{j=0}^{2i} \{a_{j,4i-j}\} \bmod 2.$$

It follows that $\rho_2 \cdot \frac{1}{2}\, \delta(\overline{W}^{2i-1}\, \overline{W}^{2i})$ contains a cocycle of the form

$$\rho^2 \sum_{j=0}^{i-1} (\{a_{2j,4i-2j}\} + \{a_{2j+1,4i-2j-1}\}) = \rho_2 \sum_{j=0}^{2i-1} \{a_{j,4i-j}\}$$

$$= \rho_2 \sum_{j=0}^{2i} \{a_{j,4i-j}\} + \rho_2 \{a_{2i,2i}\},$$

which belongs to $\overline{W}^{2i} \cup \overline{W}^{2i} + \rho_2 \overline{P}^{4i}$. Hence

$$\rho_2 \left[\overline{P}^{4i} + \frac{1}{2}\, \delta(\overline{W}^{2i-1}\, \overline{W}^{2i})\right] = \overline{W}^{2i} \cup \overline{W}^{2i}.$$

By Theorem 3, we have then

$$\rho_2 \left[l^* \overline{C}^{4i} - \overline{P}^{4i} - \frac{1}{2}\, \delta(\overline{W}^{2i-1}\, \overline{W}^{2i})\right] = 0. \tag{18}$$

By Theorem 2, we have also

$$\rho_0 \left[l^* \overline{C}^{4i} - \overline{P}^{4i} - \frac{1}{2}\, \delta(\overline{W}^{2i-1}\, \overline{W}^{2i})\right] = 0. \tag{19}$$

As $R_{n,m}$ has only torsion number 2, we have from (18) and (19)

$$l^* \overline{C}^{4i} - \overline{P}^{4i} - \frac{1}{2}\, \delta(\overline{W}^{2i-1}\, \overline{W}^{2i}) = 0, \tag{20}$$

which is the second formula of (17)$'$. The other formulas may be proved in the similar manner.

III. An intrinsic definition of Pontrjagin classes

For any m-vector bundle $\mathfrak{B}$ over a (finite) complex K we may form the Whitney product $\mathfrak{B} \cup \mathfrak{B}$ over the same base space K and turn $\mathfrak{B} \cup \mathfrak{B}$ into a "unitary bundle" as follows. Let the director space of $\mathfrak{B}$ be R^m, the fiber in $\mathfrak{B}$ over $p \in K$ be R^m_p, and the coordinate system in a coordinate neighborhood σ of K be $\xi_{\sigma p}: R^m \to R^m_p$ $(p \in \sigma)$. Take two copies R^m_i of R^m and R^m_{pi} $(i = 1, 2)$ of each R^m_p, realized by $\omega_i: R^m \to R^m_i$, $R^m_p \to R^m_{p,i}$. According to the definition of Whitney product, the director space of $\mathfrak{B} \cup \mathfrak{B}$ and the fiber over $p \in K$ in $\mathfrak{B} \cup \mathfrak{B}$ are respectively $R^{2m} = R^m_1 \oplus R^m_2$ and

$R_p^{2m} = R_{p,1}^m \oplus R_{p,2}^m$ with coordinate system $\zeta_{\sigma p}: R^{2m} \to R_p^{2m}$, $(p \in \sigma)$ in σ defined by

$$\zeta_{\sigma p}(\omega_1 \mathfrak{x}_1 + \omega_2 \mathfrak{x}_2) = \omega_1 \xi_{\sigma p}(\mathfrak{x}_1) + \omega_2 \xi_{\sigma p}(\mathfrak{x}_2),\ \mathfrak{x}_1, \mathfrak{x}_2 \in R^m,\ p \in \sigma. \quad (1)$$

Now for each complex number $c = a + ib$ (a and b are real) and any $\mathfrak{x}_1$, $\mathfrak{x}_2 \in R^m$, let us define a multiplication as follows:

$$c(\omega_1 \mathfrak{x}_1 + \omega_2 \mathfrak{x}_2) = a(\omega_1 \mathfrak{x}_1 + \omega_2 \mathfrak{x}_2) + ib(\omega_1 \mathfrak{x}_1 + \omega_2 \mathfrak{x}_2),$$
$$a(\omega_1 \mathfrak{x}_1 + \omega_2 \mathfrak{x}_2) = a\,\omega_1 \mathfrak{x}_1 + a\,\omega_2 \mathfrak{x}_2,$$
$$i(\omega_1 \mathfrak{x}_1 + \omega_2 \mathfrak{x}_2) = -\omega_1 \mathfrak{x}_2 + \omega_2 \mathfrak{x}_1.$$

Under this multiplication $R^{2m} = R_1^m \oplus R_2^m$ becomes canonically a unitary space of complex dimension m, which will be denoted by U^m. Similarly, for each $p \in K$, we may endow R_p^{2m} with a natural complex structure to turn it into a unitary space U_p^m of complex dimension m. For any two overlapping coordinate neighborhoods σ, τ of K and any point $p \in \sigma \cap \tau$, we have by (1),

$$i\,\zeta_{\tau p}^{-1}\,\zeta_{\sigma p}(\omega_1 \mathfrak{x}_1 + \omega_2 \mathfrak{x}_2) = \zeta_{\tau p}^{-1}\,\zeta_{\sigma p}\, i(\omega_1 \mathfrak{x}_1 + \omega_2 \mathfrak{x}_2),\ \mathfrak{x}_1, \mathfrak{x}_2 \in R^m.$$

It follows that, with respect to any complex coordinate system in U^m, $\zeta_{\tau p}^{-1}\,\zeta_{\sigma p}$ is a nonsingular linear transformation on m complex variables, and thus under $\zeta_{\sigma p}$, $\mathfrak{B} \cup \mathfrak{B}$ becomes an m-unitary bundle $\mathfrak{B}^*$. Evidently this bundle $\mathfrak{B}^*$ depends intrinsically on $\mathfrak{B}$ and will be called the canonical unitary bundle associated to $\mathfrak{B}$.

Now let $\mathfrak{R}_{n,m}$ and $\mathfrak{C}_{n,m}$ be respectively the universal m-vector bundle and m-unitary bundle canonically defined over the grassmannian manifolds $R_{n,m}$ and $C_{n,m}$. Suppose that $\mathfrak{B} = \alpha^* \mathfrak{R}_{n,m}$ where $\alpha: K \to R_{n,m}$ is a continuous mapping and n is sufficiently large. Then it is easy to see that $\mathfrak{B}^*$ is induced from $\mathfrak{C}_{n,m}$ by $l\alpha = \beta: K \to C_{n,m}$ where $l: R_{n,m} \to C_{n,m}$ is defined in §II, i.e. $\mathfrak{B}^* \sim \beta^* \mathfrak{C}_{n,m}$. By definition, the Pontrjagin classes $P^{4k}(\mathfrak{B})$ of the vector bundle $\mathfrak{B}$ and the Chern classes $C^{2i}(\mathfrak{B}^*)$ of the unitary bundle $\mathfrak{B}^*$ are respectively given by

$$P^{4k}(\mathfrak{B}) = \alpha^* P^{4k}, \quad (2)$$

$$C^{2i}(\mathfrak{B}^*) = \beta^* C^{2i}, \quad (3)$$

where P^{4k} and C^{2i} are respectively the P-classes of $R_{n,m}$ and C-classes of $C_{n,m}$. Similarly for the dual Pontrjagin classes and the dual Chern classes:

$$\overline{P}^{4k}(\mathfrak{B}) = \alpha^* \overline{P}^{4k}, \tag{2'}$$

$$\overline{C}^{2i}(\mathfrak{B}) = \beta^* \overline{C}^{2i}. \tag{3'}$$

The definitions of these classes are not intrinsic owing to the intervention of grassmannian manifolds. For the Chern classes we may define them intrinsically as the obstruction classes in Steenrod's sense of certain associated bundles of the given unitary bundle. The Pontrjagin classes may be defined intrinsically in a similar manner, as show by Chern and Rohlin, but with some complication. The purpose of this section is to give some relatively simple intrinsic definitions of Pontrjagin classes by reducing them to Chern classes as follows. By Theorems 2 and 4 of §II and the formulas (2), (3) above, we get easily

$$P_r^{4k}(\mathfrak{B}) = (-1)^k C_r^{4k}(\mathfrak{B}^*), \quad r = 0 \text{ or an odd prime}, \quad 0 \leq 2k \leq m, \tag{4}$$

$$P^{4k}(\mathfrak{B}) = (-1)^k C^{4k}(\mathfrak{B}^*) + \frac{1}{2}\delta[W^{2k-1}(\mathfrak{B}) \cup W^{2k}(\mathfrak{B})], \quad 0 \leq 2k \leq m. \tag{5}$$

Similar expressions may be obtained for $\overline{P}^{4k}(\mathfrak{B})$ and $\overline{P}_r^{4k}(\mathfrak{B})$. As $\mathfrak{B}^*$, $C^{2i}(\mathfrak{B}^*)$ and $W^i(\mathfrak{B})$ are all intrinsically defined from $\mathfrak{B}$, (4) and (5) furnish us the following

Intrinsic definition of Pontrjagin classes. *For a given m-vector bundle $\mathfrak{B}$, we may endow naturally the Whitney product $\mathfrak{B} \cup \mathfrak{B}$ of $\mathfrak{B}$ by itself with a unitary structure to turn it into a unitary bundle $\mathfrak{B}^*$. The Pontrjagin classes of $\mathfrak{B}$ are then intrinsically defined in terms of the Chern classes $C^{2i}(\mathfrak{B}^*)$ of $\mathfrak{B}^*$ and the Stiefel-Whitney classes $W^i(\mathfrak{B})$ of $\mathfrak{B}$ by the formulas* (4) *and* (5).

IV. Certain formulas about characteristic classes

Let A be the bundle space of an m-unitary bundle $\mathfrak{S}$, with base space a finite complex K and projection $\omega: A \to K$. In each fiber U_p^m over $p \in K$ in $\mathfrak{S}$ may be defined a (complex) grassmannian manifold $C_{m-1,1}(p)$. All of these manifolds form naturally a space B which is the bundle space of a fiber bundle $\mathfrak{R}$ associated with $\mathfrak{S}$, the base being K, the fiber over $p \in K$, $C_{m-1,1}(p)$, and projection, $\pi: B \to K$. Let A_1 be the subspace of $A \times B$ formed by all points (a, b) such that $\omega a = \pi b (= p)$, and $a, b \in C_{m-1,1}(p)$; then A_1 is the bundle space of a certain complex line bundle (or l-unitary bundle) $\mathfrak{T}$ with B as base.

Theorem 5. *In the space B we have*

$$\sum_{i=0}^{m} (-1)^i \pi^* C^{2i}(\mathfrak{S}) \cup [C^2(\mathfrak{T})]^{m-i} = 0. \tag{1}$$

Moreover, the classes $C^{2i}(\mathfrak{S})$ *are uniquely determined by* (1) *in the sense that if*

$$\sum_{i=0}^{m} (-1)^i \pi^* C'^{2i} \cup [C^2(\mathfrak{T})]^{m-i} = 0,$$

where $C'^0 = C^0(\mathfrak{S}) = 1$ *is the unit class in* K, *then* $C'^{2i} = C^{2i}(\mathfrak{S})$ *for* $i = 0, 1, \cdots, m$.

Proof. Let us consider the m-unitary bundle $\pi^*\mathfrak{S}$ over the base B. The bundle space Γ of this bundle is a subspace of $A \times B$, consisting of all points $(a, b) \in A \times B$ such that $\omega a = \pi b$, the projection $\pi'\colon \Gamma \to B$ being defined by $\tilde{\pi}(a, b) = b$. In $\pi^*\mathfrak{S}$ the fiber over $b \in B$ is an unitary space U_b^m of complex dimension m whose points (a, b) satisfy $a \in U_p^m$ and $p = \pi(b)$. Now b represents a complex line of U_p^m through its origin. The complex $(m-1)$-element in U_p^m completely orthogonal to b will be denoted by $\bar{b}$. In Γ the set of all points (a, b), where $a \in \bar{b}$, is naturally the bundle space A_2 of an $(m-1)$-unitary bundle $\bar{\mathfrak{T}}$ with B as base space. As $b \oplus \bar{b} = U_p^m (p = \pi(b))$, it turns out that $\pi^*\mathfrak{S}$ is the Whitney product of $\mathfrak{T}$ and $\bar{\mathfrak{T}}$, i.e. $\pi^*\mathfrak{S} \sim \mathfrak{T} \cup \bar{\mathfrak{T}}$. By the duality theorem for product bundles [6] we have

$$\begin{aligned} &1 + C^2(\pi^*\mathfrak{S})\, t + \cdots + C^{2m}(\pi^*\mathfrak{S})\, t^m \\ &= [1 + C^2(\mathfrak{T})] \cup [1 + C^2(\bar{\mathfrak{T}})\, t + \cdots + C^{2m-2}(\bar{\mathfrak{T}})\, t^{m-1}], \end{aligned}$$

t being an indeterminate. Multiplying both sides by $\sum_{i=0}^{\infty} (-1)^i [C^2(\mathfrak{T})]^i t^i$, we get

$$\sum_{i=0}^{m} C^{2i}(\pi^*\mathfrak{S})\, t^i \cdot \sum_{i=0}^{\infty} (-1)^i [C^2(\mathfrak{T})]^i t^i = \sum_{i=0}^{m-1} C^{2i}(\bar{\mathfrak{T}})\, t^i.$$

The comparison of coefficients of t^m gives then

$$\sum_{i=0}^{m} C^{2i}(\pi^*\mathfrak{S}) \cup (-1)^{m-i} [C^2(\mathfrak{T})]^{m-i} = 0.$$

Since $C^{2i}(\pi^*\mathfrak{S}) = \pi^* C^{2i}(\mathfrak{S})$ we get (1).

We will give $C^2(\mathfrak{T})$ in (1) another interpretation as follows. Take a fixed point $p \in K$ and consider the fiber $C_{m-1,1}(p)$ in $\mathfrak{R}$ as a complex projective space K_{m-1} of complex dimension $m-1$. Then $H^*(C_{m-1,1}(p))$ is generated by $D^2 \in H^2(C_{m-1,1}(p))$ which takes the value $+1$ on a complex line of K_{m-1}, with a single relation $(D^2)^m = 0$. On the other hand, $C_{m-1,1}(p)$ is a grassmannian manifold defined in U_p^m with a C-class C^2. It is easy to see that $C^2 = D^2$. Next, let $\mathfrak{C}_{m-1,1}(p)$ be the canonical 1-unitary bundle over $C_{m-1,1}(p)$; then $C^2 = C^2(\mathfrak{C}_{m-1,1}(p))$, while the points of the bundle space $\widetilde{C}_{m-1,1}(p)$ of $\mathfrak{C}_{m-1,1}(p)$ are of the form (a, b) where $b \in C_{m-1,1}(p)$, $a \in \bar{b} \subset U_p^m$. If $(a, b) \in \widetilde{C}_{m-1,1}(p)$, then $\pi\, b = \omega a = p$, which means $(a, b) \in A_1$. Hence we may define identical mappings $\tilde{i}: \widetilde{C}_{m-1,1}(p) \subset A_1$ and $i: C_{m-1,1}(p) \subset B$. Evidently $\tau \tilde{i} = i\, \tau_p$, where τ_p and τ are projections of $\mathfrak{C}_{m-1,1}(p)$ and $\mathfrak{L}$ respectively. Hence we have $\mathfrak{C}_{m-1,1}(p) \sim i^* \mathfrak{L}$, whence

$$i^* C^2(\mathfrak{L}) = C^2(i^* \mathfrak{L}) = C^2(\mathfrak{C}_{m-1,1}(p)) = C^2 = D^2.$$

It follows that the bundle $\mathfrak{R}$ over K has the property that its fiber is "totally nonhomologous in the bundle space B". The uniqueness of $C^{2i}(\mathfrak{S})$ occurring in (1) follows then from a theorem of Leray-Hirsch. (Cf. e.g. [9])

Let us remark that, the bundle $\mathfrak{L}$ being intrinsically associated with the given bundle $\mathfrak{S}$, the above theorem gives an intrinsic definition of $C^{2i}(\mathfrak{S})$. Now combining with (4) or (5) of §III, we get an alternative intrinsic definition of Pontrjagin classes given by

Theorem 6. *For each m-vector bundle $\mathfrak{V}$ over a finite complex let us form the Whitney product $\mathfrak{V} \cup \mathfrak{V}$ of $\mathfrak{V}$ with itself and endow it with a natural unitary structure to turn it into a unitary bundle $\mathfrak{V}^*$. Let B be the bundle space of the $C_{m-1,1}$-bundle $\mathfrak{R}$ associated with $\mathfrak{V}^*$, and $\mathfrak{L}$ the complex line bundle naturally defined over B as base space. Then the Pontrjagin classes $P^{4k}(\mathfrak{V})$ are uniquely determined by the following formulas in the space B:*

$$\sum_{k=0}^{m'} (-1)^k \pi^* P_r^{4k}(\mathfrak{V}) \cup [C_r^2(\mathfrak{L})]^{m-2k} = 0, \tag{2}$$

$$\sum_{k=0}^{m'} \pi^* [(-1)^k P^{4k}(\mathfrak{V}) + \frac{1}{2}\delta(W^{2k-1}(\mathfrak{V}) \cup W^{2k}(\mathfrak{V}))] \cup [C^2(\mathfrak{L})]^{m-2k} = 0, \tag{3}$$

in which m' denotes the greatest integer $\leq m/2$, r is either 0 or an odd prime, and π is the projection of $\mathfrak{R}$.

In the case of Stiefel-Whitney classes we have a theorem analogous to Theorem 5 as follows.

Theorem 5′. *Let $\mathfrak{V}$ be an m-vector bundle over a finite complex K. Let $\mathfrak{P}$ be a bundle associated with $\mathfrak{V}$ with m-dimensional projective space as fibers. Over the bundle space B of $\mathfrak{P}$ we may naturally define a 1-vector bundle $\mathfrak{L}$. Then the (reduced) Stiefel-Whitney classes $W^i(\mathfrak{V})$ of $\mathfrak{V}$ are uniquely determined by the following formula:*

$$\sum_{i=0}^{m'} \pi^* W^i(\mathfrak{B}) \cup [W^1(\mathfrak{S})]^{m-i} = 0, \tag{4}$$

π being the projection of the bundle $\mathfrak{P}$.

The proof of Theorem $5'$ is quite analogous to that of Theorem 5 and is therefore omitted. We remark only that both these two theorems were first discovered by G. Hirsch [9], a little before the author in 1950.

BIBLIOGRAPHY

[1] Wu Wen-tsün, *On the product of sphere bundles and the duality theorem modulo two,* Ann. of Math. (2) 49 (1948), 641–653. MR 10, 203.

[2] L. S. Pontrjagin, *Characteristic cycles on manifolds,* C. R. (Dokl.) Acad. Sci. URSS 35 (1942), 34–37. MR 4, 147.

[3] R. Thom, *Espaces fibrés en sphères et carrés de Steenrod,* Ann. Sci. École Norm. Sup. (3) 69 (1952), 109–182. MR 14, 1004.

[4] Wu Wen-tsün, *Classes caractéristiques et i-carrés d'une variété,* C. R. Acad. Sci. Paris 230 (1950), 508–511. MR 12, 42.

[5] S. Chern, *On the multiplication in the characteristic ring of a sphere bundle,* Ann. of Math. (2) 49 (1948), 362–372. MR 9, 456.

[6] Wu Wen-tsün, *Sur les classes caractéristiques des structures fibrées sphériques,* Actualités Sci. Indust., no. 1183, Hermann, Paris, 1952. MR 14, 1112.

[7] V. A. Rohlin, *Intrinsic description of Pontrjagin's characteristic cycles,* Dokl. Akad. Nauk SSSR 84 (1952), 449–452. (Russian) MR 14, 306.

[8] E. H. Spanier, *Homology theory of fiber bundles,* Proc. Internat. Congress Math. (Cambridge, Mass., 1950), vol. 2, Amer. Math. Soc., Providence, R.I., 1952, pp. 390–396. MR 13, 486.

[9] G. Hirsch, *Sur la structure multiplicative de l'anneau de cohomologie d'un espace fibré,* C. R. Acad. Sci. Paris 230 (1950), 46–48. MR 11, 379.

Amer. Math. Soc. Transl.
(2) Vol. 92, 1970

ON PONTRJAGIN CLASSES. II*

WU WEN-TSÜN

Abstract. The present paper, as well as the preceding one with the same title [6], contains the detailed proofs of the results sketched in [1], §§5 and 6. It is proved that for a closed differentiable manifold, for any odd prime p, certain cup product combinations of the Pontrjagin classes, reduced mod p, with respect to any given differential structure of the manifold are in reality independent of that structure and therefore are topological invariants of the manifold. For $p = 3$, it turns out that the mod 3 Pontrjagin classes are themselves topological invariants of the manifold. The proofs are based on the so-called "diagonal method", obtaining thus a connection between the Pontrjagin classes and the Steenrod powers, of which the latter, in the case of a differentiable manifold, are of a non-differential character, being purely topological.

Introduction

This paper contains the detailed proof of the results sketched in [1], §§5 and 6.

Let M be an m-dimensional differentiable manifold, $\mathfrak{D}$ a differential structure. The tangent planes at each point of M determine an m-Euclidean space bundle with respect to $\mathfrak{D}$, $\mathfrak{T}(\mathfrak{D})$. By definition, all classes in $\mathfrak{T}(\mathfrak{D})$ depend not only on M, but also on the differential structure $\mathfrak{D}$. Regarding Stiefel-Whitney classes, the author pointed out at the end of 1949 that $W^i(\mathfrak{T}(\mathfrak{D}))$ is related to the square (operation) in M, and proved the following formula:

$$[W^2(\mathfrak{T}(\mathfrak{D})) + W^1(\mathfrak{T}(\mathfrak{D})) \cup W^1(\mathfrak{T}(\mathfrak{D}))] \cup X = Sq^2 X,$$

$$X \in H^{m-2}(M, \bmod 2).$$

Hence we know that $W^2(\mathfrak{T}(\mathfrak{D}))$ can be determined completely by the homology properties of M. As a consequence, $W^2(\mathfrak{T}(\mathfrak{D}))$ depends only on the topological structure of M, [1)] and has nothing to do with the differential structure

* Translation of Acta Math. Sinica 4 (1954), 171–199.

1) $W^1(\mathfrak{T}(\mathfrak{D}))$ is determined by the orientability of M; $W^m(\mathfrak{T}(\mathfrak{D}))$ is determined by the Euler-Poincaré characteristic of M. Hence when $k = 1$ or m, it is obvious that $W^k(\mathfrak{T}(\mathfrak{D}))$ is a topological invariant. But when $1 < k < m$ the problem is not so simple.

$\mathfrak{D}$ in M. In other words, $W^2(\mathfrak{T}(\mathfrak{D}))$ is a topological invariant.

From the above result, Thom proved the topological invariance of all the Stiefel-Whitney classes $W^i(\mathfrak{T}(\mathfrak{D}))$, $i \geq 1$ [2, 3, 4]. The author obtained the formulas of these classes by using the cup product and square in M [5].

Whether or not the above results can be generalized to Pontrjagin classes is a natural question. In 1950–1951, the author proved that in an oriented, closed, differentiable manifold, a Pontrjagin class (mod 3) is a topological invariant. Let p be a prime ≥ 3; then after being reduced mod p, a certain structure in Pontrjagin classes, obtained by using the cup product method, is a topological invariant. 1) The proofs of these results are the main content of this paper.

In the case of Stiefel-Whitney classes, the proof of the topological invariance depends on the following formula, due to Thom:

$$Sq^i \tau^* 1 = \tau^* W^i(\mathfrak{B})$$

(see §4 for details). Thom's proof cannot be applied to Pontrjagin classes. In this paper we give a new proof of this formula. A similar proof can be used to derive a characteristic formula for Pontrjagin classes, and then to obtain some of the above-mentioned results on the topological invariance of Pontrjagin classes in differentiable manifolds.

The general problem of the topological invariance of Pontrjagin classes in differentiable manifolds has not been solved. Apparently some new tools are required in order to solve this problem completely. Nevertheless, we can point out that the nature of Pontrjagin classes lies deeper than that of Stiefel-Whitney classes. The understanding of its nature will yield a significant contribution to the development of topology.

We need the result in [6].

§1. Homology groups in locally compact spaces 2)

In this section we briefly discuss homology systems and their properties. From now on the coefficient group of all homology groups will be a fixed abelian group.

For any compact space pair (X, A); that is, a compact space X and a closed subspace A of X, $H^k_+(X, A)$ denotes the k-dimensional Čech cohomol-

1) It follows that a four-dimensional Pontrjagin class in an orientable four-dimensional differentiable manifold takes values of multiples of 3 on the oriented manifold.

2) Cf. [7]–[10].

ogy group in (X, A).

Let (X, A) be a locally compact space pair; that is, a locally compact space X and a closed subspace A of X. Take a point ∞ not belonging to X; let X^+ be the Aleksandrov compactification of X produced by adding ∞ to X. Let $A^+ = A \cup \infty$. Define

$$H^k(X, A) = H^k_+(X^+, A^+). \tag{1}$$

With this definition, U^- represents the element in $H^k(X, A)$ corresponding to $U \in H^k_+(X^+, A^+)$, and U^+ represents the element in $H^k_+(X^+, A^+)$ corresponding to $U \in H^k(X, A)$. In particular, if (X, A) is a compact space pair, X^+ and A^+ are the union of X and ∞, A and ∞ respectively. Hence $H^k(X, A) \approx H^k_+(X, A)$.

From now on we shall write $(X, A) \in \mathfrak{C}$ or $(X, A) \in \mathfrak{LC}$ depending on whether (X, A) is a compact space pair or a locally compact space pair. Let ∞ be the point outside the locally compact spaces in discussion.

Let $(X, A), (Y, B) \in \mathfrak{LC}$. If the map $f\colon (X, A) \to (Y, B)$ has the property that the inverse image of a compact set in Y is a compact set in X, then f is called an admissible map, denoted by $f \in \mathfrak{A}$. Extend f to f^+: $(X^+, A^+) \to (Y^+, B^+)$ by $f^+(X) = f$, $f^+(\infty) = \infty$. Define $f^*\colon H^k(Y, B) \to H^k(X, A)$ as f^{+*}. $H^k_+(Y^+, B^+) \to H^k_+(X^+, A^+)$. In other words,

$$f^* U = (f^{+*} U^+)^-, \quad U \in H^k(Y, B). \tag{2}$$

The above homology system H has the following properties:

$H1^\circ$. Let $(X, A) \in \mathfrak{LC}$, and define $i^+\colon (X^+, A^+) \to ((X-A)^+, \emptyset^+)$ as $i^+|(X-A) =$ identity map; $i^+(A^+) = \emptyset^+ (= \infty)$. Then i^+ is continuous and $i^{+*}\colon H^k_+((X-A)^+, \emptyset^+) \approx H^k_+(X^+, A^+)$. From this we define a homology transformation $i^*\colon H^k(X-A) \approx H^k(X, A)$. Here

$$i^* U = (i^{+*} U^+)^-, \quad U \in H^k(X-A). \tag{3}$$

Generally speaking if $(X, A), (Y, B) \in \mathfrak{LC}$, and $f\colon (X, A) \to (Y, B)$ is admissible, while $f|(X-A)$ is a homeomorphism (in brief: f is a relative homeomorphism), then $f^*\colon H^k(Y, B) \approx H^k(X, A)$.

$H2^\circ$. Suppose $X' \in \mathfrak{LC}$ and X is an open set in X'. Define k^+: $(X'^+, \emptyset^+) \to (X^+, \emptyset^+)$ as $k^+|X =$ identity map, and $k^+((X'-X) \cup \infty) = \infty$. Then $k^{+*}\colon H^k_+(X^+, \emptyset^+) \to H^k_+(X'^+, \emptyset^+)$ defines a determinant pseudo-homomorphic transformation $k^*\colon H^k(X) \to H^k(X')$ as follows:

$$k^* U = (k^{*+} U^+)^-, \ U \in H^k(X). \tag{4}$$

Let j: $X' \subset (X', X' - X)$; then obviously $k^* = j^* i^*$, where i^*: $H^k(X) \approx H^k(X', X' - X)$, as in $H1°$.

$H3°$. For every (X, A) there exists an exact sequence

$$\cdots \to H^k(X, A) \xrightarrow{j_k^*} H^k(X) \xrightarrow{i_k^*} H^k(A) \xrightarrow{\delta_k^*} H^{k+1}(X, A) \to \cdots$$

where each transformation agrees with the corresponding transformation of the exact sequence in $(X^+, A^+, \emptyset^+)$:

$$\cdots \to H_+^k(X^+, A^+) \xrightarrow{j_k^{+*}} H_+^k(X^+, \emptyset^+) \xrightarrow{i_k^{+*}} H_+^k(A^+, \emptyset^+) \xrightarrow{\delta_k^{+*}} H_+^{k+1}(X^+, A^+) \to \cdots,$$

and j_k^* and i_k^* can be regarded as induced by the inclusion relations j: $X \subset (X, A)$ and i: $A \subset X$.

$H4°$. H satisfies the Eilenberg-Steenrod axiom system for homology groups. The axioms of excision and exact sequences are mentioned in 1° and 3°.

$H5°$. If R^m is an m-dimensional Euclidean space, and G is a homology group, then

$$H^k(R^m) \approx \begin{cases} 0, & k \neq m, \\ G, & k = m. \end{cases}$$

$H6°$. If X has a cell decomposition K, [1] then $H^k(X)$ is isomorphic to the cohomology group derived from the finite cochain in K.

We shall discuss the ring structure in $H^*(X, A) = \Sigma_k H^*(X, A)$, where we assume the coefficient group to be a ring.

Suppose that $(X_1, A_1), (X_2, A_2) \in \mathfrak{C}$; then $(X, A) = (X_1, A_1) \times (X_2, A_2) = (X_1 \times X_2, A_1 \times X_2 \cup X_1 \times A_2)$. Take arbitrarily two finite open coverings $\alpha_1 = \{v_{j_1}, j_1 \in J_1\}$ and $\alpha_2 = \{v_{j_2}, j_2 \in J_2\}$, of (X_1, A_1) and (X_2, A_2) respectively. Construct a finite open covering $\alpha = \alpha_1 \times \alpha_2 = \{v_j, j \in J_1 \times J_2\}$ of (X, A), where $v_j = v_{j_1} \times v_{j_2}$, $j = (j_1, j_2)$. It is easy to see that these α's form a cofinal subsystem in the directed system formed by all finite open coverings of (X, A). Let the nerves of α_1, α_2 and α be $(K_{\alpha_1}, L_{\alpha_1})$, $(K_{\alpha_2}, L_{\alpha_2})$ and (K_α, L_α) respectively. For any simplex $\sigma \in (K_\alpha, L_\alpha)$,

1) Here the cell decomposition can be taken in the wide sense. See [7].

let its vertices be $v_{j_{10}} \times v_{j_{20}}, \cdots, v_{j_{1k}} \times v_{j_{2k}}$ $(j_{1i} \in J_1, j_{2i} \in J_2, i = 0, 1, \cdots, k)$; then $v_{j_{10}}, \cdots, v_{j_{1k}}$ determine a simplex σ_1 in K_{α_1} and $v_{j_{20}}, \cdots, v_{j_{2k}}$ determine a simplex σ_2 in K_{α_2}, while $\sigma \to \sigma_1 \times \sigma_2$ determines a cell correspondence i_α: $(K_\alpha, L_\alpha) \to (K_{\alpha_1}, L_{\alpha_1}) \times (K_{\alpha_2}, L_{\alpha_2}) = (K_{\alpha_1} \times K_{\alpha_2}, L_{\alpha_1} \times K_{\alpha_2} + K_{\alpha_1} \times L_{\alpha_2})$. Since the closure of $\sigma_1 \times \sigma_2$ possesses a simple homology, i_α induces a pseudo-homeomorphic transformation i_α^*: $H^*((K_{\alpha_1}, L_{\alpha_1}) \times (K_{\alpha_2}, L_{\alpha_2})) \to H^*(K_\alpha, L_\alpha)$.

Now let $U_1^+ \in H_+^*(X_1, A_1)$ and $U_2^+ \in H_+^*(X_2, A_2)$. Take the finite open coverings α_1 and α_2 of (X_1, A_1) and (X_2, A_2) such that U_1^+ and U_2^+ have representatives $u_{\alpha_1}^+$ and $u_{\alpha_2}^+$ in $H^*(K_{\alpha_1}, L_{\alpha_1})$ and $H^*(K_{\alpha_2}, L_{\alpha_2})$, respectively. Let $\alpha = \alpha_1 \times \alpha_2$, $u_\alpha^+ = i_\alpha^*(u_{\alpha_1}^+ \otimes u_{\alpha_2}^+) \in H^*(K_\alpha, L_\alpha)$, but $U^+ \in H_+^*(X, A)$ is the element represented by u_α^+. It is easy to prove that U^+ is independent of the choice of α_1 and α_2, and hence is determined uniquely by U_1^+ and U_2^+. We define $U^+ = U_1^+ \otimes U_2^+$.

Suppose $(X_1, A_1), (X_2, A_2) \in \mathfrak{L}\mathfrak{C}$; then $(X, A) = (X_1, A_1) \times (X_2, A_2) = (X_1 \times X_2, A_1 \times X_2 \cup X_1 \times A_2)$. Define

$$f: (X_1^+, A_1^+) \times (X_2^+, A_2^+) = (X_1^+ \times X_2^+, A_1^+ \times X_2^+ \cup X_1^+ \times A_2^+) \to (X^+, A^+)$$

as $f|X$ = identity map, and $f(\infty \times X_2^+ \cup X_1^+ \times \infty) = \infty$. We have that f is continuous and is a relative homeomorphism. By $H1°$, $f^*: H^*(X^+, A^+) \approx H^*((X_1^+, A_1^+) \times (X_2^+, A_2^+))$. For any $U_1 \in H^*(X_1, A_1)$ and $U_2 \in H^*(X_2, A_2)$, we define $U_1 \otimes U_2 \in H^*(X, A)$ as the element corresponding to $f^{*-1}(U_1^+ \otimes U_2^+)$ in $H^*(X^+, A^+)$:

$$U_1 \otimes U_2 = [f^{*-1}(U_1^+ \otimes U_2^+)]^-. \tag{5}$$

Suppose $(X, A), (X_1, A_1), (X_2, A_2) \in \mathfrak{L}\mathfrak{C}$, and every space is a subspace of the same space, and $X = X_1 \cap X_2$, $A = A_1 \cup A_2$. Let d: $(X, A) \to (X_1, A_1) \times (X_2, A_2)$ be defined as the diagonal map: $d(x) = x \times x$, $x \in X$. If d is admissible, then for any $U_1 \in H^*(X_1, A_1)$ and $U_2 \in H^*(X_2, A_2)$ we define $U_1 \cup U_2 \in H^*(X, A)$ as

$$U_1 \cup U_2 = d^*(U_1 \otimes U_2). \tag{6}$$

In particular, when $X = X_1$ or $X = X_2$, d must be admissible and is well defined on $U_1 \cup U_2$.

The above-defined tensor product $\otimes$ and cup product $\cup$ possess some familiar properties, along with the following not-so-familiar properties:

$P1°$. Let $(X_1, A_1), (X_2, A_2), (Y_1, B_1), (Y_2, B_2) \in \mathfrak{L}\mathfrak{C}$ and let

$$f_1\colon (X_1, A_1) \to (Y_1, B_1), f_2\colon (X_2, A_2) \to (Y_2, B_2)$$

be admissible. Let $(X, A) = (X_1, A_1) \times (X_2, A_2)$ and $(Y, B) = (Y_1, B_1) \times (Y_2, B_2)$. Define $f = f_1 \times f_2\colon (X, A) \to (Y, B)$ as $f(x_1, x_2) = f_1(x_1) \times f_2(x_2)$, $x_1 \in X_1$, $x_2 \in X_2$. Then f is also admissible. For any $U_1 \in H^*(Y_1, B_1)$ and $U_2 \in H^*(Y, B)$ we have

$$f^*(U_1 \otimes U_2) = (f_1 \times f_2)^*(U_1 \otimes U_2) = f_1^* U_1 \otimes f_2^* U_2. \tag{7}$$

$P2°$. Let $(X, A), (Y_1, B_1), (Y_2, B_2) \in \mathfrak{L}\mathfrak{C}$ and let $f_1\colon (X, A) \to (Y_1, B_1)$, and $f_2\colon (X, A) \to (Y_2, B_2)$ be admissable. Define $f = f_1 \cup f_2\colon (X, A) \to (Y_1, B_1) \times (Y_2, B_2)$ as $f(x) = f_1(x) \times f_2(x)$, $x \in X$. Then f is also admissible. For any $U_1 \in H^*(Y_1, B_1)$ and $U_2 \in H^*(Y_2, B_2)$ we have

$$f^*(U_1 \otimes U_2) = (f_1 \cup f_2)^*(U_1 \otimes U_2) = f_1^* U_1 \cup f_2^* U_2.$$

$P3°$. Let $X \in \mathfrak{L}\mathfrak{C}$, and let A and A' be open subsets of X with $A \subset A'$. Define k^* as in $H2°$: $H^*(A) \to H^*(A')$. Then for any $U \in H^*(A)$ and $V \in H^*(X)$ we have

$$k^*(U \cup V) = k^* U \cup V. \tag{8}$$

Proof. Since $d\colon A \to A \times X$ and $d'\colon A' \to A' \times X$ are admissible, they can be extended to $d^+\colon (A^+, \emptyset^+) \to ((A \times X)^+, \emptyset^+)$ and $d'^+\colon (A'^+, \emptyset^+) \to ((A' \times X)^+, \emptyset^+)$. Since A and $A \times X$ are open subsets of A' and $A' \times X$ respectively, by $H2°$ we may define $k^+\colon (A'^+, \emptyset^+) \to (A^+, \emptyset^+)$ and $k_1^+\colon ((A' \times X)^+, \emptyset^+) \to ((A \times X)^+, \emptyset^+)$. Also define $f\colon (A^+, \emptyset^+) \times (X^+, \emptyset^+) \to ((A \times X)^+, \emptyset^+)$ and $f'\colon (A'^+, \emptyset^+) \times (X^+, \emptyset^+) \to ((A' \times X)^+, \emptyset^+)$ as follows: $f|A \times X =$ identity map, $f(\infty \times X^+ \cup A^+ \times \infty) = \infty$, $f'|A' \times X =$ identity map, $f'(\infty \times X \cup A' \times \infty) = \infty$. From the definition of tensor product we obtain

$$(U \otimes V)^+ = f^{*-1}(U^+ \otimes V^+),$$

$$(k^* U \otimes V)^+ = f'^{*-1}((k^* U)^+ \otimes V^+) = f'^{*-1}(k^{+*-1}(k^{+*} U^+ \otimes V^+).$$

It is easy to see that $d^+ k^+ = k_1^+ d'^+$ and $k_1^+ f' = f(k^+ \times 1)$. Here $1\colon (X^+, \emptyset^+) \to (X^+, \emptyset)$ is the identity map. Hence, from the definition of cup product,

$$(U \cup V)^+ = [d^*(U \otimes V)]^+ = d^{+*}(U \otimes V)^+ = d^{+*} f^{*-1}(U^+ \otimes V^+),$$

$$(k^* U \cup V)^+ = [d'^*(k^* U \otimes V)]^+ = d'^{+*}(k^* U \otimes V)^+ = d'^{+*} f'^{*-1}(k^{+*} U^+ \otimes V^+)$$

$$= d'^{+*} f'^{*-1}(k^+ \times 1)^*(U^+ \otimes V^+) = d'^{+*} k_1^{+*} f^{*-1}(U^+ \otimes V^+)$$

$$= k^{+*} d^{+*} f^{*-1}(U^+ \otimes V^+).$$

(8) follows immediately on comparison of these two equalities.

$P4°$. Let A, X, $X' \in \mathfrak{L}\mathfrak{C}$, where A is an open subset of X', and X is a closed subset of X'. Also require that $A \subset X$ and $\bar{k}$: $X \subset X'$. Then for any $U \in H^*(A)$ and any $V' \in H^*(X')$ we have

$$U \cup \bar{k}^* V' = U \cup V'. \tag{9}$$

Proof. Let d: $A \to A \times X$ and d': $A \to A \times X'$ be diagonal maps. Let 1: $A \to A$ be the identity map. Then d, d', $1 \times \bar{k} \in \mathfrak{A}$ and $(1 \times \bar{k})d = d'$. From (6) and (7) we obtain

$$U \cup \bar{k}^* V' = d^*(U \otimes \bar{k}^* V') = d^*(1 \times \bar{k})^*(U \otimes V')$$

$$= d'^*(U \otimes V') = U \cup V'.$$

In the following let the coefficient group be the prime field of integers mod p (p is an odd prime). For any (geometric) compound pair (K, L), Steenrod has shown the existence of a set of pseudo-homomorphic transformations

$$St_p^i: H^q(K, L) \to H^{q+i}(K, L),$$

where $St_p^i = D_{k(p-1)-i}^p$. When $(X, A) \in \mathfrak{C}$, then for any $U \in H_+^*(X, A)$ take a finite open covering α of (X, A), with (K, L) as the nerve. Let u be the representative of U in $H^*(K, L)$. Then the class in (X, A) represented by $St_p^i u$ is independent of the choice of α and is defined as $St_p^i U \in H_+^{k+i}(X, A)$. When $(X, A) \in \mathfrak{L}\mathfrak{C}$ and $U \in H^*(X, A)$ we define

$$St_p^i U = (St_p^i U^+)^-. \tag{10}$$

Steenrod's power so defined has the properties $((X, A), (X', A') \in \mathfrak{L}(\mathfrak{C})$:

$S1°$. If f: $(X', A') \to (X, A)$ is admissible, then

$$f^* St_p^i = St_p^i f^*.$$

$S2°$. $St_p^i U = 0$, $i < 0$ or $i > k(p-1)$, $U \in H^k(X, A)$.

$S3°$. $St_p^{k(p-1)}U = (U)^p$, $U \in H^k(X, A)$.

$S4°$. $St_p^0 U = \lambda_k U$ and $U \in H^k(X, A)$

$$\lambda_k = (-1)^{qk(k-1)/2}(q!)^k, \quad q = (p-1)/2.$$

$S5°$. If $U \in H^k(A)$ and $\delta^*: H^k(A) \to H^{k+1}(X, A)$ is defined as in $H3°$, then $St_p^i \delta^* U = (-1)^{qk+i} q! \, \delta^* St_p^i U$.

$S6°$. $St_p^{2i+1} = (\delta/p) St_p^{2i}$.

$S7°$. If $U_1 \in H^*(X_1)$ and $U_2 \in H^*(X_2)$, then

$$St_p^{2i}(U_1 \otimes U_2) = \sum_{i_1+i_2=i} St_p^{2i_1} U_1 \otimes St_p^{2i_2} U_2.$$

$S8°$. If $U_1, U_2 \in H^*(X)$, then

$$St_p^{2i}(U_1 \cup U_2) = \sum_{i_1+i_2=i} St_p^{2i_1} U_1 \cup St_p^{2i_2} U_2.$$

$S9°$. If $X \subset X'$ and X is an open set in X', and $k^*: H^*(X) \to H^*(X')$ is defined as in $H2°$, then

$$k^* St_p^i = St_p^i k^*.$$

Proof of $S9°$. Let $U \in H^*(X)$. As in $H2°$, define $k^+: (X'^+, \emptyset^+) \to (X^+, \emptyset^+)$. From the definitions of k^* and St_p^i, i.e. (4), (10), and $S1°$, we have

$$k^* St_p^i U = k^*(St_p^i U^+)^- = (k^{+*} St_p^i U^+)^-$$

$$= (St_p^i k^{+*} U^+)^- = [St_p^i (k^* U)^+]^-$$

$$= St_p^i k^* U,$$

which is just the statement of $S9°$.

The square (operation) corresponding to $p = 2$ has similar properties with the exception of $S4°$, $7°$, $8°$, which have to be altered somewhat.

§2. Steenrod powers in Grassmann manifolds [1)]

Throughout this section p is a fixed odd prime. The coefficients of all polynomials and equations are taken from the field I_p.

Let $C_{n,m}$ be a complex Grassmann manifold, where the C-class of its integral coefficients is C^{2i}, $0 < i \leq m$. After reduction mod p, the mod p C-class is C_p^{2i}, $0 < i \leq m$. We assume that C^0 and C_p^0 have integral coef-

1) Cf. [1] and [11].

ficients and are the mod p unit class, $C^{2i} = 0$, $C_p^{2i} = 0$, $i > m$.

We know that $C_{n,m}$ does not possess torsion coefficients, and that the cohomology ring structure of its integral coefficients, induced by C^{2i} (C_p^{2i}), is similar to the mod 2 cohomology ring structure induced by W^i in the Grassmann manifold $R_{n,m}$. Similar to Theorem 2 and Lemmas 4, 5 and 6 in [12], we have the following:

Lemma 1. *The cohomology ring $H^*(C_{n,m}, I_p)$ can be induced by the* mod p *C-class C_p^{2i}, $0 < i \le m$ in $C_{n,m}$. These C-classes form a multiplicative basis for the collection of all elements with dimension $\le 2n$ in this ring. In other words, any $2k$-dimensional* mod p *cohomology class can be written as a polynomial of C_p^{2i}, $0 < i \le m$ in $C_{n,m}$ (the product is the cup product). When $k \le n$ this representation is unique.*

Lemma 2. *Let $n_1 + n_2 = n$ and $m_1 + m_2 > m$. Then there is a natural map*

$$h: C_{n_1,m_1} \times C_{n_2,m_2} \to C_{n,m}. \tag{1}$$

Let the mod p *C-classes be $C^{(j)2i}$ in C_{n_j,m_j}; then*

$$h^* C_p^{2i} = \sum_{i_1+i_2=i} C_p'^{2i_1} \otimes C_p''^{2i_2} \tag{2}$$

when $k \le n_1$ and n_2; and

$$h^*: H^{2k}(C_{n,m}, I_p) \subset H^{2k}(C_{n_1,m_1} \times C_{n_2,m_2}, I_p).$$

Now suppose $n \ge mp$. By Lemma 1, $St_p^{2r} C_p^{2s}$, $0 < s \le m$, $n \ge mp \ge r + s$, is a certain polynomial of C_p^{2i}, $0 < i \le m$, where the multiplication is the cup product; namely

$$St_p^{2r} C_p^{2s} = \phi_{r,s,m,n}(C_p^{2i}),\ 0 < s \le m,\ n \ge mp \ge r + s, \tag{3}$$

Here $\phi_{r,s,m,n}(c_i)$ is a polynomial in the c_i with weight i, $i = 1, 2, \cdots, m$, the weight of each term being $r + s$. If $n' > n \ge mp$, and $C_p'^{2i}$ is a mod p C-class in $C_{n',m}$, then with respect to the usual map $f: C_{n,m} \to C_{n',m}$ we have $f^* C_p'^{2i} = C_p^{2i}$, since

$$\begin{aligned}\phi_{r,s,m,n'}(C_p^{2i}) &= \phi_{r,s,m,n'}(f^* C_p'^{2i}) = f^* \phi_{r,s,m,n'}(C_p'^{2i}) = f^* St_p^{2r} C_p'^{2s} \\ &= St_p^{2r} f^* C_p'^{2s} = St_p^{2r} C_p^{2s} = \phi_{r,s,m,n}(C_p^{2i}).\end{aligned}$$

Hence only for n large enough ($\geq mp$) will the form of $\phi_{r,s,m,n}(c_i)$ be independent of n. We define

$$\phi_{r,s,m}(c_i) = \phi_{r,s,m,n}(c_i), \quad 0 < s \leq m, \quad n \geq mp \geq r + s; \tag{4}$$

$$\phi_{r,s,m}(c_i) = 0, \quad s > m \text{ or } mp < r + s; \tag{4$'$}$$

$$\phi_{r,s,m}(c_i) = \begin{cases} 1, & r = 0, \\ 0, & r > 0. \end{cases} \tag{4$''$}$$

Hence for n large enough, we have

$$St_p^{2r} C_p^{2s} = \phi_{r,s,m}(C_p^{2i}), \quad r, s \geq 0. \tag{3$'$}$$

By using the natural map $f\colon C_{n,m} \to C_{n',m}$ $(n' > n)$, we immediately see that the above equality is true for all n.

Now take positive integers n_1, n_2, m_1, m_2. Let $n_1 + n_2 = n$, $m_1 + m_2 = m$, $n_i \geq mp$; then with respect to the natural map h (see (1)), by Lemma 2 we have $(r + s \leq mp)$

$$h^* St_p^{2r} C_p^{2s} = St_p^{2r} h^* C_p^{2s} = St_p^{2r} \left[\sum_{s_1 + s_2 = s} C_p'^{2s_1} \otimes C_p''^{2s_2} \right]$$

$$= \sum_{s_1 + s_2 = s} St_p^{2r}(C_p'^{2s_1} \otimes C_p''^{2s_2}) = \sum_{\substack{s_1 + s_2 = s \\ r_1 + r_2 = r}} St_p^{2r_1} C_p'^{2s_1} \otimes St_p^{2r_2} C_p''^{2s_2},$$

or

$$\phi_{r,s,m}(h^* C_p^{2i})$$

$$= h^* \phi_{r,s,m}(C_p^{2i}) = \sum_{\substack{r_1 + r_2 = r \\ s_1 + s_2 = s}} \phi_{r_1,s_1,m_1}(C_p'^{2i}) \otimes \phi_{r_2,s_2,m_2}(C_p''^{2i}), \tag{5}$$

or

$$\phi_{r,s,m} \left[\sum_{i_1 + i_2 = i} C_p'^{2i_1} \otimes C_p''^{2i_2} \right]$$

$$= \sum_{\substack{r_1 + r_2 = r \\ s_1 + s_2 = s}} \phi_{r_1,s_1,m_1}(C_p'^{2i_1}) \otimes \phi_{r_2,s_2,m_2}(C_p''^{2i_2}). \tag{5$'$}$$

By hypothesis $n_i \geq mp$, $i = 1, 2$. From Lemma 1, any $2k$-dimensional mod p cohomology class in $C_{n_1, m_1} \times C_{n_2, m_2}$ can be represented as a polynomial of $C_p'^{2i_1} \otimes C_p''^{2i_2}$, $0 < i_1, i_2 \leq m_j$, (the multiplication is cup product). This representation is unique when $k \leq n_1$ and n_2, Hence from $(5)'$ we know that

$$\phi_{r,s,m}\left(\sum_{i_1+i_2=i} c'_{i_1} c''_{i_2}\right) = \sum_{\substack{r_1+r_2=r \\ s_1+s_2=s}} \phi_{r_1,s_1,m_1}(c'_{i_1})\phi_{r_2,s_2,m_2}(c''_{i_2}). \quad (6)$$

Here c'_{i_1} $(i_1 = 1, 2, \cdots, m_1)$ and c''_{i_2} $(i_2 = 1, 2, \cdots, m_2)$ are determinates with weight i_1 and i_2 respectively.

Lemma 3.

$$\phi_{r,1,m}(c_i) = \begin{cases} 0, & r \neq 0,\ p-1; \\ \lambda_2 c_1, & r = 0,\ \lambda_2 = (-1)^q (q!)^2,\ q = (p-1)/2; \\ (c_1)^p, & r = p-1. \end{cases} \quad (7)_m$$

Proof. When $s = 1$, $(5)'$ becomes (see $(4)''$)

$$\phi_{r,1,m}\left(\sum_{i_2+i_2=i} C_p'^{2i_1} \otimes C_p''^{2i_2}\right) = \phi_{r,1,m_1}(C_p'^{2i}) \otimes 1 + 1 \otimes \phi_{r,1,m_2}(C_p''^{2i}). \quad (8)$$

Now let all classes $C^{2r_1}, \cdots, C^{2r_k}$ $(0 < r_1 \leq \cdots \leq r_k \leq m)$ in $C_{n,m}$ be ordered lexicographically. For the time being let m be very large, $m > r+1$, and take $m_1 = 1$. Assume $\phi_{r,1,m}$ is not identically zero and let $C^{2i_1} \cdots \cdots C^{2i_k}$, $i_1 + \cdots + i_k = r+1$, be the term of highest degree with nonzero coefficient $a (a \in I_p)$ in $\phi_{r,1,m}(C^{2i})$. Then in the expression of $\phi_{r,1,m}$ $(\Sigma C_p'^{2i_1} \otimes C_p''^{2i_2})$ there is a term $C_p'^2 \otimes C_p''^{2i_1-2} C_p''^{2i_2} \cdots C_p''^{2i_k}$ with coefficient at, where $i_1 = \cdots = i_t \neq i_{t+1}$. If $0 < r < p-1$ and $t \leq t i_1 \leq r+1 < P$, then $at \neq 0$ $(at \in I_p)$. But the right-hand side of (8) cannot have such a term; therefore

$$\text{when } 0 < r < p-1,\ \phi_{r,1,m}(C_p^{2i}) = 0.$$

If $m \leq r+1$, we know the above equality is true.

From the properties of Steenrod powers in §2, we also know that

$$\phi_{r,1,m}(C_p^{2i}) = \begin{cases} 0, & r \neq 0, p-1; \\ \lambda_2 C_p^2, & r = 0; \text{ (for } \lambda, \text{ see §1, S4°)} \\ (C_p^2)^p, & r = p-1. \end{cases}$$

From this we have (7).

From our previous discussion it is clear that for any integer $m > 0$, $r, s \geq 0$, there exists a polynomial (in I_p) $\phi_{r,s,m}(c_i)$, $0 < i \leq m$, in c_i, each of whose terms has weight $r + s$. c_i is an indeterminant of weight i. This polynomial has the following three properties:

1°. (4) and (4)$'$ are both satisfied.

2°. (7) is satisfied.

3°. (6) is satisfied. In other words, if $m_1, m_2 > 0$ and

$$m_1 + m_2 = m, \tag{9}$$

and $c_i'(0 < i \leq m_1)$ and $c_i''(0 < i \leq m_2)$ are two sets of indeterminates with weight i, and

$$c_i = \sum_{i_1+i_2=i} c_{i_1}' c_{i_2}'' \quad (c_0' = 1,\ c_0'' = 1), \tag{10}$$

then

$$\phi_{r,s,m}(c_i) = \sum_{\substack{r_1+r_2=r \\ s_1+s_2=s}} \phi_{r_1,s_1,m_1}(c_i')\phi_{r_2,s_2,m_2}(c_i''). \tag{11}$$

Lemma 4. *Under the assumptions 1°–3°, $\phi_{r,s,m}(c_i)$ is a polynomial determined by the following method: extend the field I_p to F_p in such a way that F_p contains c_i and in this field the equation*

$$x^m - c_1 x^{m-1} + \cdots + (-1)^m c_m = 0 \tag{12}$$

has a solution $x_1, \cdots, x_m$. Then

$$\phi_{r,s,m}(c_i) = \begin{cases} 0, \text{ if } r \not\equiv 0 \mod (p-1); \\ (\lambda_2)^{s-t} \cdot S((x_1 \cdots x_t)^p x_{t+1} \cdots x_s), \text{ if } r = t(p-1). \end{cases} \tag{13$_m$}$$

Here, with respect to $x_1, \cdots, x_m$ the function $S(x_1^{\lambda_1} \cdots x_m^{\lambda_m})$ is symmetric *with* $x_1^{\lambda_1} \cdots x_m^{\lambda_m}$ *as standard terms,* [1] *and* (*see* $(7)_m$ *for* λ_2)

$$\lambda_2 = (-1)^q (q!)^2, \; q = (p-1)/2. \tag{13}'$$

Proof. Let c_i' and c_i'' be as in 3°. Extend the field F_p to a field containing c_i' and c_i', and wherein the equations

$$x^{m_j} - c_1^{(j)} x^{m_j - 1} + \cdots + (-1)^{m_j} c_{m_j}^{(j)} = 0, \; j = 1, 2,$$

have solutions $x_1^{(j)}, \cdots, x_{m_j}^{(j)}$ $(j = 1, 2)$. Equation (12) can be obtained from (9) and (10) by multiplying these two equations together. Hence $x_1, \cdots, x_m$ can be regarded as a permutation of $x_1', \cdots, x_{m_1}'$ and $x_1'', \cdots$ $\cdots, x_{m_2}''$ and

$$S(x_1 \cdots x_t)^p x_{t+1} \cdots x_s$$

$$= \sum_{\substack{s_1 + s_2 = s \\ t_1 + t_2 = t}} [S'((x_1' \cdots x_{t_1}')^p x_{t_1+1}' \cdots x_{s_1}') \cdot S''((x_1'' \cdots x_{t_2}'')^p x_{t_2+1}'' \cdots x_{s_2}'')]. \tag{14}$$

Here S' and S'' are symmetric functions with respect to x_i' and x_i''.

Now we prove (13) by induction on m. When $m = 1$, $(13)_1$ has been proved in 1° and 2°. Now assume the truth of $(13)_1, \cdots, (13)_{m-1}$. As before take m_1, m_2, c_i' and c_i'' arbitrarily, with $m_1, m_2 < m$. When $r \not\equiv 0 \bmod (p-1)$, for any r_1 and r_2 such that $r_1 + r_2 = r$, at least one of them $\not\equiv 0 \bmod (p-1)$. Hence, by induction, for any s_1 and s_2 such that $s_1 + s_2 = s$, at least one of $\phi_{r_1, s_1, m_1}(c_i')$ and $\phi_{r_2, s_2, m_2}(c_i'')$ is equal to zero. So from (11) we have $\phi_{r,s,m}(c_i) = 0$. This is the first part of $(13)_m$. Now let $r = t(p-1)$. By the induction hypothesis, along with (11) and (14) we have

$$\phi_{t(p-1),s,m}(c_i) = \sum_{\substack{t_1 + t_2 = t \\ s_1 + s_2 = s}} \phi_{t_1(p-1), s_1, m_1}(c_i'), \; \phi_{t_2(p-1), s_2, m_2}(c_i'')$$

$$= \sum_{\substack{t_1 + t_2 = t \\ s_1 + s_2 = s}} (\lambda_2)^{s_1 - t_1} S'((x_1' \cdots x_{t_1}')^p x_{t_1+1}' \cdots x_{s_2}')$$

$$\cdot (\lambda_2)^{s_2 - t_2} S''((x_1'' \cdots x_{t_2}'')^p x_{t_2+1}'' \cdots x_{s_2}'')$$

---

1) $S((x_1 \cdots x_t)^p x_{i+1} \cdots x_s) = 0$ when $s = 0$.

$$= (\lambda_2)^{s-t} S((x_1 \cdots x_t)^p x_{t+1} \cdots x_s).$$

This is the second part of $(13)_m$, and the theorem is proved.

When $s = m$, $\phi_{r,m,m}$ can be expressed explicitly as follows. Let c_i and x_i be as before. Here c_i is the elementary symmetric function of x_i. Let

$$s_k = x_1^k + \cdots + x_m^k = S(x_1^k),$$

Then, from Waring's formula

$$\begin{aligned}(-1)^k \frac{s_k}{k} &= \sum_{(\alpha)} (-1)^{\alpha_1 + \cdots + \alpha_m} \cdot \frac{(\alpha_1 + \cdots + \alpha_m - 1)}{\alpha_1! \cdots \alpha_m!} \cdot c_1^{\alpha_1} \cdots c_m^{\alpha_m} \\ &= \sum_i{}^{(k)} \frac{(-1)^i}{i} (c_1 + \cdots + c_m)^i, \end{aligned} \tag{15}$$

where $\Sigma_{(\alpha)}$ is over all m-tuples of integers $(\alpha_1, \cdots, \alpha_m)$ satisfying $\alpha_i \geq 0$, $\alpha_1 + 2\alpha_2 + \cdots + m\,\alpha_m = k$. $\Sigma^{(k)}$ stands for the partial sum of all terms with weight k in the expression. Now, letting $x_1^{p-1}, \cdots, x_m^{p-1}$ be roots, construct the following equation

$$y^m - d_1 y^{m-1} + \cdots + (-1)^m d_m = 0.$$

Denote by σ_k the sum of the kth powers of the roots of this equation. Then $\sigma_k = s_{k(p-1)}$, and from another classical formula we have

$$\begin{aligned} d_r &= S(x_1^{p-1} \cdots x_r^{p-1}) \\ &= \sum_{(\beta)} (-1)^{r+\beta_1+\cdots+\beta_m} \cdot \frac{1}{\beta_1! \cdots \beta_m!} \cdot \left(\frac{\sigma_1}{1}\right)^{\beta_1} \left(\frac{\sigma_2}{2}\right)^{\beta_2} \cdots \left(\frac{\sigma_m}{m}\right)^{\beta_m} \\ &= \sum_{(\beta)} (-1)^{\beta_1+\cdots+\beta_m} \cdot \frac{1}{\beta_1! \cdots \beta_m!} \cdot \left(-\frac{s_{p-1}}{1}\right)^{\beta_1} \left(+\frac{s_{2(p-1)}}{2}\right)^{\beta_2} \cdots \\ &\qquad \cdots \left((-1)^m \cdot \frac{s_{m(p-1)}}{m}\right)^{\beta_m}, \end{aligned}$$

or

$$d_r = \Sigma_j^{(r)} \frac{(-1)^j}{j!}\left[-\frac{s_{p-1}}{1}+\frac{s_{2(p-1)}}{2}-\cdots+(-1)^m\cdot\frac{s_{m(p-1)}}{m}\right]^j, \tag{16}$$

where $\Sigma_{(\beta)}$ is over all m-tuples of integers $(\beta_1, \cdots, \beta_m)$ satisfying $\beta_i \geq 0$ and $\beta_1 + 2\beta_2 + \cdots + m\beta_m = r$. $\Sigma^{(r)}$ stands for the partial sum of terms with weight $r(p-1)$ (the weight of s_k is k). In particular, when $p = 3$,

$$d_r = (-1)^r \sum_{i+j=2r} (-1)^{ij} c_i c_j \quad (p = 3). \tag{17$_3$}$$

From (15) and (16) we have

$$d_r = S\, x_1^{p-1} \cdots x_r^{p-1} = \Sigma_j^{(r)} \frac{(-1)^j}{j!}\left[\Sigma_i^{[p-1]} \frac{(-1)^i}{i}(c_1+\cdots+c_m)^i\right]^j, \tag{17}$$

where $\Sigma^{(r)}$ is the same as before and $\Sigma^{[p-1]}$ denotes the partial sum of all terms of weight divisible by $p-1$. From $(13)_m$, when $r = t(p-1)$,

$$\phi_{r,m,m}(c_i) = S(x_1, \cdots x_t)^p x_{t+1} \cdots x_m = x_1 \cdots x_m \cdot S(x_1^{p-1} \cdots x_t^{p-1}) = c_m d_t,$$

or

$$\phi_{t(p-1),m,m}(c_i) = c_m \cdot \Sigma_j^{(t)} \frac{(-1)^j}{j!}\left[\Sigma_i^{[p-1]} \frac{(-1)^i}{i}(c_1+\cdots+c_m)^i\right]^j. \tag{18}$$

Now define $D_p^{2r} \in H^{2r}(C_{n,m}, I_p)$, $0 < r \leq m(p-1)$, as

$$D_p^{2r} = \begin{cases} 0, & r \not\equiv 0 \bmod (p-1) \\ \Sigma_j^{(r)} \dfrac{(-1)^j}{j!}\left[\Sigma_i^{[p-1]} \dfrac{(-1)^i}{i}(C_p^2+\cdots+C_p^{2m})^i\right]^j, & r = t(p-1); \end{cases} \tag{19}$$

$\Sigma^{[p-1]}$ means the partial sum of all terms whose dimension in the expression is a multiple of $2(p-1)$. $\Sigma^{(r)}$ stands for the partial sum of all terms with dimension $2r$. Hence from (18) and $(3)'$ we have

$$St_p^{2r} C_p^{2m} = \begin{cases} 0, & r \not\equiv 0 \bmod (p-1); \\ C_p^{2m} D_p^{2r}, & r = t(p-1),\ 0 < t \leq m. \end{cases} \tag{20}$$

When $p = 3$, by $(17)_3$, D_p^{2r} becomes

$$D_3^{2r} = \begin{cases} 0, \; r \not\equiv 0 \mod 2; \\ (-1)^k \sum\limits_{i+j=2k} (-1)^{ij} C_3^{2i} C_3^{2j}, \; r = 2k. \end{cases} \tag{21$_3$}$$

§3. Gysin's exact sequence in euclidean space bundles [1)]

Let $\mathfrak{B} = \{A, \pi, K, R^m, G\}$ be an m-Euclidean space bundle, where A is a bundle space, π is a projection, K is a base space, and the structure group G is an orthogonal group $\underline{O}_m$ or a subgroup of one. There is an m-dimensional closed cell-bundle $\overline{\mathfrak{B}}$ in K and, associated with $\mathfrak{B}$, an $(m-1)$-dimensional sphere bundle $\mathfrak{S}$. Let their bundle spaces be $\overline{A}$ and E and their projections be $\overline{\pi}$ and ω respectively. We can suppose that $\overline{A}$ is an ω mapping cylinder, $A = \overline{A} - E$, $\omega = \overline{\pi}|\overline{E}$, $\pi = \overline{\pi}|A$, and for any $p \in K$, $\omega^{-1}(p)$ is the side of $\overline{\pi}^{-1}(p)$, and that $\pi^{-1}(p) = \overline{\pi}^{-1}(p) - \omega^{-1}(p)$.

We suppose that K is a finitely decomposable space. Then A, $\overline{A}$ and E are the same. Take any cell-decomposition of K; let it be K also. Then A has a cell-decomposition also; let it be A; and so on. Let σ be an arbitrary cell in K. Then A is composed of all cells of the form $\tilde{\sigma} = \pi^{-1}(\sigma)$. Obviously $\dim \tilde{\sigma} = \dim \sigma + m$.

In the sequel, H^k is the k-dimensional cohomology group, as mentioned in §1. The coefficient group is understood. For convenience we only discuss the following two cases:

1. The coefficient group is I_2; the structure group is $G = O_m$.
2. The coefficient group is I_0; the structure group is cyclic.

In other words, $\mathfrak{B}, \overline{\mathfrak{B}}, \mathfrak{S}$ are orientable bundles. Now we suppose that all fibers are oriented, and form an oriented bundle. Also we assume that for any $p \in K$, the orientations of $\pi^{-1}(p)$, $\overline{\pi}^{-1}(p)$ and $\omega^{-1}(p)$ are agreeable.

For the space pair $(\overline{A}, E)$, according to $H1°$ and $H3°$ in §1, there is an exact sequence

$$\cdots \longrightarrow H^{r-1}(E) \xrightarrow{\alpha^*} H^r(A) \xrightarrow{\beta^*} H^r(\overline{A}) \xrightarrow{r^*} H^r(E) \longrightarrow \cdots \tag{1}$$

and β^* can be regarded as the pseudo-homeomorphism corresponding to k^* when A is an open set in $\overline{A}$.

---

1) Cf. [4], [15] and [16].

Since $K \subset \bar{A}$ and K is a kernel of deformation retraction in $\bar{A}$, on letting $\iota: K \to \bar{A}$ be the identity map we obtain

$$\begin{cases} \bar{\pi}^*: & H^*(K) \approx H^*(\bar{A}), \\ \iota^*: & H^*(\bar{A}) \approx H^*(K), \\ \iota^* = \bar{\pi}^{*-1}. \end{cases} \tag{2}$$

Furthermore we define the homeomorphic transformation

$$\tau^*: H^r(K) \approx H^{r+m}(A) \tag{3}$$

as follows. Let $u \in C^r(K)$. Define $\tau^{\#}(u) \in C^{r+m}(A)$ as

$$\tau^{\#} u(\tilde{\sigma}) = u(\sigma). \tag{4}$$

In the second case, the orientation of $\tilde{\sigma}$ can be derived from the orientations of the fibers R^m and σ, since $\tilde{\sigma}$ and $R^m \times \sigma$ are homeomorphic. Hence $\tau^{\#}: C^r(K) \approx C^{r+m}(A)$ is a homeomorphic transformation. Obviously $\tau^{\#}\delta = \delta\tau^{\#}$. Hence $\tau^{\#}$ determines (3).

τ^* was introduced by Thom; hence we call it a Thom transformation. The transformation has the following properties:

1°. Let K be another finitely decomposable space, and let $f: K' \to K$ be a continuous map. Let $\mathfrak{B}' = f^*\mathfrak{B}$, $\bar{\mathfrak{B}}' = f^*\bar{\mathfrak{B}}$ and $\mathfrak{S}' = f^*\mathfrak{S}$, have A', $\bar{A}'$ and E' as space bundles, and so on. Let $\tilde{f}: A' \to A$ be the bundle map satisfying $\pi\tilde{f} = f\pi'$. Choose the orientation of the bundles in such a way that f preserves the orientation of the fibers. Let τ^*, τ'^* be the Thom transformations corresponding to these orientations (with respect to suitable decompositions of K and K'). Then

$$\tilde{f}^*\tau^* = \tau'^* f^*. \tag{5}$$

Proof. It is easy to see that $\tilde{f}$ is an admissible map. Choose the decompositions of K and K', denote them again by K and K', and let f be a cell map. Then so is $\tilde{f}$, and from (4) we have $\tilde{f}^{\#}\tau^{\#} = \tau'^{\#} f^{\#}$, thus obtaining (5). In general we can take a suitably small decomposition K_1' in K' and change f to a cell map f_1 from K_1' to K. By the homotopy covering theorem of fiber bundles we know that there is a bundle map $\tilde{f}_1: A' \to A$ such that $\tilde{f}_1 \simeq \tilde{f}$, $\pi\tilde{f}_1 = f_1\pi'$. Hence we obtain $\tilde{f}_1^*\tau^* = \tau_1'^* f_1^* = \tau'^* f_1^*$. This is (5).

From 1°, we know that

2°. The Thom transformation τ^* is independent of the cell decomposition of K.

3°. For any $U_1, U_2 \in H^*(K)$ we have

$$\tau^*(U_1 \cup U_2) = \tau^* U_1 \cup \bar{\pi}^* U_2. \tag{6}$$

Proof. Define $\pi_1\colon A \times K \to K \times K$ by $\pi_1(p, q) = (\pi(p), q)$, $p \in A$, $q \in K$. Let $d\colon K \to K \times K$ be defined as the diagonal map. Define $\tilde{d}\colon A \to A \times K$ by $\tilde{d}(p) = (p, \pi(p))$, $p \in A$. Then

$$\pi_1 \tilde{d} = d\pi. \tag{7}$$

It is easy to see that $\tilde{d}$ is admissible. Obviously $A \times K$ is a bundle space in an m-Euclidean space bundle $\mathfrak{B}_1$ in $K \times K$, π_1 is the bundle projection in $\mathfrak{B}_1$, $\pi^{-1}(p) \times q$ is the fiber of $(p, q) \in K \times K$, and $\tilde{d}$ is the bundle map from $\mathfrak{B}$ to $\mathfrak{B}_1$. Thus from (7)

$$d^* \mathfrak{B}_1 \sim \mathfrak{B}. \tag{8}$$

The cells σ_1 and σ_2 in K, where $\pi_1^{-1}(\sigma_1 \times \sigma_2) = \pi^{-1}(\sigma_1) \times \sigma_2 = \tilde{\sigma}_1 \times \tilde{\sigma}_2$, form a decomposition. Orient $\mathfrak{B}_1$ in such a way that $\tilde{d}$ keeps the orientation of the corresponding fiber. Then the orientation of each cell $\tilde{\sigma}_1 \times \sigma_2$ is determined by the orientation of $\tilde{\sigma}_1$ and σ_2, which have already been determined. Let $\tau_1^{\#}$ and τ_1^* be the Thom transformations with respect to $\mathfrak{B}_1$. Then for any $u_1, u_2 \in C^*(K)$, we obtain from (4)

$$\begin{aligned} \tau_1^{\#}(u_1 \otimes u_2)(\tilde{\sigma}_1 \times \sigma_2) &= (u_1 \otimes u_2)(\sigma_1 \times \sigma_2) = u_1(\sigma_1) \cdot u_2(\sigma_2) \\ &= \tau^{\#} u_1(\tilde{\sigma}_1) \cdot u_2(\sigma_2) = (\tau^{\#} u_1 \otimes u_2)(\tilde{\sigma}_1 \times \sigma_2). \end{aligned}$$

Hence $\tau_1^{\#}(u_1 \otimes u_2) = \tau^{\#} u_1 \otimes u_2$ and

$$\tau_1^*(U_1 \otimes U_2) = \tau^* U_1 \otimes U_2,\ U_1, U_2 \in H^*(K). \tag{9}$$

From (5), (7), (8) and (9) we have $(U_1, U_2 \in H^*(K))$

$$\begin{aligned} \tau^*(U_1 \cup U_2) &= \tau^* d^*(U_1 \otimes U_2) = \tilde{d}^* \tau_1^*(U_1 \otimes U_2) \\ &= \tilde{d}^* (\tau^* U_1 \otimes U_2). \end{aligned}$$

Now let $\tilde{d}'\colon A \to A \times \bar{A}$ be the diagonal map; then $\tilde{d} = (1 \times \bar{\pi})\tilde{d}'$, and

$\tilde{d}^* = \tilde{d}'^*(1 \times \bar{\pi})^*$, since $1: A \to A$ is the identity map. By (6) and (7) in §1, we obtain from the above

$$\tau^*(U_1 \cup U_2) = \tilde{d}^*(\tau^* U_1 \otimes U_2) = \tilde{d}'^*(1 \times \bar{\pi})^*(\tau^* U_1 \otimes U_2)$$

$$= \tilde{d}'^*(\tau^* U_1 \otimes \bar{\pi}^* U_2) = \tau^* U_1 \cup \bar{\pi}^* U_2,$$

as desired.

Now define

$$\phi^* = \tau^{*-1}\alpha^*, \tag{10}$$

$$\psi^* = \bar{\pi}^{*-1}\beta^*\tau^* = \iota^*\beta^*\tau^*. \tag{11}$$

Then, since $\omega^* = \gamma^*\pi^*$, (1) becomes the so-called Gysin's exact sequence

$$\cdots \to H^{r-1}(E) \xrightarrow{\phi^*} H^{r-m}(K) \xrightarrow{\psi^*} H^r(K) \xrightarrow{\omega^*} H^r(E) \to \cdots. \tag{1$'$}$$

From (6) (1 = unit class, $U \in H^*(K)$)

$$\psi^* U = \iota^*\beta^*\tau^* U = \iota^*\beta^*\tau^*(1 \cup U) = \iota^*\beta^*(\tau^* 1 \cup \bar{\pi}^* U).$$

Using $P3°$ in §1 on $A \subset \bar{A} \subset \bar{A}$, we have

$$\beta^*(\tau^* 1 \cup \bar{\pi}^* U) = \beta^*\tau^* 1 \cup \bar{\pi}^* U.$$

Hence the previous equality becomes

$$\psi^* U = \iota^*(\beta^*\tau^* 1 \cup \bar{\pi}^* U) = \iota^*\beta^*\tau^* 1 \cup \iota^*\bar{\pi}^* U,$$

which is the same as

$$\psi^* U = \psi^* 1 \cup U. \tag{12}$$

§4. Generalizations of Thom's theorem and another proof [1)]

Let $\mathfrak{B} = \{A, \pi, K, R^m, G\}$, $\bar{\mathfrak{B}} = \{\bar{A}, \bar{\pi}, \bar{K}, \bar{R}^m, G\}$ and $\mathfrak{S} = \{E, \omega, K, S^{m-1}, G\}$ be defined as at the beginning of §3, where $\bar{R}^m$ is an m-dimensional closed cell. In the second case of §3, every bundle has been oriented. As in §3, let $W_2^i(\mathfrak{B})$ be their mod 2 W-classes. Thom [4] proved that

1) Cf. [1] and [4].

$$Sq^i\tau^* 1 = \tau^* W_2^i(\mathfrak{B}). \tag{1}$$

where $\tau^* : H^i(K, I_2) \approx H^{i+m}(A, I_2)$ is a Thom transformation and 1 is a mod 2 unit class. Furthermore, we have

$$\pi^{*-1}\beta^*\tau^* 1 = \psi^* 1 = -c^m(\mathfrak{S}), \tag{2}$$

where, in the second case, $c^m(\mathfrak{S}) \in H^m(K, \pi_{m-1}(S^{m-1})) \approx H^m(K, I_0)$ is the cross-section class defined by Steenrod. Its homomorphic transformation $\pi_{m-1}(S^{m-1}) \approx I_0$ is determined by the orientation of the fibers in the oriented bundle $\mathfrak{S}$, and 1 is the integral coefficient unit class. In the first case, every class in (2) is assumed to be reduced mod 2.

Now we give the proof of (1).

Let $K = R_{n,m}$ be a Grassmann manifold and $\mathfrak{B} = \mathfrak{N}_{n,m}$ the model m-Euclidean space bundle in $R_{n,m}$. Let W_2^i be the mod 2 W-class in $R_{n,m}$. Let $\tau^*_{n,m}$ be the transformation corresponding to τ^* with respect to $\mathfrak{N}_{n,m}$. Let $A_{n,m}$ be the space corresponding to A, and so on. By §4 of [11], we have

$$Sq^i W_2^m = W_2^i W_2^m. \tag{3}$$

By (2) and (12) in §3, we have

$$W_2^m = \psi^*_{n,m} 1_{n,m}, \tag{4}$$

$$XW_2^m = \psi^*_{n,m}X, \; X \in H^*(K, I_2) \tag{5}$$

Hence $\psi^*_{n,m} = \pi^{*-1}_{n,m}\beta^*_{n,m}\tau^*_{n,m}$, and $1_{n,m}$ is the mod 2 unit class in $R_{n,m}$. Since (see $S9°$ in §1 and §3 for notation)

$$\bar{\pi}^{*-1}_{n,m}Sq^i = \iota^*_{n,m}Sq^i = Sq^i\iota^*_{n,m} = Sq^i\bar{\pi}^{*-1}_{n,m},$$

$$\beta^*_{n,m}Sq^i = Sq^i\beta^*_{n,m},$$

we have, from (3), (4) and (5),

$$\bar{\pi}^{*-1}_{n,m}\beta^*_{n,m}Sq^i\tau^*_{n,m}1_{n,m} = \bar{\pi}^{*-1}_{n,m}\beta^*_{n,m}\tau^*_{n,m}W_2^i. \tag{6}$$

From Lemmas 4 and (5) in [11] we know that

$$\psi^*_{n,m}: H^k(R_{n,m}, I_2) \subset H^{k+m}(R_{n,m}, I_2), \; k + m \leq n.$$

From this we have

$$\beta^*_{n,m}\colon H^k(A_{n,m}, I_2) \subset H^k(\overline{A}_{n,m}, I_2),\ k \leq n.$$

Hence when n is sufficiently large $(n \geq 2m)$ we obtain from (6)

$$Sq^i \tau^*_{n,m} 1 = \tau^*_{n,m} W^i_2. \tag{7}$$

Now let bundles $\mathfrak{B}$, $\overline{\mathfrak{B}}$, and $\mathfrak{S}$ be as before. Then we have $\lambda\colon K \to R_{n,m}$ such that $\mathfrak{S} = \lambda^* \mathfrak{N}_{n,m}$ (n sufficiently large). Let $\overline{\lambda}\colon A \to A_{n,m}$ be its bundle map. Then from (5) in §3, we have

$$Sq^i \tau^* 1 = Sq^i \tau^* \lambda^* 1_{n,m} = Sq^i \tilde{\lambda}^* \tau^*_{n,m} 1_{n,m} = \tilde{\lambda}^* Sq^i \tau^*_{n,m} 1_{n,m}$$

$$= \tilde{\lambda}^* \tau^*_{n,m} W^i_2 = \tau^* \lambda^* W^i_2 = \tau^* W^i_2(\mathfrak{B}).$$

This is (1).

The above proof can be generalized to obtain a formula similar to (1) for the Pontrjagin class.

First look at the model m-unitary space bundle $\mathfrak{C}_{n,m}$ in $C_{n,m}$. According to §2,

$$St_p^{2t(p-1)}\, C_p^{2m} = C_p^{2m} D_p^{2t(p-1)},\quad (0 < t \leq m). \tag{8}$$

From [13], [14] and (2) we obtain

$$C^{2m} = c^{2m}(\mathfrak{C}_{n,m}), \tag{9}$$

$$c^{2m}(\mathfrak{C}_{n,m}) = (-1)^m c^{2m}(\mathfrak{C}'_{n,m}), \tag{10}$$

$$\psi^*_{n,m} 1_{n,m} = -c^{2m}(\mathfrak{C}'_{n,m}), \tag{11}$$

where c^{2m} is the cross-section class of the equivalent bundle according to Steenrod's definition, $\mathfrak{C}'_{n,m}$ is the oriented bundle determined by the natural orientation of the fibers when $\mathfrak{C}_{n,m}$ is regarded as the $2m$-Euclidean space bundle, and $c^{2m}(\mathfrak{C}_{n,m})$, $c^{2m}(\mathfrak{C}'_{m,n}) \in H^{2m}(C_{n,m}, \pi_{2m-1}(S^{2m-1})) \approx H^{2m}(C_{n,m}, I_0)$. The coefficient group $\pi_{2m+1}(S^{2m-1}) \approx I_0$ is determined by the natural orientation of the fibers S^{2m-1}, $\psi_{n,m}$ is the pseudo-homomorphic transformation equivalent to ψ in $\mathfrak{S}'_{n,m}$, and $1_{n,m}$ is the integral unit class in $C_{n,m}$ etc.

After reducing these equalities mod p, we obtain from (12) in §3

$$\psi^*_{n,m} X = (-1)^{m+1} C_p^{2m} X, \ X \in H^{2k}(C_{n,m}, I_p). \tag{12}$$

Substituting in (8), we obtain

$$St_p^{2t(p-1)} \psi^*_{n,m} 1_{n,m} = \psi^*_{n,m} D_p^{2t(p-1)} \tag{13}$$

From Lemma 1 in §2, if n is sufficiently large,

$$\psi^*_{n,m} = \overline{\pi}^{*-1}_{n,m} \beta^*_{n,m} \tau^*_{n,m}: \ H^{2k}(C_{n,m}, I_p) \subset H^{2k+2m}(C_{n,m}, I_p).$$

Hence $\beta^*_{n,m}: \ H^{2k}(A_{n,m}, I_p) \subset H^{2k}(\overline{A}_{n,m}, I_p)$, $n \geq k$. Here $A_{n,m}$ and $\overline{A}_{n,m}$ are the spaces corresponding to A and $\overline{A}$ in $\mathfrak{C}'_{n,m}$. As in the proof of (1), (13) yields

$$St_p^{2t(p-1)} \tau^*_{n,m} 1_{n,m} = \tau^*_{n,m} D_p^{2t(p-1)}, \ n \geq 2mp. \tag{14}$$

Now on K let $\mathfrak{B}$ be any m-Euclidean space oriented bundle, and let $\mathfrak{B} = \lambda^* \mathfrak{R}_{n,m}$, λ: $K \to R_{n,m}$ (n sufficiently large). Let $\mathfrak{B}^*$ be its m-unitary space bundle determined by $\mathfrak{B} \cup \mathfrak{B}$. Then, from §5 in [6]

$$\mathfrak{B}^* \sim (\iota \lambda)^* \mathfrak{C}_{n,m},$$

where l: $R_{n,m} \to C_{n,m}$ is the natural map. This yields

$$\mathfrak{B} \cup \mathfrak{B} \sim (l\lambda)^* \mathfrak{C}'_{n,m}.$$

Using (5) in §4, let $\widetilde{\lambda}$ be the bundle map from $\mathfrak{B} \cup \mathfrak{B}$ to $(l\lambda)^* \mathfrak{C}'_{n,m}$, and let τ^* be the Thom transformation with respect to $\mathfrak{B} \cup \mathfrak{B}$. From (14) we obtain

$$\begin{aligned} St_p^{2t(p-1)} \tau^* 1 &= St_p^{2t(p-1)} \tau^* (l\lambda)^* 1_{n,m} = St_p^{2t(p-1)} \widetilde{\lambda}^* \tau^*_{n,m} 1_{n,m} \\ &= \widetilde{\lambda}^* St_p^{2t(p-1)} \tau^*_{n,m} 1_{n,m} = \widetilde{\lambda}^* \tau^*_{n,m} D_p^{2t(p-1)} \\ &= \tau^* (l\lambda)^* D_p^{2t(p-1)} = \tau^* \lambda^* l^* D_p^{2t(p-1)}. \end{aligned}$$

Define

$$l^* D_p^{2t(p-1)} = Q_p^{2t(p-1)}, \tag{15}$$

$$Q_p^{2t(p-1)}(\mathfrak{B}) = \lambda^* Q_p^{2t(p-1)}. \tag{15$'$}$$

Then from the previous equality

$$St_p^{2t(p-1)}\tau^* 1 = \tau^* Q_p^{2t(p-1)}(\mathfrak{B}). \tag{16}$$

From §5 in [6], we have

$$l^* C_p^{2i} = \begin{cases} 0, & i \text{ is odd;} \\ (-1)^{i/2} P_p^{2i}, & i \text{ is even;} \end{cases} \tag{17}$$

where P_p^{4k} is the mod$p$ P-class in $R_{n,m}$. From (19) and (21) in §2, we obtain ($q = (p-1)/2$)

$$Q_p^{2t(p-1)}(\mathfrak{B})$$

$$= \sum_j^{(t(p-1))} \frac{(-1)^j}{j!}\left[\sum_i^{[p-1]} \frac{(-1)^i}{i}\left[-P_p^4(\mathfrak{B}) + P_p^8(\mathfrak{B}) - \cdots\right]^i\right]^j,$$

$$(0 < t \leq m), \tag{18}$$

$$Q_3^{4k}(\mathfrak{B}) = \sum_{i+j=k} P_3^{4i}(\mathfrak{B})\, P_3^{4j}(\mathfrak{B}) \tag*{$(18)_3$}$$

See §2 for the notation $\Sigma^{(\ )}$ and $\Sigma^{[\ ]}$.

§5. Pontrjagin classes in differentiable manifolds

Let M be an m-dimensional differentiable manifold. Choose a differe - tial structure $\mathfrak{D}$ arbitrary, where $\mathfrak{D}$ is determined by a set of coordinate neighborhoods V_j, $j \in J$, and a set of coordinate functions ϕ_j: $R^m \to V_j$ (ϕ_j is a topological map). Set $\mathfrak{D} = \{M, V_j, \phi_j, j \in J\}$. With respect to $\mathfrak{D}$, the tangent vectors of M forman m-Euclidean space bundle $\mathfrak{T}(\mathfrak{D})$, called the tangent bundle of M under $\mathfrak{D}$. The structure group of $\mathfrak{T}(\mathfrak{D})$ is a linear group, but by a theorem of Ehresmann, in $\mathfrak{T}(\mathfrak{D})$ there is an m-Euclidean space subordonné bundle, with an orthogonal O_m as structure group. With respect to O_m its bundle homomorphic classes are completely determined by $\mathfrak{T}(\mathfrak{D})$. Hence we can assume the structure group of $\mathfrak{T}(\mathfrak{D})$ is O_m.

For any integer $n > 1$, construct the topological product $M^n = M_1 \times \cdots \cdots \times M_n$. Here M_i is a model of M. Let ω_i: $M^n \to M_i$ be the projection of M^n onto its ith component. Thus, in M^n, $\mathfrak{D}$ induces a differential structure $\mathfrak{D}^n = \{M^n, V_{(j_1,\cdots,j_n)}, \phi_{(j_1,\cdots,j_n)}, (j_1, \cdots, j_n) \in J^n\}$. Here $V_{(j_1,\cdots,j_n)} =$

$V_{j_1} \times \cdots \times V_{j_n}$, and $\phi_{(j_1,\cdots,j_n)}$: $R^{mn} = \underbrace{R^m \oplus \cdots \oplus R^m}_{n} \to V_{(j_1,\cdots,j_n)}$ is defined as $\phi_{(j_1,\cdots,j_n)}(x_1, \cdots, x_n) = (\phi_{j_1}(x_1), \cdots, \phi_{j_n}(x_n))$. Obviously $\mathfrak{D}^n$ is a differential structure in M^n with the same order as $\mathfrak{D}$.

We know that $\mathfrak{D}$ has a countable Riemann quantity ds^2. Let ds_i^2 be the quantity corresponding to ds^2 in M_i. Then $\widetilde{ds}^2 = ds_1^2 + \cdots + ds_n^2$ is a Riemann quantity in M^n consistent with $\mathfrak{D}^n$. Let d: $M \to M^n$ be a diagonal map. Then with respect to $\widetilde{ds}^2$, all tangent vectors at every point of $d(M)$ and perpendicular to $d(M)$ naturally form an $m(n-1)$-Euclidean space bundle $\mathfrak{N}(\mathfrak{D})$, with M as base space. As before, we may assume the structure group of $\mathfrak{N}(\mathfrak{D})$ to be the orthogonal group $O_{m(n-1)}$. We have

Lemma 5. $\mathfrak{N}(\mathfrak{D}) \sim \underbrace{\mathfrak{T}(\mathfrak{D}) \cup \cdots \cup \mathfrak{T}(\mathfrak{D})}_{n-1}$. *Here* $\cup$ *is the Whitney product in the Euclidean space bundle.*

Proof. Let $\tilde{p} = d(p)$, $p \in M$. Let $T^n(\tilde{p}, \mathfrak{D}^n)$ be the plane tangent to M^n at the point $\tilde{p}$ with respect to $\mathfrak{D}^n$, and let $T(p, \mathfrak{D})$ be the tangent plane to M at the point p with respect to $\mathfrak{D}$. Then ω_i: $M^n \to M$ induces a linear map $\bar{\omega}_i$: $T^n(\tilde{p}, \mathfrak{D}^n) \to T(p, \mathfrak{D})$, $i = 1, \cdots, n$. Let $\mathrm{a} \in T^n(\tilde{p}, \mathfrak{D}^n)$, $\bar{\omega}_i(\mathrm{a}) = \mathrm{a}_i \in T(p, \mathfrak{D})$ and $p = d(\tilde{p})$. Then $\mathrm{a} \longleftrightarrow (\mathrm{a}_1, \cdots, \mathrm{a}_n)$ forms a linear topological correspondence between $T^n(\tilde{p}, \mathfrak{D}^n)$ and $\underbrace{T(p, \mathfrak{D}) \oplus \cdots \oplus T(p, \mathfrak{D})}_{n}$. Under this correspondence we can easily prove

1°. The necessary and sufficient condition that a is tangent to $d(M)$ is $\mathrm{a}_1 = \cdots = \mathrm{a}_n$.

2°. With respect to $\widetilde{ds}^2$, the necessary and sufficient condition that a is perpendicular to $d(M)$ is $\mathrm{a}_1 + \cdots + \mathrm{a}_n = 0$.

Let $N^n(p)$ be the perpendicular plane to M^n at $\tilde{p} \in d(M)$ with respect to $\widetilde{ds}^2$. In $N^n(\tilde{p})$ let $N_k^n(\tilde{p})$, $k = 2, \cdots, n$ be the m-plane formed by the tangent vectors $\mathrm{a} = (\mathrm{a}_1, \cdots, \mathrm{a}_n) \in N^n(\tilde{p})$ and satisfying the following condition:

$$(C_k) \quad \mathrm{a}_{k+1} = \cdots = \mathrm{a}_n = 0, \; -(k-1)\mathrm{a}_1 = \mathrm{a}_3 = \cdots = \mathrm{a}_k.$$

Obviously $N^n(\tilde{p}) = N_2^n(\tilde{p}) \oplus \cdots \oplus N_n^n(\tilde{p})$. Naturally the set $\mathbf{U}_{\tilde{p} \in d(M)} N_k^n(\tilde{p}) = N_k^n$ forms a topological space. Let λ_k: $N_k^n \to M$ be defined as $\lambda_k(\mathrm{a}) = p$, where $\mathrm{a} \in N_k^n(\tilde{p})$, and $\tilde{p} = d(p)$. Hence N_k^n forms a space bundle in an m-Euclidean space bundle $\mathfrak{N}_k^n$ in M, with λ_k as projection, and

$$\mathfrak{N}(\mathfrak{D}) \sim \mathfrak{N}_2^n \cup \cdots \cup \mathfrak{N}_n^n. \tag{1}$$

Now for $a_p \in T(p, \mathfrak{D})$, $p \in M$, let

$$h_k(a_p) = ((k-1)\,a_p, \underbrace{-a_p, \cdots, -a_p}_{k-1}, \underbrace{0, \cdots, 0}_{n-k}) \in N_k^n(p),\ \tilde{p} = d(p).$$

Then h_k is obviously a map from $\mathfrak{T}(\mathfrak{D})$ to $\mathfrak{N}_k^n$, and hence $\mathfrak{N}_k^n \sim \mathfrak{T}(\mathfrak{D})$. Thus from (1) we obtain our theorem.

Now let M be an orientable closed manifold, where the order of $\mathfrak{D} \geq 3$. Then M^n is also an orientable closed manifold, and the order of $\mathfrak{D}^n$ is also ≥ 3. For any $a \in N^n(\tilde{p})$, $a \neq 0$, $\tilde{p} = d(p)$, $p \in M$, let $g(a)$ be the shortest (with respect to $\widetilde{ds^2}$) line starting at $\tilde{p}$ and tangent at a. Let $q(a, \epsilon)$ be those points on $g(a)$ such that the shortest arc length from $\tilde{p}$ to $q(a, \epsilon)$ is equal to ϵ. Since M and M^n are closed, there must be an $\eta < 0$ such that if $a \neq a'$ ($a \in N^n(\tilde{p})$, $a' \in N^n(\tilde{p}\,')$, $a, a' \neq 0$) or $\epsilon \neq \epsilon'$ ($\eta \geq \epsilon, \epsilon' > 0$), then $q(a, \epsilon) \neq q(a', \epsilon')$, and all points $q(a, \epsilon)$ such that $a \in N^n(\tilde{p})$, $\tilde{p} = d(p)$, $p \in M$, $\eta \geq \epsilon \geq 0$ form a neighborhood of $d(M)$ in M^n. Call it A. Then all points $q(a, \epsilon)$ such that $\eta \geq \epsilon \geq 0$ form a closure of A in M^n, written $\bar{A}$. Hence A can be regarded as the bundle space of the bundle $\mathfrak{N}(\mathfrak{D})$. Its projection is $\pi(q(a, \epsilon)) = p$, where $a \in N^n(\tilde{p})$, $\tilde{p} = a(p)$. $\bar{A}$ can be regarded as the bundle space of the associated closed cell bundle of $\mathfrak{N}(\mathfrak{D})$. Its projection is $\bar{\pi}(q(a, \epsilon)) = p$, where a and p are as before and $\bar{\pi}|A = \pi$.

Choose an orientation for M arbitrarily. Since $M^n = \underbrace{M \times \cdots \times M}_{n}$, the orientation of A is determined. Next for any $p \in M$ choose the orientation of $\pi^{-1}(p) \subset A$ to be such that in the vicinity of p the orientation of A determined by $\pi^{-1}(p)$ and M in this order is the same as before. Let

$$\tau^*:\ H^k(M) \approx H^{k+m(n-1)}(A)$$

be the Thom transformation with respect to the oriented bundle $\mathfrak{N}(\mathfrak{D})$.[1] Its coefficient group is I_p, where p is an odd prime. According to $H2°$ in §1, since A is an open set in M^n we have the transformation k^*: $H^k(A) \to H^k(M^n)$. Define

$$\bar{\tau}^* = k^*\tau^*:\ H^k(M) \to H^{k+m(n-1)}(M^n). \tag{2}$$

It is easy to see that $\bar{\tau}^*$ does not have any relation to the magnitude of η

1) By a classical theorem, we know that M must have a cell decomposition consistent with $\mathfrak{D}$; hence τ^* is well defined, and according to 2° in §3 we know τ^* has nothing to do with the choice of its cell decomposition.

and ds^2, but we shall prove the following theorem.

Theorem 1. *$\bar{\tau}^*$ is independent of the choice of the differential structure $\mathfrak{D}$ of M. In other words, $\bar{\tau}^*$ is a topologically invariant transformation between M and M^*. Furthermore,*

$$\bar{\tau}^*: H^*(M) \subset H^*(M^n).$$

Proof. Let $\mathfrak{D}'$ be a differential structure in M with order ≥ 3. We use the same symbol to denote all objects derived from $\mathfrak{D}'$ and equivalent to those in $\mathfrak{D}$, but put a prime in the upper right-hand corner to denote that the object is derived from $\mathfrak{D}'$. Also when we define A' and $\bar{A}'$, we shall choose η' small enough so that $\bar{A}' \subset A$. This is possible, as M is closed. According to $H2^\circ$ in §1, there exists a transformation

$$j^*: H^*(A') \to H^*(A)$$

and

$$k^*j^* = k'^*: H^*(A') \to H^*(M^n). \tag{3}$$

We know [1] that $H^m(M)$, $H^{mn}(M^n)$, $H^{mn}(A)$ and $H^{mn}(A)$ are $\approx I_p$. Corresponding to the orientations of M, M^n, A and A' these groups have certain generating elements. Let them be M_0, M_0^n, A_0' and A_0'. Then

$$k^*(A_0) = M_0^n,\ k'^*(A_0') = M_0^n,\ j^*(A_0') = A_0. \tag{4}$$

Choose any decomposition of M; call it M also. According to §3, we can construct a corresponding decomposition of A; call this A also. Then, according to the definition of τ^* and the choice of the orientations of M and A, it is easy to see that

$$\tau^*(M_0) = A_0. \tag{5}$$

Or, from (6) in §3,

$$\tau^*\, 1 \cup \bar{\pi}^* M_0 = A_0. \tag{6}$$

Simultaneously,

$$\tau'^*\, 1 \cup \bar{\pi}'^* M_0 = A_0'. \tag{6'}$$

---

1) See [7], especially §7.

Let $\bar{j}$: $\bar{A}' \subset \bar{A}$, $\iota: M \to \bar{A}$ and $\iota': M \to \bar{A}'$ be defined as diagonal maps d. Then $\bar{j}: \iota' = \iota$, and $\bar{\pi}^* = \iota^{*-1}$. Hence

$$\bar{j}^* \bar{\pi}^* = \bar{\pi}'^* . \tag{7}$$

Using $P3^\circ$ and $P4^\circ$ in §1, we have

$$j^*(\tau'^* 1 \cup \bar{\pi}'^* M_0) = j^*(\tau'^* 1 \cup \bar{j}^* \bar{\pi}^* M_0) = j^*(\tau'^* 1 \cup \bar{\pi}^* M_0)$$

$$= j^* \tau'^* 1 \cup \bar{\pi}^* M_0.$$

From (4) and $(6)'$ we obtain

$$j^* \tau'^{*-1} \cup \bar{\pi}^* M_0 = A_0. \tag{8}$$

Since $H^{m(n-1)}(A) \approx H^0(M) \approx I_p$, we see that $\tau^* 1$, $j^* \tau'^* 1 \in H^{m(n-1)}(A)$ differ only by an integral multiple mod p. By comparing (6) and (8), it must follow that

$$\tau^* 1 = j^* \tau'^* 1. \tag{9}$$

From (2), (3) and (9) we obtain

$$\bar{\tau}^* 1 = \bar{\tau}'^* 1. \tag{10}$$

In other words, $\bar{\tau}^* 1$ is a class defined invariantly in M^n, and independent of the choice of $\mathfrak{D}$.

Next, let $U \in H^*(M)$ be arbitrary. Then, from (6) in §3, and (7) and (9) in this section, and $P3^\circ$ and $P4^\circ$ in §1, we have

$$j^* \tau'^* U = j^*(\tau'^* 1 \cup \bar{\pi}'^* U) = j^*(\tau'^* 1 \cup j^* \bar{\pi}^* U) = j^*(\tau'^* 1 \cup \bar{\pi}^* U)$$

$$= j^* \tau'^* 1 \cup \bar{\pi}^* U = \tau^* 1 \cup \bar{\pi}^* U = \tau^* U.$$

Hence

$$j^* \bar{\tau}'^* = \bar{\tau}^* . \tag{11}$$

From (3), (11) and (2) we obtain $\bar{\tau}'^* = k'^* \tau'^* = k^* \tau^* = \tau^*$ or

$$\bar{\tau}'^* = \bar{\tau}^* . \tag{12}$$

In other words, $\bar{\tau}^*$ is independent of the choice of the differential structure

$\mathfrak{D}$. This proves the first part of the theorem.

Now let $V \in H^*(M^n)$, and let $\bar{k}$: $\bar{A} \subset M^n$. Then $\bar{k}\iota = d$, $\iota^* \bar{k}^* = d^*$ and $\bar{k}^* = \iota^{*-1} d^* = \bar{\pi}^* d^*$. Hence from (2), $P3°$ and $P4°$ in §1, and (6) in §3, we have

$$\bar{\tau}^* d^* V = k^* \tau^* d^* V = k^*(\tau^* 1 \cup \bar{\pi}^* d^* V) = k^*(\tau^* 1 \cup \bar{k}^* V)$$
$$= k^*(\tau^* 1 \cup V) = k^* \tau^* 1 \cup V,$$

or [1)]

$$\bar{\tau}^* d^* V = \bar{\tau}^* 1 \cup V. \tag{13}$$

Finally, let $U \in H^*(M)$ be arbitrary with $U \neq 0$. We can choose $U_1 \in H^*(M)$ such that $U \cup U_1 = M_0$. Because of the covering property of d^*, we can choose $V, V_1 \in H^*(M)$ such that $d^* V = U$, $d^* V_1 = U_1$. From (2), (4), (5) and (13) we have

$$\bar{\tau}^* U \cup V_1 = \bar{\tau}^* d^* V \cup V_1 = (\bar{\tau}^* 1 \cup V) \cup V_1 = \bar{\tau}^* 1 \cup (V \cup V_1)$$
$$= \bar{\tau}^* d^*(V \cup V_1) = \bar{\tau}^*(U \cup U_1) = \bar{\tau}^* M_0 = M_0^n \neq 0.$$

Hence $\bar{\tau}^* U \neq 0$ and $\bar{\tau}^*$: $H^*(M) \subset H^*(M^n)$. The proof of the theorem is completed.

Now let $n = 3$. By Lemma 5, $\mathfrak{N}(\mathfrak{D}) \sim \mathfrak{T}(\mathfrak{D}) \cup \mathfrak{T}(\mathfrak{D})$. Hence by (16) in §4, we have

$$St_p^{2r(p-1)} \tau^* 1 = \tau^* Q_p^{2t(p-1)}(\mathfrak{T}(\mathfrak{D})). \tag{14}$$

Apply k^* to both sides. Then, by $S9°$ in §1, we obtain

$$St_p^{2t(p-1)} \bar{\tau}^* 1 = \bar{\tau}^* Q_p^{2t(p-1)}(\mathfrak{T}(\mathfrak{D})). \tag{15}$$

Here $\bar{\tau}^*$ is defined as in (2).

In (15), we suppose that the order of $\mathfrak{D}$ is ≥ 3. Now let the order of $\mathfrak{D}$ be arbitrary. Then, according to [17], M has a differential structure $\mathfrak{D}'$ with order ≥ 3, and $\mathfrak{D}' \sim \mathfrak{D}$. From this we have $\mathfrak{T}(\mathfrak{D}) \sim \mathfrak{T}(\mathfrak{D}')$, and hence

1) Up to now we have used neither the fact that I_p is the coefficient group, nor the dual theorems in $H^*(M)$, nor the relations between $H^*(M)$ and $H^*(M^n)$ such as the covering property of d^*. In fact, for any differentiable closed manifold which is imbedded differentially in a differentiable closed manifold N, all of the results are still true: the mapping $\bar{\tau}^*(M) \to H^*(N)$ exists and is a topological invariant of (N, M) (arbitrary coefficient group).

$Q_p^{2t(p-1)}(\mathfrak{T}(\mathfrak{D})) = Q_p^{2t(p-1)}(\mathfrak{T}(\mathfrak{D}'))$. With respect to $\mathfrak{D}$, define $\bar{\tau}^* = \bar{\tau}'^*$ as in (12). Then for this $\mathfrak{D}$ with arbitrary order, (15) is still true. Since the definition of $\bar{\tau}^*$ is independent of the choice of $\mathfrak{D}$ by Theorem 1, and $\bar{\tau}^*$: $H^*(M) \subset H^*(M^n)$, (15) can be written as

$$Q_p^{2t(p-1)}(\mathfrak{T}(\mathfrak{D})) = \bar{\tau}^{*-1} St_p^{2t(p-1)} \bar{\tau}^* 1. \qquad (15)_p'$$

As a conclusion, we have the following theorem:

Theorem 2. *If M is a differentiable, orientable, closed manifold, then for any differential structure $\mathfrak{D}$ in M, the* mod p *Q-classes $Q_p^{2t(p-1)}(\mathfrak{T}(\mathfrak{D}))$ (p is an odd prime), in the tangent bundle determined by $\mathfrak{D}$, are independent of the choice of $\mathfrak{D}$. Hence M is a topological invariant. These classes can be determined by $(15)_p'$.*

According to $(19)_3$ in §4, $P_3^{4k}(\mathfrak{T}(\mathfrak{D}))$ can be determined by $Q_3^{4k}(\mathfrak{T}(\mathfrak{D}))$, $i \leq k$. Hence from Theorem 2 we have

Theorem 3. *If M is a differentiable, orientable, closed manifold, then for any differential structure $\mathfrak{D}$ in M, the* mod 3 *Pontrjagin classes in the tangent bundle $\mathfrak{T}(\mathfrak{D})$ determined by $\mathfrak{D}$ are independent of the choice of $\mathfrak{D}$. Hence M is a topological invariant. Also, these classes can be determined by $(15)_p'$ and $(19)_3$ in §4.*

BIBLIOGRAPHY

[1] Wu Wen-tsün, *Sur les puissances de Steenrod,* Colloq. de Topologie de Strasbourg, 1951, no. 9, La Bibliothèque Nationale et Universitaire de Strasbourg, 1952. MR 14, 491.

[2] R. Thom, *Classes caractéristiques et i-carrés,* C. R. Acad. Sci. Paris 230 (1950), 427–429. MR 12, 42.

[3] ———, *Variétés plongées et i-carrés,* C. R. Acad. Sci. Paris 230 (1950), 507–508. MR 12, 42.

[4] ———, *Espaces fibrés en sphères et carrés de Steenrod,* Ann. Sci. École Norm. Sup. (3) 69 (1952), 109–182. MR 14, 1004.

[5] Wu Wen-tsün, *Classes caractéristiques et i-carrés d'une variété,* C. R. Acad. Sci. Paris 230 (1950), 508–511. MR 12, 42.

[6] ———, *On Pontrjagin classes.* I, Acta Math. Sinica 3 (1953), 291–315; English transl., Amer. Math. Soc. Transl. (2) 92 (1970), 49–62. MR 17, 521.

[7] H. Cartan, *Méthodes modernes en topologie algébrique,* Comment. Math. Helv. 18 (1945), 1–15. MR 7, 138.

[8] S. Eilenberg and N. Steenrod, *Foundations of algebraic topology*, Princeton Univ. Press, Princeton, N. J., 1952. MR 14, 398.

[9] S. Lefschetz, *Algebraic topology*, Amer. Math. Soc. Colloq. Publ., vol. 27, Amer. Math. Soc., Providence, R. I., 1942. MR 4, 84.

[10] N. Steenrod, *Reduced powers of cohomology classes*, Ann. of Math. (2) 56 (1952), 47–67. MR 13, 966.

[11] A. Borel and J.-P. Serre, *Groupes de Lie et puissances réduites de Steenrod*, Amer. J. Math. 75 (1933), 409–448. MR 15, 338.

[12] Wu Wen-tsün, *On squares in Grassmann manifolds*, Acta Sci. Sinica 2 (1953), 203–229; English transl., Amer. Math. Soc. Transl. (2) 38 (1964), 235–258. MR 17, 396.

[13] S. Chern, *Characteristic classes of Hermitian manifolds*, Ann. of Math. (2) 47 (1946), 85–121. MR 7, 470.

[14] Wu Wen-tsün, *Sur les classes caractéristiques des structures fibrées sphériques*, Actualités Sci. Indust., no. 1183, Hermann, Paris, 1952. MR 14, 1112.

[15] W. Gysin, *Zur Homologietheorie der Abbildungen und Faserungen von Mannigfaltigkeiten*, Comment. Math. Helv. 14 (1942), 61–122. MR 3, 317.

[16] S.-S. Chern and E. Spanier, *The homology structures of sphere bundles*, Proc. Nat. Acad. Sci. U.S.A. 36 (1950), 248–255. MR 12, 42.

[17] H. Whitney, *Differentiable manifolds*, Ann. of Math. (2) 37 (1936), 645–680.

Translated by:
B. B. Murphy

Amer. Math. Soc. Transl.
(2) Vol. 92, 1970

ON PONTRJAGIN CLASSES. IV* †

WU WEN-TSÜN

Abstract. The aim of this paper is to give a second proof that the Pontrjagin classes reduced mod 3 of a closed differentiable manifold are topological invariants of that manifold. The proof is different from our preceding one in that Steenrod powers do not come explicitly into the consideration. A comparison of the two proofs which exhibits some general aspects about Steenrod powers will be given later.

Introduction

In the previous paper [2], we used Steenrod powers to prove the topological invariant property of the mod 3 Pontrjagin classes in a differentiable manifold. In this paper we introduce another method for proving the same result that does not use Steenrod powers. The same method can be applied to Stiefel-Whitney classes, thus proving the topological invariance of these classes without using Steenrod's square.

The method used in this paper can be described briefly as follows. Let M be an m-dimensional closed differentiable manifold. Set $M^p = \underbrace{M \times \cdots \times M}_{p}$, where p is an odd prime. Starting from a differential structure $\mathfrak{D}$, a 'symmetric' Riemann quantity $d\tilde{s}^2$ can be defined in M^p. With respect to $d\tilde{s}^2$, let $\tilde{N}_\epsilon$ be the space formed by all points at a distance ϵ (ϵ sufficiently small) from the diagonal manifold in M^p. In M^p, there is a topological transform with period p that does not possess a fixed point:

$$t\colon (a_1, \cdots, a_p) \to (a_2, \cdots, a_p, a_1);$$

t is simultaneously a topological transform in $\tilde{N}_\epsilon$. Let N_ϵ be the modular space of $\tilde{N}_\epsilon$ under t. Then N_ϵ can be regarded as the bundle space of the fiber bundle in M. Its fibers are a certain special lens space. Its projection is π. According to the theory of P. A. Smith [3, 4], a system of cohomology classes $U^i \in H^i(N_\epsilon, I_p)$ can be defined in N_ϵ. Hence in N_ϵ

$$\prod_{i=1}^{q} \sum_{k=0}^{m'} (-1)^k i^{m-2k} U^{2m-4k} \cup \pi^* P_p^{4k}(\mathfrak{D}) = 0 \tag{1}$$

---

* Translation of Acta Math. Sinica 5 (1955), 37–63

† For a translation of part III see Amer. Math. Soc. Transl. (2), 11 (1959), 155–172.

where $q = (p-1)/2$, m' is the largest integer $\leq m/2$, and $\mathfrak{T}$ is the tangent bundle determined by the differential structure $\mathfrak{D}$ of M. The above equality can also be written as

$$\sum_{k=0}^{m'q} U^{2qm-4k} \cup \pi^* R_p^{4k}(\mathfrak{T}) = 0,$$

where

$$R_p^{4k}(\mathfrak{T})$$
$$= (-1)^k \sum_{k_1+\cdots+k_q=k} 1^{m-2k_1}\cdot 2^{m-2k_2}\cdots q^{m-2k_q} P_p^{4k_1}(\mathfrak{T}) P_p^{4k_2}(\mathfrak{T})\cdots P_p^{4k_q}(\mathfrak{T});$$

Hence we can prove that for two tangent bundles $\mathfrak{T}$ and $\mathfrak{T}'$, determined by arbitrary differential structures in M, that

$$R_p^{4p}(\mathfrak{T}) = R_p^{4k}(\mathfrak{T}').$$

In other words, $R_p^{4k}(\mathfrak{T})$ is a topological invariant in M. In particular, when $p = 3$,

$$R_3^{4k}(\mathfrak{T}) = R_3^{4k}(\mathfrak{T}').$$

Hence we prove the topological invariant properties of the mod 3 Pontrjagin classes, the main theorem of this paper.

In §1, we shall state the general theory of Smith, especially the part concerning the Smith transform μ_i. Some properties of μ_i were obtained by the author [5]. In §2, we shall discuss a certain special lens space. In §3 and §4 we shall prove that, corresponding to any Euclidean space bundle, a special lens space can be constructed for which there exists a relation similar to (1) in its bundle space. In §5 we shall use this to prove the topological invariant properties of the mod 3 Pontrjagin classes and state briefly the case of Stiefel-Whitney classes.

The author obtained the above results in 1950 but did not publish them. In the next paper we shall compare this proof with that in [2] and mention their similarities.

§1. Spaces under periodic transforms*

Let $\tilde{K}$ be an abstract complex, and T a multiplicative group on $\tilde{K}$. In other words, corresponding to every $t \in T$ there is a one-to-one mapping of

---

* Cf. [3]–[5].

$\tilde{K}$ onto itself. Call this mapping t. This relationship has the property that $t_2(t_1\tilde{\sigma}) = (t_2 t_1)\tilde{\sigma}$ and $1\tilde{\sigma} = \tilde{\sigma}$ $(\tilde{\sigma} \in \tilde{K}, t_1, t_2 \in T)$, where 1 is the identity element in T. If the following are satisfied, $(\tilde{K}, T)$ is called a primitive system $(t \in T; \tilde{\sigma}, \tilde{\tau} \in \tilde{K})$:

1°. When $t \neq 1$, $t\tilde{\sigma}$ and $\tilde{\sigma}$ are never the same.

2°. $\dim (t\tilde{\sigma}) = \dim (\tilde{\sigma})$.

3°. If $\tilde{\tau}$ is a face of $\tilde{\sigma}$, then $t\tilde{\tau}$ is a face of $t\tilde{\sigma}$.

4°. If $\dim \tilde{\sigma} = \dim \tilde{\tau} + 1$, the incidence numbers of $t\tilde{\sigma}$ and $t\tilde{\tau}$ are equal to those of $\tilde{\sigma}$ and $\tilde{\tau}$ respectively.

Now we define a concept of equivalence. Let $\tilde{\sigma} \sim \tilde{\tau}$ mean that there exists $t \in T$ such that $t\tilde{\sigma} = \tilde{\tau}$. In any equivalent system choose $\tilde{\sigma} \in \sigma$. The collection of these $\tilde{\sigma}$'s is called a fundamental domain of $\tilde{K}$. Those with dimension r are denoted by $\{\tilde{\sigma}_i^r\}$, $i \in \mathfrak{J}^r$. For an arbitrary coefficient group G, let $G(T)$ be the group of all elements of the form $\Sigma g_i t_i$ $(g_i \in G, t_i \in T)$ under the usual addition. Assume $n(\Sigma g_i t_i) = \Sigma g_i \in G$. Thus any chain in $C_r(\tilde{K}, G)$ can be denoted by $\Sigma_{i \in \mathfrak{J}^r} \alpha_i \tilde{\sigma}_i^r$, $\alpha_i \in G(T)$, and the boundary relation in $\tilde{K}$ can be written as

$$\partial \tilde{\sigma}_i^r = \sum_{j \in \mathfrak{J}^{r-1}} \alpha_{ij}^r \tilde{\sigma}_j^{r-1}, \quad i \in \mathfrak{J}^r, \tag{1}$$

where $\alpha_{ij}^r \in I(T)$ and I is the additive group of integers. By the natural multiplication of elements in $I(T)$ and $G(T)$, we can obtain those elements in $G(T)$. Hence for any $\alpha \in I(T)$

$$\alpha\Big[\sum_{i \in \mathfrak{J}^r} \alpha_i \tilde{\sigma}_i^r\Big] = \sum_{i \in \mathfrak{J}^r} (\alpha\alpha_i)\tilde{\sigma}_i^r, \quad (\alpha_i \in G(T)).$$

This defines a chain transform with coefficient group G in $\tilde{K}$. Let the homology group transform induced by α be α_*.

Now let K be a collection of equivalent systems σ in $\tilde{K}$. From 2°, the dimension of any $\tilde{\sigma} \in \sigma$ is independent of the choice of $\tilde{\sigma}$. We define it to be the dimension of σ. If for any two $\sigma, \tau \in K$ there is a $\tilde{\sigma} \in \sigma$ and a $\tilde{\tau} \in \tau$ such that $\tilde{\tau}$ is the face of $\tilde{\sigma}$, then define τ to be the face of σ. In K, define a boundary relation ∂ as

$$\partial \sigma_i^r = \sum_{j \in \mathfrak{J}^{r-1}} n(\alpha_{ij}^r)\sigma_j^{r-1}, \quad i \in \mathfrak{J}^r. \tag{2}$$

Here $\alpha_{ij}^r \in I(T)$ as before, and σ_i^r is the equivalent system containing $\tilde{\sigma}_i^r$. Thus K becomes a complex, which we shall call the modular complex in $\tilde{K}$ under T, and denote by $K = \tilde{K}_T$. The natural correspondence

$\pi: \tilde{\sigma} \to \sigma$ $(\tilde{\sigma} \in \sigma \in K)$ is called the projection of $(\tilde{K}, T)$, $\pi: \tilde{K} \to K$. The chain transform and homology group transform induced by π are denoted by π and π_*.

Now let T be a finite group. Then $s = \sum_{t \in T} t \in I(T)$ is meaningful. Suppose that for $x \in C_r(K, G)$ there is $\tilde{x} \in C_r(\tilde{K}, G)$ such that $\pi\tilde{x} = x$. It is easy to see that $s\tilde{x} \in C_r(\tilde{K}, G)$ depends only on x. Let it be called $\bar{\pi}x$. Then $\bar{\pi}$ defines a chain transform $\bar{\pi}: C_r(K, G) \to C_r(\tilde{K}, G)$. Obviously

$$\bar{\pi}\pi = s, \quad \pi\bar{\pi} = n(s) \cdot 1. \tag{3}$$

We shall let π', $\bar{\pi}'$ and α' $(\alpha \in I(T))$ be the cochain transforms dual to π, $\bar{\pi}$ and α, and π^*, $\bar{\pi}^*$ and α^* the cohomology group transforms induced by them, and so on.

Now let T be arbitrary, and let $(\tilde{K}_1, T)$ be another primitive system. Let $K_1 = (\tilde{K}_1)_T$ be its modular complex, and let its projection be π. Suppose that the set transform $\tilde{S}: \tilde{K} \to \tilde{K}_1$ and chain transform $\tilde{f}: \tilde{K} \to \tilde{K}_1$ are such that for any $t \in T$ we have $t\tilde{S} = \tilde{S}t$ and $t\tilde{f} = \tilde{f}t$. Then we call $\tilde{S}$ and $\tilde{f}$ T-transforms, denoted by $\tilde{S}: (\tilde{K}, T) \to (\tilde{K}_1, T)$ and $\tilde{f}: (\tilde{K}, T) \to (\tilde{K}_1, T)$. Now $\tilde{S}$ induces a set transform $S: K \to K_1$ such that $\pi\tilde{S} = S\pi$. It is easy to see that for any $x \in C_r(K, G)$, when we choose any $\tilde{x} \in C_r(\tilde{K}, G)$ such that $\pi\tilde{x} = x$, the projection $\pi\tilde{f}\tilde{x} \in C_r(K_1, G)$ depends only on x and is independent of the choice of $\tilde{x}$. We define $\pi\tilde{f}\tilde{x}$ as fx; hence

$$\pi\tilde{f} = f\pi. \tag{4}$$

In particular, when T is a finite group, we can define $\bar{\pi}: C_r(K, G) \to C_r(\tilde{K}, G)$ and $\bar{\pi}: C_r(K_1, G) \to C_r(\tilde{K}_1, G)$. Now for any $x \in C_r(K, G)$, choose arbitrarily an $\tilde{x} \in C_r(\tilde{K}, G)$ such that $\pi\tilde{x} = x$. Then from (4) we have $\pi\tilde{f}\tilde{x} = f\pi\tilde{x} = fx$, and so $\bar{\pi}fx = s\tilde{f}\tilde{x} = \tilde{f}s\tilde{x} = \tilde{f}\bar{\pi}x$, and so

$$\tilde{f}\bar{\pi} = \bar{\pi}f. \tag{4$'$}$$

If $\tilde{K}_1$ is a decomposition of $\tilde{K}$,* and the set transform $\tilde{S}: \tilde{K} \to \tilde{K}_1$ and chain transforms $\tilde{f}: C_r(\tilde{K}) \to C_r(\tilde{K}_1)$ and $\tilde{g}: C_r(\tilde{K}_1) \to C_r(\tilde{K}_1)$ which define the decomposition are T-transforms, then $\tilde{K}_1$ is called a T-decomposition of $\tilde{K}$. Now it is easy to see that the set transforms S and the chain transforms f and g induced by $\tilde{S}$, $\tilde{f}$ and $\tilde{g}$ also define a T-decomposition of K; that is, K_1.

---

* See S. Lefschetz, *Algebraic topology*, Amer. Math. Soc. Colloq. Publ., vol. 27, Amer. Math. Soc., Providence, R. I., 1942, pp. 162–163, for the definition of a decomposition.

We shall suppose that p is a fixed prime and that $T \approx I_p$, the additive group of integers mod p, and let t be a fixed generating element. We shall occasionally write $(\tilde{K}, T)$ and $(\tilde{K}, t)$ as $\tilde{K}_T$ and $\tilde{K}_t$. Let $(i \geq 0)$

$$s_i = \begin{cases} 1 + t + \cdots + t^{p-1} \in I(T), & i \text{ is even} \\ 1 - t \in I(T), & i \text{ is odd}. \end{cases} \tag{5}$$

Here s_{2i} is the s above. According to P. A. Smith we have

A. Let $\tilde{x} \in C_r(\tilde{K}, G)$ (or $\tilde{x}' \in C^r(\tilde{K}, G)$). If $s_i \tilde{x} = 0$ (or $s_i' \tilde{x}' = 0$), then there must be $\tilde{y} \in C_r(\tilde{K}, G)$ (or $\tilde{y}' \in C^r(\tilde{K}, G)$) such that $\tilde{x} = s_{i+1} \tilde{y}$ (or $\tilde{x}' = s_{i+1}' \tilde{y}'$).

Now let $X \in H_r(K, I_p)$. Choose any $x_0 \in X$, and $\tilde{x}_0 \in C_r(\tilde{K}, I_p)$ such that $\pi\tilde{x}_0 = x_0$. By (3) we have $s_0 \partial\tilde{x}_0 = \partial s_0 \tilde{x}_0 = \partial\bar{\pi}\tilde{x}_0 = \bar{\pi}\partial x_0 = 0$. Thus from A we have $\tilde{x}_1 \in C_{r-1}(\tilde{K}, I_p)$ such that $\partial\tilde{x}_0 = s_1 \tilde{x}_1$. Since $s_1 \partial\tilde{x}_1 = \partial s_1 \tilde{x}_1 = \partial\partial\tilde{x}_0 = 0$, from A there is an $\tilde{x}_2 \in C_{r-2}(\tilde{K}, I_p)$ such that $\partial\tilde{x}_1 = s_2 \tilde{x}_2$. Continuing this reasoning, we have $\tilde{x}_i \in C_{r-i}(\tilde{K}, I_p)$ such that

$$\pi\tilde{x}_0 = x_0 \in X, \quad \partial\tilde{x}_i = s_{i+1} \tilde{x}_{i+1}, \quad r > i \geq 0. \tag{6}$$

It is easy to see that for any $\pi\tilde{x}_i$ there is a mod p closed chain. If there is another system $y_0 \in X$, $\tilde{y}_i \in C_{r-i}(\tilde{K}, I_p)$ such that $\pi\tilde{y}_0 = y_0 \in X$ and $\partial\tilde{y}_i = s_{i+1} \tilde{y}_{i+1}$, $r > i \geq 0$, then by (3), (A), and induction, there must exist $\tilde{z}_i \in C_{r-i-1}(\tilde{K}, I_p)$, $r > i \geq 0$ such that $\tilde{y}_i - \tilde{x}_i = \partial\tilde{z}_i + s_{i+1} \tilde{z}_{i+1}$, $r > i \geq 0$. Thus $\pi\tilde{y}_i \sim \pi\tilde{x}_i$. In other words the mod p homology classes determined by $\pi\tilde{x}_i$ are independent of the choices of $x_0, \tilde{x}_0, \cdots, \tilde{x}_i$. We define it as $\mu_{i,t} X$, $r \geq i \geq 0$, or $\mu_i X$ when there is no confusion. From this we obtain a system of pseudo-homomorphic transforms

$$\mu_i \colon \; H_r(K, I_p) \to H_{r-i}(K, I_p), \quad r \geq i \geq 0,$$

where μ_0 is obviously the identity transform.

Similarly, let $X^* \in H^r(K, I_p)$. We may choose $x_0' \in X^*$ and $\tilde{x}_0' \in C^{r+i}(\tilde{K}, I_p)$, $i \geq 0$ such that

$$\bar{\pi}'\tilde{x}_0' = x_0' \in X^*, \quad \delta\tilde{x}_i' = s_{i+1}' \tilde{x}_{i+1}', \quad i \geq 0. \tag{6$'$}$$

Thus $\bar{\pi}'\tilde{x}_i'$ is a mod p closed cochain whose mod p cohomology classes are independent of the choices of $x_0', \tilde{x}_0', \cdots, \tilde{x}_i'$. Define them as $\mu_{i,t}^* X^*$ or $\mu_i^* X^*$. Then we obtain a system of pseudo-homomorphic transformations ($\mu_0^* =$ identity transform)

$$\mu_i^* \colon \; H^r(K, I_p) \to H^{r+i}(K, I_p), \quad i \geq 0.$$

Let 1 be the mod p unit class in K; then we call

$$\mu_i^* 1 = U^i(\widetilde{K}, t), \quad i \geq 0 \tag{7}$$

the (mod p) class with respect to the generating element t or (K, t) in $(\widetilde{K}, T)$. For the same $t \in T$, the transforms μ_i, μ_i^* and classes U^i have the following properties:

A1°. $\mu_i \mu_j = \mu_j \mu_i$, $\mu_{2i} = (\mu_2)^i$ and $\mu_{2i+1} = \mu_1(\mu_2)^i$. When $p = 2$ we have in addition $(\mu_1)^2 = \mu_2$, and when $p > 2$, $(\mu_1)^2 = 0$.

A2°. If $\widetilde{f}$ is the chain T transform from the primitive system $(\widetilde{K}, T)$ to $(\widetilde{K}_1, T)$, and f is the chain transform from K to K_1 induced by $\widetilde{f}$, then $f_* \mu_i = \mu_i f_*$ and $\mu_i^* f^* = f^* \mu_i^*$.

A3°. If $(\widetilde{K}_1, T)$ is the T-decomposition of $(\widetilde{K}, T)$, defined by the set transforms $\widetilde{s}$ and the chain transforms $\widetilde{f}(\widetilde{K} \to \widetilde{K}_1)$ and $\widetilde{g}(\widetilde{K}_1 \to \widetilde{K})$, then for any chain transforms $f(K \to K_1)$ and $g(K_1 \to K)$ induced by $\widetilde{f}$ and $\widetilde{g}$ we have $\mu_i^* f^* = f^* \mu_i^*$ and $g^* \mu_i^* = \mu_i^* g^*$, etc.

A4°. Let $K = \widetilde{K}_T$, and let K and $\widetilde{K}$ be simple complexes. In other words, for any $\sigma \in K$ and $\widetilde{\sigma} \in \widetilde{K}$, $\mathrm{cl}\sigma$ and $\mathrm{cl}\widetilde{\sigma}$ have simple homology solutions. Now we can define the cup product $\cup$ on K and $\widetilde{K}$. Thus

$$\mu_i^* X^* = \mu_i^* 1 \cup X^* = U^i(\widetilde{K}, t) \cup X^*, \quad X^* \in H^r(K, I_p) \tag{8}$$

(see [5], §1).

Proof. First let K and $\widetilde{K}$ be simple complexes, and arrange the vertices of K and $\widetilde{K}$ in a sequence such that the projection preserves the order of the vertices. Now use Whitney's method to define the cup product of these orders among the cochains in K, and in $\widetilde{K}$. Thus for any $\widetilde{y} \in C^r(\widetilde{K}, I_p)$ and $\widetilde{z} \in C^s(\widetilde{K}, I_p)$ we should have

$$\left.\begin{aligned} \bar{\pi}'(\widetilde{y} \cup s'\widetilde{z}) &= \bar{\pi}'\widetilde{y} \cup \bar{\pi}'\widetilde{z}, \\ s_i'(\widetilde{y} \cup s'\widetilde{z}) &= s_i'\widetilde{y} \cup s'\widetilde{z}, \quad i \geq 0. \end{aligned}\right\} \tag{9}$$

The proof of the above equality goes as follows. Let $(a_0 \cdots a_{r+s})$ be any $(r+s)$-dimensional simplex in K; according to the chosen order we have $a_0 < \cdots < a_{r+s}$. In $\widetilde{K}$ choose an $(r+s)$-dimensional simplex $(\widetilde{a}_0 \cdots \widetilde{a}_{r+s})$ such that $\pi\widetilde{a}_i = a_i$, $i = 0, \cdots, r+s$. Since π preserves the order of the vertices, $\widetilde{a}_0 < \cdots < \widetilde{a}_{r+s}$. Let $t^j\widetilde{a}_i = \widetilde{a}_i^j$; then

$$\bar{\pi}'(\widetilde{y} \cup s'\widetilde{z})(a_0 \cdots a_{r+s}) = (\widetilde{y} \cup s'\widetilde{z})(\bar{\pi}(a_0 \cdots a_{r+s}))$$

$$= (\widetilde{y} \cup s'\widetilde{z})(s(\widetilde{a}_0 \cdots a_{r+s})) = \sum_{j=0}^{p-1} (\widetilde{y} \cup s'\widetilde{z})(\widetilde{a}_0^j \cdots \widetilde{a}_{r+s}^j) =$$

$$= \sum_{j=0}^{p-1} [\tilde{y}(\tilde{a}_0^j \cdots \tilde{a}_r^j) \cdot s'\tilde{z}(\tilde{a}_r^j \cdots \tilde{a}_{r+s}^j)]$$

$$= \sum_{j=0}^{p-1} [\tilde{y}(\tilde{a}_0^j \cdots \tilde{a}_r^j) \cdot \tilde{z}(s(\tilde{a}_r^j \cdots \tilde{a}_{r+s}^j))]$$

$$= \sum_{j=0}^{p-1} [\tilde{y}(\tilde{a}_0^j \cdots \tilde{a}_r^j) \cdot \tilde{z}(s(\tilde{a}_r \cdots \tilde{a}_{r+s}))]$$

$$= [\sum_{j=0}^{p-1} \tilde{y}(\tilde{a}_0^j \cdots \tilde{a}_r^j)] \cdot \tilde{z}(s(\tilde{a}_r \cdots \tilde{a}_{r+s}))$$

$$= \tilde{y}(s(\tilde{a}_0 \cdots \tilde{a}_r)) \cdot \tilde{z}(s(\tilde{a}_r \cdots \tilde{a}_{r+s}))$$

$$= \tilde{y}(\bar{\pi}(a_0 \cdots a_r)) \cdot \tilde{z}(\bar{\pi}(\tilde{a}_r \cdots \tilde{a}_{r+s}))$$

$$= \bar{\pi}'\tilde{y}(a_0 \cdots a_r) \cdot \bar{\pi}'\tilde{z}(a_r \cdots a_{r+s})$$

$$= (\bar{\pi}'\tilde{y} \cup \bar{\pi}'\tilde{z})(a_0 \cdots a_{r+s}).$$

Since $(a_0 \cdots a_{r+s})$ is arbitrary, the first equality in (9) is proven. Similarly for the second one.

Now choose $\tilde{u}_i \in C^i(\tilde{K}, I_p)$ $(i \geq 0)$ such that (1 = mod p unit cochain)

$$\bar{\pi}'\tilde{u}_0 = 1, \quad \delta\tilde{u}_i = s'_{i+1}\tilde{u}_{i+1}, \quad i \geq 0. \tag{10}$$

Take $x'_0 \in X^*$ and $\tilde{x}'_i \in C^{r+i}(\tilde{K}, I_p)$ such that $(6)'$ is satisfied. From (9)

$$\bar{\pi}'(\tilde{u}_0 \cup s'\tilde{x}'_0) = \bar{\pi}'\tilde{u}_0 \cup \bar{\pi}'\tilde{x}'_0 = 1 \cup \bar{\pi}'\tilde{x}'_0 = \bar{\pi}'\tilde{x}'_0 = x'_0 \in X^*,$$

$$\begin{aligned}\delta(\tilde{u}_i \cup s'\tilde{x}'_0) &= \delta\tilde{u}_i \cup s'\tilde{x}'_0 + (-1)^i\tilde{u}_i \cup \delta s'\tilde{x}'_0 \\ &= s'_{i+1}\tilde{u}_{i+1} \cup s'\tilde{x}'_0 + (-1)^i\tilde{u}_i \cup s'\delta\tilde{x}'_0 \\ &= s'_{i+1}(\tilde{u}_{i+1} \cup s'\tilde{x}'_0) + (-1)^i\tilde{u}_i \cup s's'_1\tilde{x}'_1 \\ &= s'_{i+1}(\tilde{u}_{i+1} \cup s'\tilde{x}'_0),\end{aligned}$$

$$\bar{\pi}'(\tilde{u}_i \cup s'\tilde{x}'_0) = \bar{\pi}'\tilde{u}_i \cup \bar{\pi}'\tilde{x}'_0 = \bar{\pi}'\tilde{u}_i \cup x'_0.$$

But from the definition $\bar{\pi}'\tilde{u}_i \in \mu_i^* 1 = U^i(\tilde{K}, t)$ and $\bar{\pi}'(\tilde{u}_i \cup s'\tilde{x}'_0) \in \mu_i^* X^*$. The theorem follows from this last equality.

In general, we can take $\tilde{K}$ to be the derived complex of $\tilde{K}$. Now T is applied on $\tilde{K}_1$ in such a way that it becomes a primitive system $(\tilde{K}_1, T)$. It is a simple matter to see that $K_1 = (\tilde{K}_1)_T$ is the derived complex of K. Since derivatives do not affect the homology group and cup products, the theorem follows from A3° and the previous case.

From A1° and A4°, we know that the following relations hold among the classes $U^i = U^i(\tilde{K}, t)$ in $(\tilde{K}, t)$:

A5°. If $p = 2$, then $U^i = (U^1)^i$. If $p > 2$, then $(U^1)^2 = 0$, $U^{2i} = (U^2)^i$ and $U^{2i+1} = U^1(U^2)^i$.

A6°. If $\tilde{L}$ is another complex and T is applied on $\tilde{K} \times \tilde{L}$ in such a way that $t(\tilde{\sigma} \times \tilde{\tau}) = t\tilde{\sigma} \times \tilde{\tau}$ $(\tilde{\sigma} \in \tilde{K}, \tilde{\tau} \in \tilde{L})$, then $(\tilde{K} \times \tilde{L}, T)$ forms a primitive system. For t, let the transforms corresponding to μ_i and μ_i^* be $\bar{\mu}_i$ and $\bar{\mu}_i^*$. Then

$$\bar{\mu}_i(X \otimes Y) = \mu_i X \otimes Y, \quad \bar{\mu}_i^*(X^* \otimes Y^*) = \mu_i^* X^* \otimes Y^*.$$

Here

$$X \in H(K, I_p),\ Y \in H(\tilde{L}, I_p),\ X^* \in H^*(K, I_p),\ Y^* \in H^*(\tilde{L}, I_p).$$

Now let $\tilde{K}$ be an m-dimensional orientable (finite) associated manifold having fixed orientation. Let the r-dimensional cell in $\tilde{K}$ be $\tilde{\sigma}^r_\lambda$, where λ is an incidence number related to r. For the orientation of $\tilde{K}$, the dual of $\tilde{\sigma}^r_\lambda \in \tilde{K}$ is denoted by $\tilde{\tau}^{m-r}_\lambda$. Hence the boundary relations in $\tilde{K}$ and $\tilde{K}^*$ can be written as

$$\left.\begin{aligned} \partial\tilde{\sigma}^r_\lambda &= \sum_\mu a^r_{\lambda\mu}\tilde{\sigma}^{r-1}_\mu, \\ \partial\tilde{\tau}^{m-r+1}_\mu &= (-1)^r \sum_\lambda a^r_{\lambda\mu}\tilde{\tau}^{m-r}_\lambda \end{aligned}\right\} \quad (a^r_{\lambda\mu} = \pm 1). \tag{11}$$

For any coefficient group G and chain $x = \Sigma_\lambda g_\lambda \tilde{\sigma}^r_\lambda \in C_r(\tilde{K}, G)$, define $\tilde{d}x \in C^{m-r}(\tilde{K}^*, G)$ such that $\tilde{d}x(\tilde{\tau}^{m-r}_\lambda) = g_\lambda$. Then $\tilde{d}$ defines a homomorphic transform

$$\tilde{d}\colon\ C_r(\tilde{K}, G) \approx C^{m-r}(\tilde{K}^*, G).$$

We know that

$$\delta\tilde{d}x = (-1)^r \tilde{d}\delta x, \qquad x \in C_r(\tilde{K}, G). \tag{12}$$

Hence $\tilde{d}$ induces the homomorphic transformation

$$\tilde{D}\colon\ H_r(\tilde{K}, G) \approx H^{m-r}(\tilde{K}^*, G).$$

Let the associated manifold $\tilde{K}$ have a group T applied on it in such a way that $(\tilde{K}, T)$ is a primitive system. Let $t\tilde{\sigma}^r_\lambda = \tilde{\sigma}^r_{t(\lambda)}$, where $t \in T$ is arbitrary. We define $t\tilde{\tau}^{m-r}_\lambda = \tilde{\tau}^{m-r}_{t(\lambda)}$. Thus T is also applied on $\tilde{K}^*$. From the first equality in (11), we have

$$\partial t\tilde{\sigma}^r_\lambda = \partial\tilde{\sigma}^r_{t(\lambda)} = \sum_\mu a^r_{t(\lambda),\,t(\mu)}\,\tilde{\sigma}^{r-1}_{t(\mu)},$$

$$t\partial\tilde{\sigma}^r_\lambda = t\sum_\mu a^r_{\lambda\mu}\,\tilde{\sigma}^{r-1}_\mu = \sum_\mu a^r_{\lambda\mu}\,t\tilde{\sigma}^{r-1}_\mu = \sum_\mu a^r_{\lambda\mu}\tilde{\sigma}^{r-1}_{t(\mu)}.$$

Since $(\tilde{K}, T)$ is a primitive system, $\partial t = t\partial$, and from the above two equalities we have $a^r_{\lambda\mu} = a^r_{t(\lambda),\,t(\mu)}$. Similarly, from the second part of (11) we obtain

$$\partial t\tilde{\tau}^{m-r+1}_\mu = \partial\tilde{\tau}^{m-r+1}_{t(\mu)} = (-1)^r\sum_\lambda a^r_{t(\lambda),\,t(\mu)}\,\tilde{\tau}^{m-r}_{t(\lambda)},$$

$$t\partial\tilde{\tau}^{m-r+1}_\mu = (-1)^r\,t\sum_\lambda a^r_{\lambda\mu}\,\tilde{\tau}^{m-r}_\lambda = (-1)^r\sum_\lambda a^r_{\lambda\mu}\,\tilde{\tau}^{m-r}_{t(\lambda)}.$$

From $a^r_{\lambda\mu} = a^r_{t(\lambda),\,t(\mu)}$ we obtain $\partial t\tilde{\tau}^{m-r+1}_\mu = t\partial\tilde{\tau}^{m-r+1}_\mu$. Hence $(\tilde{K}^*, T)$ satisfies the condition corresponding to 4°. As 1°–3° are obvious, (K^*, T) is also a primitive system.

Now let $T \approx I_p$, and let the generator $t \in I_p$ be an orientation-preserving transform. Then $K = \tilde{K}_T$ and $K^* = \tilde{K}^*_T$ are also dual decompositions of an orientable m-dimensional (finite) associated manifold, and the orientation of $\tilde{K}$ induces the orientation of K in such a way that the projection $\pi\colon \tilde{K} \to K$ is orientation preserving. For this orientation of K, we can define the dual transforms as before:

$$d\colon\ C_r(K, G) \approx C^{m-r}(K^*, G),$$
$$D\colon\ H_r(K, G) \approx H^{m-r}(K^*, G).$$

When $G \approx I_p$, we have the following relations; their proof is omitted.

$$\pi'\tilde{d} = d\pi, \tag{13}$$
$$(t')^{-1}\tilde{d} = \tilde{d}t, \tag{14}$$
$$\tilde{d}s_i = s_i'\bar{s}_i'\tilde{d}. \tag{15}$$

Here

$$\bar{s}_i' = \begin{cases} 1, & i \text{ is even} \\ 1 + t' + \cdots + t'^{p-2}, & i \text{ is odd} \end{cases} \tag{15$'$}$$

$$\bar{\pi}'\bar{s}_i' = (-1)^i\bar{\pi}'. \tag{16}$$

A7°. For the primitive systems $(\tilde{K}, t)$, $(\tilde{K}^*, t)$, define

$$\mu_i\colon\ H_r(K, I_p) \to H_{r-i}(K, I_p),$$
$$\mu_i^*\colon\ H^r(K^*, I_p) \to H^{r+i}(K^*, I_p)$$

as before. Then

$$D\mu_i X = (-1)^{i(r+1)}\mu_i^* DX, \quad X \in H_r(\widetilde{K}, I_p). \tag{17}$$

Proof. Choose $x_0 \in X$, and $\widetilde{x}_i \in C_{r-i}(\widetilde{K}, I_p)$ such that (6) is satisfied. From (6) and (12)–(16) we obtain

$$\pi' \widetilde{d}\widetilde{x}_0 = d\pi\widetilde{x}_0 = dx_0 \in DX,$$

$$\delta(\bar{s}_1' \cdots \bar{s}_i' \widetilde{d}\widetilde{x}_i) = \bar{s}_1' \cdots \bar{s}_i' \delta\widetilde{d}\widetilde{x}_i = (-1)^{r-i}\bar{s}_1' \cdots \bar{s}_i' \widetilde{d}\,\partial\widetilde{x}_i$$

$$= (-1)^{r-i}\bar{s}_1' \cdots \bar{s}_i' \widetilde{d}s_{i+1}\widetilde{x}_{i+1} = (-1)^{r-i}s_{i+1}'(\bar{s}_1' \cdots \bar{s}_i' \bar{s}_{i+1}' \widetilde{d}\widetilde{x}_{i+1}),$$

$$\pi'(\bar{s}_1' \cdots \bar{s}_i' \widetilde{d}\widetilde{x}_i) = (-1)^{1+2+\cdots+i}\bar{\pi}'\widetilde{d}\widetilde{x}_i = (-1)^{1+2+\cdots+i} d\pi\widetilde{x}_i.$$

By definition $\pi\widetilde{x}_i \in \mu_i X$, and $\mu_i^* DX$ contains

$$(-1)^{r+(r-1)+\cdots+(r-i+1)}\bar{\pi}'(\bar{s}_1' \cdots \bar{s}_i' \widetilde{d}\widetilde{x}_i) = (-1)^{i(r+1)} d\pi\widetilde{x}_i.$$

Hence the assertion.

Now let $\widetilde{M}$ be a compact space, T the group of topological transforms in $\widetilde{M}$, and let each $t \in T$ be such that when $t \neq 1$, t does not have a fixed point. Then we call $(\widetilde{M}, T)$ a primitive system. Given another primitive system $(\widetilde{M}_1, T)$, a continuous map $\widetilde{f}: \widetilde{M} \to \widetilde{M}_1$ such that $\widetilde{f}t = t\widetilde{f}$ for each $t \in T$ is called a T-map and is denoted by $\widetilde{f}: (\widetilde{M}, T) \to (\widetilde{M}_1, T)$. Identifying, as one point, all those points in $\widetilde{M}$ which can be transformed to each other, we obtain a space M, called the modular space of $(\widetilde{M}, T)$, which is denoted by $M = \widetilde{M}_T$. Its natural projection is $\pi: \widetilde{M} \to M$. Hence every T-map must induce an $f: M \to M_1$, such that $f\pi = \pi f$. Let $\widetilde{V}_\alpha = \{\widetilde{V}_{i\alpha}\}$ be a finite open covering of $\widetilde{M}$ such that $t\widetilde{V}_{i\alpha} \in \widetilde{V}_\alpha$ whenever $\widetilde{V}_{i\alpha} \in \widetilde{V}_\alpha$, and every $\widetilde{V}_{i\alpha}$ and every $t\widetilde{V}_{i\alpha}$ $(t \neq 1)$ are disjoint (here $t \in T$). Then $\widetilde{V}_\alpha$ is said to be simple. Now $V_\alpha = \{V_{i\alpha} = \pi\widetilde{V}_{i\alpha}\}$ is an open covering of M. Let $\widetilde{K}_\alpha$ be the nerve of $\widetilde{V}_\alpha$ and let K_α be the nerve of V_α. Then in $\widetilde{K}_\alpha$ t will induce a simple transform, t_α. Hence the group $T_\alpha = \{t_\alpha \mid t \in T\}$ acts on $\widetilde{K}_\alpha$ and makes $(\widetilde{K}_\alpha, T_\alpha)$ a primitive system, and $K_\alpha = (\widetilde{K}_\alpha)_{T_\alpha}$. Now let $T \approx I_p$ and $t \in T$ be a fixed generating element. According to P. A. Smith, there exists a simple open covering, and the totality of these open coverings and the totality of all open coverings are cofinal. Let $\widetilde{V}_\alpha$ and $\widetilde{V}_\beta$ be simple, with $\widetilde{V}_\alpha \subset \widetilde{V}_\beta$; then we can choose a simple map $\widetilde{f}_{\alpha\beta}: \widetilde{K}_\alpha \to \widetilde{K}_\beta$ such that $t_\beta \widetilde{f}_{\alpha\beta} = \widetilde{f}_{\alpha\beta} t_\alpha$. Now $\widetilde{f}_{\alpha\beta}$ induces $f_{\alpha\beta}: K_\alpha \to K_\beta$. By A2°, we have

$$\mu_i^* f_{\alpha\beta}^* = f_{\alpha\beta}^* \mu_i^*. \tag{18}$$

Now let $X \in H^r(M, I_p)$. Take a $\widetilde{V}_\alpha$ such that X has a representative X_α in K_α. Then, by (18), the representative of $\mu_i^* X_\alpha$ in $H^{r+1}(M, I_p)$ is independent of the choice of α. Define it to be $\tilde{\mu}_i^* X$. Similarly define μ_i: $H_r(M, I_p) \to H_{r-1}(M, I_p)$ and the classes of $(\tilde{M}, T)$ as

$$U^i(\tilde{M}, t) = \mu_i^* 1 \in H^i(M, I_p), \quad i \geq 0. \tag{19}$$

Naturally the definitions of μ_i, μ_i^* and U^i depend on the choice of the generating element t.

If $\tilde{M}$ is also the space of a complex $\tilde{K}$, and the cell transforms in $\tilde{K}$ induced by t make $(\tilde{K}, t)$ a primitive system, then from A2° and A3° we know that for $(\tilde{K}, t)$ and $(\tilde{M}, t)$ (i. e. $(\tilde{M}, T)$), the defining μ_i and μ_i^* coincide. In particular $U^i(\tilde{M}, t) = U^i(\tilde{K}, t)$. μ_i and μ_i^* have unchanging meaning, and hence not only A3°, but also A1°–A6° can be generalized to be properties of (M, t).

§2. Lens spaces

From now on, p is a fixed odd prime, $q = (p-1)/2$, and $\epsilon = e^{2\pi i/p}$ unless otherwise specified.

Let U^n be a complex n-dimensional unitary space with $(z_1, \cdots, z_n)$ as complex coordinate system. Let $\tilde{L}$ be the unit ball in U^n: $\sum_{i=1}^n z_i \bar{z}_i = 1$, where $\bar{z}_i$ is the conjugate of z_i. Let l_i, m_i, $i = 1, \cdots, n$ be a system of integers relatively prime to p, with $l_i m_i \equiv 1 \bmod p$. Then

$$t: (z_1, \cdots, z_n) \to (\epsilon^{m_1} z_1, \cdots, \epsilon^{m_n} z_n) \tag{1}$$

is a topological orientation-preserving transform in $\tilde{L}$ with period p and without a fixed point. Thus $(\tilde{L}, t)$ is a primitive system whose modular space $L = \tilde{L}_t$ is called an $(l_1, \cdots, l_n)$ type lens space defined in U^n with respect to the coordinates $(z_1, \cdots, z_n)$. $\tilde{L}$ and L are orientable manifolds with natural orientations. Define $\tilde{S}_k$, $\tilde{I}_{k,j} \subset \tilde{L}$ and $\tilde{P}_{k,j} \in \tilde{L}$ as follows ($k = 1, 2, \cdots, n$; j an arbitrary integer):

$$\tilde{S}_k: \ z_1 = \cdots = z_{k-1} = 0,\ z_{k+1} = \cdots = z_n = 0,\ |z_k| = 1;$$

$$\tilde{I}_{k,j}: z_1 = \cdots = z_{k-1} = 0,\ z_{k+1} = \cdots = z_n = 0,\ |z_k| = 1,$$

$$\frac{2\pi}{p} \cdot j \leq \arg z_k \leq \frac{2\pi}{p} \cdot (j+1);$$

$$\tilde{P}_{k,j}: z_1 = \cdots = z_{k-1} = 0,\ z_{k+1} = \cdots = z_n = 0,\ |z_k| = 1,$$

$$\arg z_k = \frac{2\pi}{p} \cdot j.$$

Define $[k]_j^* \subset \tilde{L}$ ($k = 0, 1, \cdots, n-1$; j an arbitrary integer) as follows ($\circ$ is one kind of join; its meaning is self-explanatory):

$$[2k+1]_j^* = \widetilde{S}_1 \circ \cdots \circ \widetilde{S}_k \circ I_{k+1,\, j}, \; [2k]_j^* = \widetilde{S}_1 \circ \cdots \circ \widetilde{S}_k \circ \widetilde{P}_{k+1,\, j} \tag{2}$$

or

$$\left.\begin{aligned} &[2k+1]_j^*: \frac{2\pi}{p}\cdot j \le \arg z_{k+1} \le \frac{2\pi}{p}\cdot (j+1), \\ &z_{k+2} = \cdots = z_n = 0, \quad \sum_{i=1}^n z_i \bar{z}_i = 1, \\ &[2k]_j^*: \quad \arg z_{k+1} = \frac{2\pi}{p}\cdot j, \\ &z_{k+1} = \cdots = z_n = 0, \quad \sum_{i=1}^n z_i \bar{z}_i = 1. \end{aligned}\right\} \tag{2$'$}$$

Then $[k]_j = [k]_j^* - \Sigma_{i=1}^p [k-1]_i^*$ is a k-dimensional open cell and the collection of $[k]_j$ $(k = 0, 1, \cdots, n-1;\ j = 1, \cdots, p)$ forms a decomposition $\widetilde{K}$ of $\widetilde{L}$. Obviously $[k]_j^*$ is the closure of, and a pseudo-manifold of, $[k]_j$, oriented naturally by (2). The natural orientations of $[2n-1]_i^*$ are naturally agreeable, that is the natural orientation of $\widetilde{L}$. For these $[k]_j$ with natural orientation, we have

$$\left.\begin{aligned} \partial[2k]_j &= [2k-1]_1 + \cdots + [2k-1]_p, \\ \partial[2k+1]_j &= [2k]_{j+1} - [2k]_j. \end{aligned}\right\} \tag{3}$$

Let t also be the cell transform in $\widetilde{K}$ induced by t; then

$$\left.\begin{aligned} t[2k+1]_j &= [2k+1]_{j+m_{k+1}}, \\ t[2k]_j &= [2k]_{j+m_{k+1}}. \end{aligned}\right\} \tag{4}$$

or

$$\left.\begin{aligned} t^{l_{k+1}}[2k+1]_j &= [2k+1]_{j+1}, \\ t^{l_{k+1}}[2k]_j &= [2k]_{j+1}. \end{aligned}\right\} \tag{4$'$}$$

From (4) or (4)$'$ we know that $(\widetilde{K}, t)$ is a primitive system whose modular complex $K = \widetilde{K}t$ is a decomposition of L. We call $\widetilde{K}$ and K the standard decompositions of $\widetilde{L}$ and L with respect to the coordinate $(z_1, \cdots, z_n)$. Let the projection be $\lambda: \widetilde{K} \to K$, $\widetilde{L} \to L$, and let

$$\lambda[k]_j = [k], \qquad (k = 0, 1, \cdots, 2n-1). \tag{5}$$

Then from (3) we have

$$\left.\begin{aligned} \partial[2k] &= p[2k-1], \\ \partial[2k+1] &= 0. \end{aligned}\right\} \tag{6}$$

From (6), we known that for any k, $[k]$ is a mod p closed chain with the naturally oriented pseudomanifold $[k]^* = \lambda[k]_j^*$ as its geometric representative. The homology class of $[k]$ is a generating element of $H_k(L, I_p) \approx I_p$. But these generating elements and the decomposition K of L depend on the choices of the coordinates $(z_1, \cdots, z_n)$.

Now let the $(2n-1)$-dimensional fundamental closed cochain determined by the naturally oriented L be x_0, and let

$$t_j = 1 + t + \cdots + t^{l_j - 1}, \quad j = 1, \cdots, n, \tag{7}$$

$$\left.\begin{aligned} \tilde{x}_{2k-1} &= (-1)^k t_n t_{n-1} \cdots t_{n-k+1} [2n-2k]_1, \\ \tilde{x}_{2k} &= (-1)^k t_n t_{n-1} \cdots t_{n-k+1} [2n-2k-1]_1; \end{aligned}\right\} (k = 0, 1, \cdots, n-1) \tag{8}$$

Then from (3) and (4)$'$ we obtain (see §1 (5) for the notation s_i)

$$\partial\tilde{x}_i = s_i \tilde{x}_{i+1}, \tag{9}$$

$$\left.\begin{aligned} \lambda\tilde{x}_{2k-1} &= (-1)^k l_n l_{n-1} \cdots l_{n-k+1} [2n-2k], \\ \lambda\tilde{x}_{2k} &= (-1)^k l_n l_{n-1} \cdots l_{n-k+1} [2n-2k-1] \end{aligned}\right\} (k = 0, 1, \cdots, n-1). \tag{10}$$

From (9), (10) and the definition of the μ_i, if X is the $(2n-1)$-dimensional mod p fundamental homology class in L with natural orientation, then we have the following results:

B1°. $\mu_{2n-2k-1} X$ and $\mu_{2n-2k} X$ contain mod p closed chains

$$(-1)^{n-k} l_{k+1} \cdots l_n [2k] \text{ and } (-1)^{n-k} l_{k+1} \cdots l_n [2k-1].$$

Since l_i and p are relatively prime,

B2°. $\mu_k X$, $k = 0, 1, \cdots, 2n-1$ becomes an additive basis of $H(L, I_p)$.

Now for the natural orientation of L, let $D: H_r(L, I_p) \approx H^{2n-1-r}(L, I_p)$ be its dual transform. Let 1 be the unit class in L, and let $U^j = U^j(\tilde{L}, t) = \mu_j^* 1$ be the mod p classes of $(\tilde{L}, t)$. Then from A4°, 5°, 7° and the above B2°, we obtain

B3°. In the lens space L, the classes $U^i = U^i(\tilde{L}, t)$, $i = 0, 1, \cdots, 2n-1$ form an additive basis of the cohomology ring $H^*(L, I_p)$. Speaking of the ring structure, $H^*(L, I_p)$ is generated by U^1, U^2 where U^1 and U^2 satisfy

$$(U^1)^2 = 0, \ (U^2)^n = 0, \ U^1 U^2 = U^2 U^1; \tag{11}$$

$$U^{2i}\,(U^2)^i,\ U^{2i+1} = U^1(U^2)^i, \quad i = 0, 1, \cdots, n-1; \tag{11$'$}$$

and

$$U^k = D\mu_k X, \quad k = 0,\ 1, \cdots, 2n-1. \tag{12}$$

In this paper, one special type of lens space possesses a very important property. Its definition follows: let R^m be m-dimensional Euclidean space. In R^m choose p vectors $\mathfrak{x}_1, \cdots, \mathfrak{x}_p$ satisfying

$$\sum_{i=1}^{p} |\mathfrak{x}_i|^2 = 1, \tag{13}$$

$$\sum_{i=1}^{p} \mathfrak{x}_i = 0. \tag{13$'$}$$

These (ordered) vector systems $(\mathfrak{x}_1, \cdots, \mathfrak{x}_p)$ naturally form a space $\tilde{S}$ homeomorphic to the $(2qm-1)$-dimensional sphere. In $\tilde{S}$ define an orientation-preserving topological transform:

$$t\colon\ (\mathfrak{x}_1, \cdots, \mathfrak{x}_p) \longrightarrow (\mathfrak{x}_2, \cdots, \mathfrak{x}_p, \mathfrak{x}_1). \tag{14}$$

Then t has period p and has no fixed point. Let the modular space of the primitive system $(\tilde{S}, t)$ be L^p_m. Then L^p_m is a lens space. The assertion follows from this.

Let $(x_1, \cdots, x_m)$ be a system of coordinates in R^m and U^m be an m-dimensional unitary space as before. For any $\mathfrak{x} = (x_1, \cdots, x_m) \in R^m$, define $\tilde{f}(\mathfrak{x}) = \mathfrak{z} = (z_1, \cdots, z_m) \in U^m$ as $z_i = x_i$, $i = 1, 2, \cdots, m$. Hence we obtain a map $\tilde{f}: R^m \longrightarrow U^m$. For any vector system $(\mathfrak{x}_1, \cdots, \mathfrak{x}_p) \in \tilde{S}$, $(\mathfrak{x}_i \in R^m)$ let

$$\mathfrak{x}'_i = \sum_{j=1}^{p} \epsilon^{(j-1)i}\, \bar{f}(\mathfrak{x}_j), \quad i = 0, 1, \cdots, p-1; \tag{15}$$

then $\mathfrak{z}'_i$ and $\mathfrak{z}'_{p-i}$ are conjugate complex vectors in U^m:

$$\mathfrak{z}'_{p-i} = \bar{\mathfrak{z}}'_i, \quad i = 1, \cdots, q, \tag{16}$$

and

$$p\bar{f}(\mathfrak{x}_j) = \sum_{i=0}^{p-1} \epsilon^{-(j-1)i}\, \mathfrak{z}'_i, \quad j = 1, \cdots, p. \tag{17}$$

From (15) we know (13)$'$ is equivalent to

$$\mathfrak{z}'_0 = 0. \tag{18}$$

From (13), (16) and (18) we have

$$2\sum_{i=1}^{q}|\mathfrak{z}_i'|^2=\sum_{i=0}^{p-1}\mathfrak{z}_i'\bar{\mathfrak{z}}_i'=p\sum_{j=1}^{p}|\mathfrak{x}_i|^2=p. \tag{19}$$

Hence we can let

$$\mathfrak{z}_i=\sqrt{\frac{2}{p}}\,\mathfrak{z}_i', \tag{20}$$

and thus

$$\sum_{i=1}^{q}|\mathfrak{z}_i|^2=1. \tag{21}$$

Let $\tilde{S}'$ be the space naturally determined by all (ordered) vector systems $(\mathfrak{z}_1,\cdots,\mathfrak{z}_p)$ $(\mathfrak{z}_i\in U^m)$ satisfying (21). Then clearly (17), (19) and (20) define a topological transform from $\tilde{S}$ to $\tilde{S}'$,

$$h:\ (\mathfrak{x}_1,\cdots,\mathfrak{x}_p)\to(\mathfrak{z}_1,\cdots,\mathfrak{z}_q). \tag{22}$$

Under h, t becomes a topological transform in $\tilde{S}'$:

$$t'=hth^{-1}:\ (\mathfrak{z}_1,\cdots,\mathfrak{z}_q)\to(\epsilon^{-1}\mathfrak{z}_1,\cdots,\epsilon^{-i}\mathfrak{z}_i,\cdots,\epsilon^{-q}\mathfrak{z}_q). \tag{23}$$

Let $U_q^m=\underbrace{U^m\otimes\cdots\otimes U^m}_{q}$. In U_q^m take coordinates

$$(z_1^{(1)},\cdots,z_m^{(1)},z_1^{(2)},\cdots,z_m^{(2)},\cdots,z_1^{(q)},\cdots,z_m^{(q)}),$$

Here $(z_1^{(i)},\cdots,z_m^{(i)})=\tilde{\mathfrak{z}}_i$ and $(\mathfrak{z}_1,\cdots,\tilde{\mathfrak{z}}_q)\in U_q^m$. We see that $(\tilde{S}',t')$, the modular space L_m^p of $(\tilde{S},\tilde{t})$, is a lens space. Its form is

$$(1',\cdots,1',\cdots,i',\cdots,i',\cdots,q',\cdots,q'),$$

where i' satisfies

$$ii'\equiv-1 \bmod p,\quad i=1,\cdots,q,$$

and every i' appears m times. This is what we wanted to prove.

The spaces L_m^p will be called the special spaces defined in $R_p^m=\underbrace{R^m\oplus\cdots\oplus R^m}_{p}$ or U_q^m.

§3. A lens space bundle in a complex Grassmann manifold

In this section p, q and ϵ are the same as in §2. All chains, homology relations and homology classes refer to the coefficient group I_p.

Let U^{n+m} be an $(n+m)$-dimensional complex unitary space, with

$(z_1, \cdots, z_{n+m})$ as a system of complex coordinates and U^k as the complex k-face defined by $z_1 = \cdots = z_{n+m-k}$. Let $C_{n,m}$ be the complex Grassmann manifold defined in U^{n+m}, where the chain $[a_1, \cdots, a_m]_c$ and pseudomanifold $[a_1, \cdots, a_m]_c^*$ $(0 \le a_1 \le \cdots \le a_m \le n)$ are defined as in §3 of [1], with respect to the standard cell decomposition determined by the sequence $U^1 \subset \cdots \subset U^{n+m-1}$. Let $\widetilde{C}_{n,m}$ be the bundle space of the model sphere fiber bundle $\mathfrak{C}_{n,m}$ in $C_{n,m}$. Its points can be expressed in the form $(\mathfrak{z}, Z)$, where $Z \in C_{n,m}$ and $\mathfrak{z} \in Z \subset U^{n+m}$, $|\mathfrak{z}| = 1$. Construct the qth Whitney power $\mathfrak{C}^q_{n,m} = \underbrace{\mathfrak{C}_{n,m} \cup \cdots \cup \mathfrak{C}_{n,m}}_{q}$ in $\mathfrak{C}_{n,m}$, with bundle space $\widetilde{\Delta}$. Then the points in $\widetilde{\Delta}$ are representable as $\{\mathfrak{z}_1, \cdots, \mathfrak{z}_q; Z\}$, where $Z \in C_{n,m}$, $\mathfrak{z}_i \in Z \subset U^{n+m}$, $i = 1, \cdots, q$ and $\Sigma_{i=1}^q |\mathfrak{z}_i|^2 = 1$.

Let $\widetilde{S}$ be the unit sphere of $U_q^{n+m} = (U^{n+m})^q$. Its points can be expressed as $(\mathfrak{z}_1, \cdots, \mathfrak{z}_q)$, where $\mathfrak{z}_i \in U^{n+m}$ and $\Sigma_{i=1}^q |\mathfrak{z}_i|^2 = 1$. In $\widetilde{S}$ the transform

$$t: \quad (\mathfrak{z}_1, \cdots, \mathfrak{z}_q) \to (\epsilon^{-1} \mathfrak{z}_1, \cdots, \epsilon^{-q} \mathfrak{z}_q)$$

is a topological transform with period p, and no fixed points. The modular space $\widetilde{S}_t$ is the special lens space L^p_{n+m} mentioned in §2. Similarly, t induces a topological transformation $\{\mathfrak{z}_1, \cdots, \mathfrak{z}_q; Z\} \to \{\epsilon^{-1} \mathfrak{z}_1, \cdots, \epsilon^{-q} \mathfrak{z}_q; Z\}$ in $\widetilde{\Delta}$ with period p and no fixed points. Call this transformation t also. Obviously the modular space $\Delta = \widetilde{\Delta}_t$ is a bundle space of a bundle $\mathfrak{L}$. Its base space is $C_{n,m}$. If $Z \in C_{n,m}$ is a fiber, then it is the special lens space L^p_m defined in $Z_q = \underbrace{Z \times \cdots \times Z}_{q}$. Let $\pi: \Delta \to C_{n,m}$ be its projection; let $U^i = U^i(\widetilde{\Delta}, t)$ be the class in Δ for t; let C_p^{2i} be the (mod p) C-class in $C_{n,m}$ The purpose of this section is to prove the validity of the equality

$$\prod_{i=1}^{q} \sum_{k=0}^{m} i^{m-k} U^{2m-2k} \cup \pi^* C_p^{2k} = 0. \tag{1}$$

For this, let $\widetilde{\Lambda} = \widetilde{S} \times C_{n,m}$. Its points can be expressed as $\{\mathfrak{z}_1, \cdots, \mathfrak{z}_q; Z\}$, where $(\mathfrak{z}_1, \cdots, \mathfrak{z}_q) \in \widetilde{S}$ and $Z \in C_{n,m}$. Let $\widetilde{\Delta}_i$ $(i = 1, 2, \cdots, n)$ be the subspace of $\widetilde{\Lambda}$ determined by the condition that $\mathfrak{z}_i \in Z$. Obviously t induces topological transformations in $\widetilde{\Lambda}$ and in $\widetilde{\Delta}_i$ with period p, and no fixed points. Also call these transforms t. Let

$$\Lambda = \widetilde{\Lambda}_t = L^p_{n+m} \times C_{n,m}, \quad \Delta_i = (\widetilde{\Delta}_i)_t, \tag{2}$$

Then Λ and Δ are manifolds, the Δ_i's are pseudomanifolds and

$$\Delta = \Delta_1 \cap \cdots \cap \Delta_q. \tag{3}$$

Also

$$\begin{cases} \text{Dim } L^p_{n+m} = 2q(n+m) - 1, \\ \text{Dim } \Lambda = 2nm + 2q(n+m) - 1, \\ \text{Dim } \Delta = 2nm + 2qm - 1, \\ \text{Dim } \Delta_i = \text{Dim } \Lambda - 2n. \end{cases}$$

Let δ_i be the (mod p) closed chain in Λ determined by the natural orientation of Δ_i. We shall first prove

$$\delta_i \sim \sum_{k=0}^{n} i^{n-k} \xi_{2n-2k} \times [n-k, n, \cdots, n]_c, \quad i = 1, \cdots, q. \tag{4}$$

Here $\xi_j \in \mu_i X$ and X is the $[2q(n+m)-1]$-dimensional fundamental homology class determined by the natural orientation of L^p_{n+m}. If we construct a standard decomposition of L^p_{n+m} according to §2 without taking the coordinates, then from B1° in §2 $(0 \le k \le n)$

$$[2n-2k] \in (q!)^{n+m} \mu_{2q(n+m)-1-2n+2k} X, \tag{5}$$

and from B3° in §2 and dual theorems

$$\mu_{2n-2k} X \cdot \mu_{2q(n+m)-1-2n+2k} X = \mu_{2q(n+m)-1} X,$$

where $\xi_{2n-2k} \cdot [2n-2k] \sim [0]$. From this we know that (4) is equivalent to the following two formulas in the naturally oriented Λ:

$$\delta_i \cdot ([2n-2k] \otimes [a_1, \cdots, a_m]_c) = \begin{cases} 0, & a_m < k, \\ i^{n-k}, & a_m = k, \end{cases} \tag{4$'$}$$

where $\Sigma_{i=0}^m \; a_i = k$, $k = 0, 1, \cdots, n$. Now we prove (4)$'$.

Case 1: $a_m < k$.

Let $\lambda: \widetilde{S} \to L^p_{n+m}$ be the projection on $(\widetilde{S}, t)$, and let $[2n-2k]^*_0$ be the pseudomanifold formed by all points $\lambda(\mathfrak{z}_1, \cdots, \mathfrak{z}_q) \in L^p_{n+m}$ determined by the following conditions:

$$\left.\begin{aligned} &\mathfrak{z}_j = (z^{(j)}_1, \cdots, z^{(j)}_{n+m}), \quad \mathfrak{z}_j = 0, \quad j \ne i, \\ &z^{(i)}_{n-k+2} = \cdots = z^{(i)}_{n+m} = 0, \ \arg z^{(i)}_{n-k+1} = 0. \end{aligned}\right\} \tag{6}$$

From §2 we know that $[2n-2k]^*_0$ represents the following chain:

$$[2n-2k]_0 \in (-i')^{n-k} (q!)^{n+m} \mu_{2q(n+m)-1-2n-2k} X.$$

Comparing this with (5), we obtain

$$[2n-2k]_0 \sim (-i')^{n-k}[2n-2k]. \tag{7}$$

According to §3 in [1], any element Z in $[a_1, \cdots, a_m]^*_c$ must satisfy the condition that $Z \subset U^{a_m+m} \subset U^{m+k-1}$. Let $\eta^* = [2n-2k]^*_0 \times [a_0, \cdots, a_m]^*_c$. If there is a $\{\mathfrak{z}_1, \cdots, \mathfrak{z}_q; Z\} \in \eta^* \cap \Delta_i$, we should have $\mathfrak{z}_i \in Z \subset U^{m+k-1}$. Then

$$z_1^{(i)} = \cdots = z_{n-k+1}^{(i)} = 0.$$

Hence from (6) we obtain that $\mathfrak{z}_1 = \cdots = \mathfrak{z}_q = 0$, contradicting $\Sigma_{i=1}^q |\mathfrak{z}_i|^2 = 1$. Hence η^* and Δ_i have no common point and we obtain the first formula in (4)′.

Case 2: $a_m = k$.

Now $[a_1, \cdots, a_m]_c = [0, \cdots, 0, k]_c$. Take $[2n-2k]^*_0$ as before and $[0, \cdots, 0, k]^*_c$ as the pseudomanifold defined by

$$U^{m-1} \subset Z \subset U^{m+k}. \tag{8}$$

Thus by (7) the (mod p) closed chain $\eta = [2n-2k]_0 \otimes [0, \cdots, 0, k]_c$ and $[2n-2k] \otimes [0, \cdots, 0, k]_c$ represented by the pseudomanifold $\eta^* = [2n-2k]^*_0 \times [0, \cdots, 0, k]_c$ satisfies the following relation:

$$[2n-2k] \otimes [0, \cdots, 0, k]_c \sim i^{n-k}\eta. \tag{7$'$}$$

Now let $\{\lambda(\mathfrak{z}_1, \cdots, \mathfrak{z}_q); Z\} \in \eta^* \cap \Delta_i$. Then from (8) we have $\mathfrak{z}_i \in Z \subset U^{m+k}$. Hence $z_1^{(i)} = \cdots = z_{n-k}^{(i)} = 0$. Further, from (6), $\mathfrak{z}_i$ must be

$$\mathfrak{z}_i^0: \ z_1^{(i)} = \cdots = z_{n-k}^{(i)} = 0, \ z_{n-k+2}^{(i)} = \cdots = z_{n+m}^{(i)} = 0, \ z_{n-k+1}^{(i)} = 1.$$

From (8) and $\mathfrak{z}_i \in Z$ we know Z must be the following complex m-plane:

$$Z_0: \quad z_1 = \cdots = z_{n-k} = z_{n-k+2} = \cdots = z_{n+1} = 0.$$

Hence let $\mathfrak{z}_j^0 = 0$ for $j \neq i$. Then η^* and Δ_i have only one point in common, namely $\{\lambda(\mathfrak{z}_1^0, \cdots, \mathfrak{z}_q^0); Z_0\}$.

In Λ this point $\{\lambda(\mathfrak{z}_1^0, \cdots, \mathfrak{z}_q^0); Z_0\}$ has a neighborhood $V \times W$, where V is the subset formed from all $\lambda(\mathfrak{z}_1, \cdots, \mathfrak{z}_q)$. Here $\mathfrak{z}_k = (z_1^{(k)}, \cdots, z_{n+m}^{(k)})$ is the following:

$$\mathfrak{z}_j: \quad z_r^{(j)} = \lambda_r^{(j)} + i\mu_r^{(j)}, \ j \neq i, \quad r = 1, \cdots, n+m,$$

$$\mathfrak{z}_i: \begin{cases} z_r^{(i)} = \lambda_r^{(i)} + i\mu_r^{(i)}, \quad r \neq n-k+1, \\ z_{n-k+1}^{(i)} = 1 - \lambda_{n-k+1}^{(i)} + i\mu_{n-k+1}^{(i)}, \end{cases}$$

where λ and μ take on all real values satisfying

$$\sum_{(r,\,j)\neq(n-k+1,\,i)} [(\lambda_r^{(j)})^2 + (\mu_r^{(j)})^2] + (\mu_{n-k+1}^{(i)})^2 + (1-\lambda_{n-k+1}^{(i)})^2 = 1,$$
$$1-\lambda_{n-k+1}^{(i)} > 0.$$

W is formed from all complex m-spaces Z determined by the equation

$$Z:\ z_r = \alpha_{r1} z_{n-k+1} + \sum_{s=2}^{m} \alpha_{rs} z_{n+s},\ r = 1, \cdots, n-k, n-k+2, \cdots, n+1,$$

where α takes on all complex values.

Obviously the natural orientations of V and W determine the natural orientation of Λ. From the definition of η^* and Δ_i we see that $\eta^* \cap (V \times W) = \eta'^*$ and $\Delta_i \cap (V \times W) = \Delta_i'$ are defined by the following equation systems:

$$\eta'^*: \begin{cases} \lambda_r^{(j)} = \mu_r^{(j)} = 0, & j \neq i, \\ \lambda_r^{(i)} = \mu_r^{(i)} = 0, & r = n-k+2, \cdots, n+m, \\ \mu_{n-k+1}^{(i)} = 0, & \\ \alpha_{rs} = 0,\ 2 \leq s \leq m,\ r = n-k+2, \cdots, n+1, & \text{or} \\ 1 \leq s \leq m,\ r = 1, \cdots, n-k. & \end{cases}$$

$$\Delta_i':\ \lambda_r^{(i)} + i\mu_r^{(i)} = \alpha_{r1}(1-\lambda_{n-k+1}^{(i)} + i\mu_{n-k+1}^{(i)})$$
$$+ \sum_{s=2}^{m} \alpha_{rs}(\lambda_{n+s}^{(i)} + i\mu_{n+s}^{(i)}),\quad r = 1, \cdots, n-k,\ n-k+2, \cdots, n+1.$$

Let $\alpha_{rs} = \beta_{rs} + i\gamma_{rs}$, where β and γ are real numbers. Then the above becomes

$$\Delta_i': \begin{cases} \lambda_r^{(i)} = \beta_{r1}(1-\lambda_{n-k+1}^{(i)}) - \gamma_{r1}\mu_{n-k+1}^{(i)} + \sum_{s=2}^{m} (\beta_{rs}\lambda_{n+s}^{(i)} - \gamma_{rs}\mu_{n+s}^{(i)}), \\ \mu_r^{(i)} = \beta_{r1}\mu_{n-k+1}^{(i)} + \gamma_{r1}(1-\lambda_{n-k+1}^{(i)}) + \sum_{s=2}^{m} (\beta_{rs}\mu_{n+s}^{(i)} + \gamma_{rs}\lambda_{n+s}^{(i)}), \\ r = 1, \cdots, n-k,\ n-k+2, \cdots, n+1. \end{cases}$$

The only intersections of η'^* and Δ_i', that is the only intersections of η^* and Δ_i (the above mentioned $0 = \{\lambda(\mathfrak{z}_1^0, \cdots, \mathfrak{z}_q^0); Z_0\}$) correspond to all $\lambda_r^{(j)}, \mu_r^{(j)}, \alpha_{rs} = 0$. If we want to determine the number of intersections at 0, we may first translate η'^* to the position of the following $\eta_\epsilon'^*$:

$$\eta_\epsilon'^*:\quad \begin{cases} \lambda_r^{(j)} = \mu_r^{(j)} = 0, \quad j \neq i, \\ \lambda_r^{(i)} = \mu_r^{(i)} = 0, \quad r = n-k+2, \cdots, n+m, \\ \mu_{n-k+1}^{(i)} = \epsilon, \\ \alpha_{rs} = 0, \ 2 \le s \le m, \ r = n-k+2, \cdots, n+1 \quad \text{or} \\ 1 \le s \le m, \ r = 1, \cdots, n-k, \end{cases}$$

where $\epsilon < 1$ is sufficiently small. It follows easily that $\eta_\epsilon'^*$ and Δ_i' only intersect at the following O_ϵ:

$$O_\epsilon:\quad \begin{cases} \lambda_r^{(j)} = \mu_r^{(j)} = 0, \quad (r, j) \neq (n-k+1, i), \\ \lambda_{n-k+1}^{(i)} = 1 - \sqrt{1-\epsilon^2}, \quad \mu_{n-k+1}^{(i)} = \epsilon, \\ \alpha_{rs} = 0, \ 1 \le s \le m, \ r = 1, \cdots, n-k, \ n-k+2, \cdots, n+1. \end{cases}$$

Also, the number of intersections at O_ϵ of $\eta_\epsilon'^*$ and Δ_i' is the number of intersections at O of η'^* and Δ_i.

The tangent plane of Δ_i' at O_ϵ is

$$T_i:\quad \begin{cases} d\lambda_r^{(i)} = \sqrt{1-\epsilon^2}\, d\beta_{r1} - \epsilon\, d\gamma_{r1}, \\ \qquad\qquad (r = 1, \cdots, n-k, \ n-k+2, \cdots, n+1), \\ d\mu_r^{(i)} = \epsilon\, d\beta_{r1} + \sqrt{1-\epsilon^2}\, d\gamma_{r1}, \end{cases}$$

and the tangent plane of $\eta_\epsilon'^*$ at O_ϵ is

$$T_\epsilon:\quad \begin{cases} d\lambda_r^{(j)} = d\mu_r^{(j)} = 0, \quad j \neq i, \\ d\lambda_r^{(i)} = d\mu_r^{(i)} = 0, \quad r = n-k+2, \cdots, n+m, \\ d\mu_{n-k+1}^{(i)} = 0, \\ d\beta_{rs} = d\gamma_{rs} = 0, \ 2 \le s \le m, \ r = n-k+2, \cdots, n+1, \quad \text{or} \\ 1 \le s \le m, \ r = 1, \cdots, n-k. \end{cases}$$

Obviously T_i and T_ϵ are linearly independent, and their number of intersections is 1. That is, $\Delta_i' \cdot \eta_\epsilon'^* = 1$ or $\delta_i \cdot \eta = 1$. Hence from $(7)'$ we obtain the last part of $(4)'$, and so (4) is proven.

From (4) we have

$$\delta_1 \cdots \delta_q \sim \prod_{i=0}^{q} \sum_{k=0}^{n} i^{n-k} \cdot \xi_{2n-2k} \otimes [n-k, n, \cdots, n]_c. \tag{9}$$

By (3) and the dimensionality relation, there exists an integer a such that*

$$\delta_1 \cdots \delta_q \sim a\delta. \tag{10}$$

Here δ is the closed chain determined by the natural orientation in Δ. But from (9)

$$\delta_1 \cdots \delta_q \sim (q!)^n \xi_{2qn} \otimes [n, \cdots, n]_c + \cdots \nsim 0.$$

Hence

$$a \not\equiv 0 \bmod p. \tag{10$'$}$$

Now let

$$\bar{\delta}_i = \sum_{j=0}^{m} i^{m-j} \xi_{2m-2j} \otimes [\underbrace{n-1, \cdots, n-1}_{j}, \underbrace{n, \cdots, n}_{m-j}]_c, \tag{11}$$

$$\bar{\delta} = \bar{\delta}_1 \cdots \bar{\delta}_q. \tag{12}$$

Since in $C_{n,m}$,

$$\sum_{j+k=r} [\underbrace{n-1, \cdots, n-1}_{i}, \underbrace{n, \cdots, n}_{m-j}]_c \cdot [n-k, n, \cdots, n]_c$$

$$\sim \begin{cases} 0, & r > 0, \\ [n, \cdots, n]_c, & r = 0; \end{cases}$$

we have

$$\delta_i \cdot \bar{\delta}_i \sim \sum_{\substack{m \geq j \geq 0 \\ n \geq k \geq 0}} i^{n+m-j-k} \xi_{2m-2j} \xi_{2n-2k}$$

$$\otimes [\underbrace{n-1, \cdots, n-1}_{j}, n, \cdots, n]_c \cdot [n-k, n, \cdots, n]_c$$

$$\sim \sum_{r \geq 0} \Big[i^{n+m-r} \xi_{2n+2m-2r}$$

$$\otimes \Big(\sum_{j+k=r} \Big[\underbrace{n-1, \cdots, n-1}_{j}, n, \cdots, n]_c \cdot [n-k, n, \cdots, n]_c \Big\} \Big]$$

$$\sim i^{n+m} \xi_{2n+2m} \otimes [n, \cdots, n]_c,$$

$$a\delta \cdot \bar{\delta} \sim (q!)^{n+m} (\xi_{2n+2m})^q \otimes [n, \cdots, n]_c \sim 0.$$

---

* See S. Lefschetz, *Topology*, Amer. Math. Soc. Colloq. Publ., Vol. 12, Amer. Math. Soc., Providence, R. I., 1930, Chapter IV, §6.

By (10)′, we have

$$\delta \cdot \bar{\delta} \sim 0. \tag{13}$$

Now $\tilde{\Lambda}$ is the bundle space in the sphere bundle $\tilde{\mathfrak{S}}^q_{n,m}$ in $C_{n,m}$. Let $\tilde{\pi}$ be its projection. In $\tilde{\Lambda}$ take arbitrary fibers $\tilde{S}_0$. Obviously t is also a topological transform in $\tilde{S}_0$ with period p and no fixed point. Let $L_0 = (\tilde{S}_0)_t$; then L_0 is a fiber in the bundle space Δ in the L^p_m bundle of $C_{n,m}$. Let $U^i_0 = U^i(\tilde{S}_0, t)$, $j: L_0 \subset \Delta$. Then U^i_0 generates $H^*(L_0)$ and $j^*U^i = U^i_0$. Hence, by a theorem of Leray and Hirsch, $H^*(\Delta)$ has additive basis $\{U^i \cup \pi^* A_p\}$ where $0 \leq i \leq 2qm - 1$, and $\{A_p\}$ is an additive basis of $H^*(C_{n,m})$. By A6° in §1, $U^i(\tilde{\Lambda}, t) = V^i \times 1$, where 1 is the (mod p) unit class in $C_{n,m}$ and $V^i = U^i(\tilde{S}, t)$.

Now let $\kappa: \Delta \subset \Lambda = L^p_{n+m} \times C_{n,m}$ and let $\omega: \Lambda \to L^p_{n+m}$ and $\omega': \Lambda \to C_{n,m}$ be its projections. Similarly let $\tilde{\kappa}: \tilde{\Delta} \subset \tilde{\Lambda} = \tilde{S} \times C_{n,m}$, and let $\tilde{\omega}: \tilde{\Lambda} \to \tilde{S}$ and $\tilde{\omega}': \tilde{\Lambda} \to C_{n,m}$ be its projections. Then $(\tilde{\omega}\tilde{\kappa})t = t(\tilde{\omega}\tilde{\kappa})$ and $(\omega\kappa)^* V^i = U^i$, and, since $\omega'\kappa = \pi$, we have

$$\kappa^*(V^i \otimes A_p) = \kappa^*(\omega^* V^i \cup \omega'^* A_p) = \kappa^*\omega^* V^i \cup \kappa^*\omega'^* A_p = U^i \cup \pi^* A_p. \tag{14}$$

whence $\kappa^* H^*(\Lambda) = H^*(\Delta)$. By duality we have

$$\kappa_*: H(\Delta) \subset H(\Lambda). \tag{15}$$

Now let

$$D: H_r(\Lambda) \approx H_{d(\Lambda)-r}(\Lambda), \quad (d(\Lambda) = \dim \Lambda)$$

and

$$D': H_r(\Delta) \approx H_{d(\Delta)-r}(\Delta), \quad (d(\Delta) = \dim \Delta)$$

be the dual transforms for the natural orientation. Let the fundamental closed chain in Δ represented by the natural orientation of Δ be δ_0. Then by (13) and the general formulas for the relation of intersections and cup product, we have

$$\begin{aligned} \kappa_* D'^{-1} \kappa^* D\{\bar{\delta}\} &= \kappa^*(\kappa^* D\{\bar{\delta}\} \cap \{\delta_0\}) = D\{\bar{\delta}\} \cap \kappa_*\{\delta_0\} \\ &= \{\bar{\delta}\} \cdot \kappa_*\{\delta_0\} = \{\bar{\delta}\} \cdot \{\delta\} = 0. \end{aligned}$$

From this and (15)

$$\kappa^* D\{\bar{\delta}\} = 0. \tag{16}$$

But from (11) and A7° in §1

$$D\{\bar{\delta}_i\} = \sum_{k=0}^{m} i^{m-k} D\mu_{2m-2k} X \otimes D\{[\underbrace{n-1, \cdots, n-1, n, \cdots, n}_{k}]_c\}$$

$$= \sum_{k=0}^{m} i^{m-k} V^{2m-2k} \otimes C_p^{2k},$$

whence

$$D\{\bar{\delta}\} = D\{\bar{\delta}_1\} \cdots D\{\bar{\delta}_q\} = \prod_{i=1}^{q} \sum_{k=0}^{m} i^{m-k} V^{2m-2k} \otimes C_p^{2k},$$

and from (14), we obtain

$$\kappa^* D\{\bar{\delta}\} = \prod_{i=1}^{q} \sum_{k=0}^{m} i^{m-k} U^{2m-2k} \cup \pi^* C_p^{2k}.$$

Thus from (16) follows (1), the desired result.

§ 4 A lens space bundle corresponding to a Euclidean space bundle

Let $R_{n,m}$ be the (real) Grassmann manifold defined in $(n+m)$-dimensional Euclidean space R^{n+m}, and let $\Re_{n,m}$ be the model m-Euclidean space bundle in $R_{n,m}$, with $\tilde{R}_{n,m}$ as its bundle space. Let the space determined by all systems $\{\mathfrak{x}_1, \cdots, \mathfrak{x}_p; X\}$ with

$$\mathfrak{x}_i \in X \in R_{n,m}, \quad \sum_{i=1}^{p} |\mathfrak{x}_i|^2 = 1, \quad \sum_{i=1}^{p} \mathfrak{x}_i = 0 \tag{1}$$

be $\tilde{\Xi}$. $\tilde{\Xi}$ is then obviously a bundle space of a $(2qm-1)$-sphere fiber bundle $\tilde{\mathfrak{X}}$ corresponding to $\Re_{n,m}$ with $R_{n,m}$ as base space. In $\tilde{\Xi}$, there is a topological transform

$$t': \{\mathfrak{x}_1, \cdots, \mathfrak{x}_p; X\} \to \{\mathfrak{x}_2, \cdots, \mathfrak{x}_p, \mathfrak{x}_1; X\}. \tag{2}$$

with period p and no fixed point.

Let $\Xi = \tilde{\Xi}_{t'}$. Since t' is a fiber preserving transform, Ξ is a bundle space of a bundle $\mathfrak{X}$ corresponding to $\tilde{\Xi}$ and hence to $\Re_{n,m}$. Let $\tilde{S}_X$ be a fiber sphere in $X \in R_{n,m}$ in $\tilde{\mathfrak{X}}$. Then, from § 2, the collection of fibers $(\tilde{S}_X)_{t'}$ in $\mathfrak{X}$ is the special lens space L_m^p defined in the Euclidean space $\underbrace{X \oplus \cdots \oplus X}_{p} = (X)^p$.

Now let $C_{n,m}$ be the complex Grassmann manifold defined in the unitary space U^{n+m}, and let Δ be the bundle space defined on the lens space bundle $\mathfrak{L}$ in $C_{n,m}$ as in §3. After choosing coordinates in R^{n+m} and U^{n+m}, we can define natural maps $\tilde{l}: R^{n+m} \to U^{n+m}$ and $l: R^{n+m} \to C_{n,m}$ (as in §5 of [1]). Let

$$h:\quad (\mathfrak{x}_1, \cdots, \mathfrak{x}_p) \to (\mathfrak{z}_1, \cdots, \mathfrak{z}_q)$$

be the map defined as in (22) of §2. Then

$$\{\mathfrak{x}_1, \cdots, \mathfrak{x}_p;\ X\} \to \{h(\mathfrak{x}_1, \cdots, \mathfrak{x}_p);\ l(X)\}$$

defines a map $\tilde{\tilde{l}}: \tilde{\tilde{\Xi}} \to \tilde{\Delta}$. As in §3, let

$$t:\ \{\mathfrak{z}_1, \cdots, \mathfrak{z}_q;\ X\} \to \{\epsilon^{-1}\mathfrak{z}_1, \cdots, \epsilon^{-q}\mathfrak{z}_q;\ Z\}$$

be the topological transform in $\tilde{\Delta}$ with period p. Then clearly $t\tilde{\tilde{l}} = \tilde{\tilde{l}}t'$, and $\tilde{\tilde{l}}$ induces a map $\tilde{l}: \Xi \to \Delta$. By A2° in §1, we have $\tilde{l}^* U^i(\tilde{\Delta}, t) = U^i(\tilde{\tilde{\Xi}}, t')$.

Next let π and π' be the projections in $\mathfrak{L}$ and $\mathfrak{X}$ respectively. Then, since $\tilde{l}$ is a fiber preserving map satisfying $\pi\tilde{l} = l\pi'$, we have $\mathfrak{X} \sim l^*\mathfrak{L}$. According to §5 in [1],

$$l^* C_p^{2i} = \begin{cases} 0, & i \text{ is odd}, \\ (-1)^{i/2} P_p^{2i}, & i \text{ is even}. \end{cases}$$

Hence, on letting $U^i = U^i(\tilde{\tilde{\Xi}}, t)$, (1) in §3 yields

$$\prod_{i=1}^{q} \sum_{k=0}^{m} (-1)^k i^{m-2k} U^{2m-4k} \cup \pi'^* P_p^{4k} = 0. \tag{3}$$

Now let $\mathfrak{B} = \{V, \pi, K, R^m, O_m\}$ be any m-Euclidean space bundle, and $\mathfrak{B} \sim g^* \mathfrak{R}_{n,m}$. Let R_x be the fiber of any $x \in K$ in $\mathfrak{B}$. Choose all systems $\{\mathfrak{x}_1, \cdots, \mathfrak{x}_p;\ R_x\}$ satisfying

$$x \in K,\quad \mathfrak{x}_i \in R_x,\quad \sum_{i=1}^{p} |\mathfrak{x}_i|^2 = 1,\quad \sum_{i=1}^{p} \mathfrak{x}_i = 0$$

which form a space $\tilde{\tilde{\Xi}}(\mathfrak{B})$. Let t be the following topological transform in $\tilde{\tilde{\Xi}}(\mathfrak{B})$ with period p and no fixed point:

$$t:\ \{\mathfrak{x}_1, \cdots, \mathfrak{x}_p;\ R_x\} \to \{\mathfrak{x}_2, \cdots, \mathfrak{x}_p, \mathfrak{x}_1;\ R_x\}.$$

Let $\Xi(\mathfrak{B}) = (\tilde{\tilde{\Xi}}(\mathfrak{B}))_t$; then $\Xi(\mathfrak{B})$ naturally becomes a lens space bundle $\mathfrak{X}(\mathfrak{B})$ corresponding to $\mathfrak{B}$, and it is easy to see that $\mathfrak{X}(\mathfrak{B}) \sim g^*\mathfrak{X}$. Let

$$U^i = U^i(\tilde{\tilde{\Xi}}(\mathfrak{B}), t), \tag{4}$$

and let ϕ be the bundle projection of $\mathfrak{X}(\mathfrak{B})$. Then from (3)

$$\prod_{i=0}^{q} \sum_{k=0}^{m'} (-1)^k i^{m-2k} U^{2m-4k} \cup \phi^* P_p^{4k}(\mathfrak{B}) = 0, \tag{5}$$

or

$$\sum_{k=0}^{m'q} U^{2qm-4k} \cup \phi^* R_p^{4k}(\mathfrak{B}) = 0. \tag{5$'$}$$

Here m' is the largest integer $\leq m/2$ and

$$R_p^{4k}(\mathfrak{B}) = (-1)^k \sum_{k_1+\cdots+k_q=k} 1^{m-2k_1} \cdot 2^{m-2k_2} \cdots q^{m-2k_q} P_p^{4k_1}(\mathfrak{B}) \cdot P_p^{4k_2}(\mathfrak{B}) \cdots P_p^{4k_q}(\mathfrak{B}). \tag{6}$$

In particular, when $p = 3$ and $q = 1$,

$$R_3^{4k}(\mathfrak{B}) = (-1)^k P_3^{4k}, \qquad 0 \leq k \leq m'. \tag{7}$$

§5. Pontrjagin classes in differentiable manifolds

Let M be an m-dimensional closed analytic manifold with Riemann element ds^2 consistent with its analytic structure $\mathfrak{D}$. Let $\tilde{M} = \underbrace{M \times \cdots \times M}_{p}$, and let $d\colon M \longrightarrow \tilde{M}$ be the diagonal map. Then $\tilde{M}$ is also an analytic manifold and has the natural 'corresponding' Riemann element $\tilde{ds}^2 = ds_1^2 + \cdots + ds_p^2$, where ds_i^2 is the Riemann element of the ith component M in $\tilde{M}$ corresponding to ds^2, and p is any integer greater than one. Now for any $\tilde{a}, \tilde{b} \in \tilde{M}$ (or $a, b \in M$) define the distance $\tilde{d}(\tilde{a}, \tilde{b})$ (or $d(a, b)$) to be the greatest lower bound of the arclengths of the arcs joining $\tilde{a}$ and $\tilde{b}$ (or a and b) with respect to $\tilde{ds}^2$ (or ds^2). We also define $\tilde{d}(\tilde{a})$ as the greatest lower bound of the distances from $\tilde{a}$ to points $\tilde{b}$ in $d(M)$. Call this the distance from $\tilde{a}$ to $d(M)$. Since M is closed, there exists $\eta > 0$ such that the following assertions are valid:

1°. If $\tilde{a}, \tilde{b} \in \tilde{M}$ and $\tilde{d}(\tilde{a}, \tilde{b}) \leq \eta$, then there is exactly one shortest line $\tilde{g}_{\tilde{a}\tilde{b}}$ joining $\tilde{a}$ and $\tilde{b}$ (with respect to $\tilde{ds}^2$). The length of $\tilde{g}_{\tilde{a}\tilde{b}}$ with respect to $\tilde{ds}^2$ is $\tilde{d}(\tilde{a}, \tilde{b})$.

2°. If $a, b \in M$ and $d(a, b) \leq \eta$, then there is exactly one shortest line $g_{ab} \subset M$ joining a to b with respect to ds^2, and the length of g_{ab} with respect to ds^2 is $d(a, b)$.

3°. If $\tilde{a} \in \tilde{M}$ and $\tilde{d}(\tilde{a}) \leq \eta$, there is exactly one shortest line $\tilde{g}_{\tilde{a}}$ with respect to $\tilde{ds}^2$ which passes through $\tilde{a}$ and is perpendicular to M. The length of $\tilde{g}_{\tilde{a}}$ with respect to $\tilde{ds}^2$ is $\tilde{d}(\tilde{a})$. The foot of the perpendicular is denoted by $d\tilde{\phi}(\tilde{a})$, $\tilde{\phi}(\tilde{a}) \in M$.

Now let $0 < \epsilon \leq \eta$. Define $\tilde{N}_\epsilon, \tilde{V}_\epsilon \subset \tilde{M}$ by

$$\tilde{N}_\epsilon = \{\tilde{a} \mid \tilde{a} \in \tilde{M},\ \tilde{d}(\tilde{a}) = \epsilon\}, \quad \tilde{V}_\epsilon = \{\tilde{a} \mid \tilde{a} \in \tilde{M},\ \tilde{d}(\tilde{a}) \leq \eta\}.$$

Obviously $\tilde{N}_\epsilon$ is a bundle space in a sphere bundle $\tilde{\mathfrak{N}}_\epsilon$ in M. Let $\mathfrak{T}$ be the tangent sphere bundle formed from all the tangent vectors with unit length (with respect to ds^2), with T as bundle space and T_a as the fiber sphere in $a \in M$. Now let $\mathrm{a}_1, \cdots, \mathrm{a}_p$ be the tangent vectors of M at the same point $a \in M$ with respect to $\mathfrak{D}$, satisfying

$$\sum_{i=1}^{n} \mathrm{a}_i = 0, \quad \sum_{i=1}^{p} |\mathrm{a}_i|^2 = 1.$$

Then from the proof of Lemma 5 in §5 of [2] we know that $\mathrm{a} = (\mathrm{a}_1, \cdots, \mathrm{a}_p)$ can be regarded as a unit normal vector of $\tilde{M}$ which is perpendicular to $d(M)$ in $d(\mathrm{a})$ (with respect to $\tilde{ds}^2$). These vectors form a bundle space $\tilde{N}$ in the sphere bundle $\tilde{\mathfrak{N}}$ of M. Clearly $\tilde{\mathfrak{N}} \sim \tilde{\mathfrak{N}}_\epsilon$, and a bundle map $h: \tilde{N} \to \tilde{N}_\epsilon$ can be defined as follows. Through $d(\mathrm{a})$ construct the shortest line $g(\mathrm{a})$ with respect to $\tilde{ds}^2$ tangent to the normal vector a in $d(\mathrm{a})$. In $g(\mathrm{a})$ choose one point whose distance from $d(M)$ is ϵ. Then this point is $h(\mathrm{a})$. Comparison with the definition of $\tilde{\mathfrak{X}}(\mathfrak{B})$ in §4 shows that when p is an odd prime, $\tilde{\mathfrak{N}} \sim \tilde{\mathfrak{X}}(\mathfrak{T})$, and hence

$$\tilde{\mathfrak{N}}_\epsilon \sim \tilde{\mathfrak{X}}(\mathfrak{T}), \qquad (p \text{ is an odd prime}).$$

Now in $\tilde{M}$ let t be the following topological transform with period p:

$$t:\ (a_1, \cdots, a_p) \to (a_2, \cdots, a_p, a_1), \quad a_i \in M.$$

Since $\tilde{ds}^2$ is 'symmetric', $t(\tilde{N}_\epsilon) \subset \tilde{N}_\epsilon$, and t is also a topological transform with period p in $\tilde{N}_\epsilon$ and no fixed point in $\tilde{N}_\epsilon$. Hence $(\tilde{N}_\epsilon, t)$ forms a primitive system with modular space $N_\epsilon = (\tilde{N}_\epsilon)_t$. Obviously N_ϵ is a bundle space in the bundle $\mathfrak{N}_\epsilon$ of M. Its projection ϕ is defined as $\phi\lambda(a_1, \cdots, a_p) = \tilde{\phi}(a_1, \cdots, a_p)$. Here λ is the projection of $(\tilde{N}_\epsilon, t)$ and $\tilde{\phi}$ is the projection of $\tilde{N}_\epsilon$ (see 3°).

Obviously, under $h: \tilde{N} \to \tilde{N}_\epsilon$, t also induces a topological transform

$$hth^{-1} = \bar{t}:\ (\mathrm{a}_1, \cdots, \mathrm{a}_p) \to (\mathrm{a}_2, \cdots, \mathrm{a}_p, \mathrm{a}_1).$$

Let $N = \widetilde{N}_t$; then, according to §4, when p is an odd prime, N is a bundle space of the lens space bundle $\mathfrak{X}(\mathfrak{T})$. From this it follows easily that

$$\mathfrak{N}_\epsilon \sim \mathfrak{X}(\mathfrak{T}), \qquad (p \text{ is an odd prime}) \tag{1}$$

Now let $\mathfrak{D}'$ be another analytic structure of M, and let $d\widetilde{s}'^2$, η', ϵ', $N'_{\epsilon'} = (\widetilde{N}'_{\epsilon'})_{t'}$, ϕ' and $\mathfrak{N}'_{\epsilon'}$ be defined as before. Here $\epsilon, \epsilon' \leq \min(\eta, \eta')$. Define a map (see Chapter IV in [6]) $\widetilde{f}: \widetilde{N}_\epsilon \to \widetilde{N}'_{\epsilon'}$ as follows. Let $\widetilde{a} \in \widetilde{N}_\epsilon$, and let $g_{\widetilde{a}}$ be the shortest normal from $\widetilde{a}$ to $d(M)$ (with respect to $d\widetilde{s}^2$), whose foot is $d\widetilde{\phi}'(\widetilde{a}) \in d(M)$. On $g_{\widetilde{a}}$ choose one point $\widetilde{a}'$ such that under $d\widetilde{s}'^2$ the distance $\widetilde{d}'(\widetilde{a}', d\widetilde{\phi}'(\widetilde{a})) = \epsilon'$. Hence $\widetilde{a}' \in \widetilde{N}'_{\epsilon'}$. We define $\widetilde{f}$ as $\widetilde{f}(\widetilde{a}) = \widetilde{a}'$.

LEMMA. $\widetilde{\phi} \simeq \widetilde{\phi}'\widetilde{f}$.

Proof. For any $\widetilde{a} \in \widetilde{V}_\eta \cap \widetilde{V}'_{\eta'}$ with $\widetilde{d}(\widetilde{a}) \leq \eta$ and $\widetilde{d}'(\widetilde{a}) \leq \eta'$, the functions $\widetilde{\phi}(\widetilde{a}), \widetilde{\phi}'(\widetilde{a}) \in M$ are defined and $d(\widetilde{\phi}(\widetilde{a}), \widetilde{\phi}'(\widetilde{a}))$ is a real continuous function on the set $\widetilde{V}_\eta \cap \widetilde{V}'_{\eta'}$. In particular, $d(\widetilde{\phi}(\widetilde{a}), \widetilde{\phi}'(\widetilde{a})) = 0$ when $\widetilde{a} \in d(M)$. Since M and $\widetilde{M}$ are closed, $\widetilde{V}_\eta \cap \widetilde{V}'_{\eta'}$ is closed and we can choose $\epsilon, \epsilon' \leq \min(\eta, \eta')$ such that $d(\widetilde{\phi}(\widetilde{a}), \widetilde{\phi}'(\widetilde{a})) \leq \eta$ for any $\widetilde{a} \in \widetilde{V}_\epsilon \cap \widetilde{V}'_{\epsilon'}$. We suppose that ϵ and ϵ' have been so chosen. Obviously this assumption does not affect the conclusion. From 3°, with respect to ds^2, there exists exactly one shortest line $g_{\widetilde{a}}$ joining $\widetilde{\phi}(\widetilde{a})$ and $\widetilde{\phi}'(\widetilde{a})$. Moving along $g_{\widetilde{a}}$, we obtain $\widetilde{\phi} \simeq \widetilde{\phi}''$. Here $\widetilde{\phi}'': \widetilde{N}_\epsilon \to N$ is defined as $\widetilde{\phi}''(\widetilde{a}) = \widetilde{\phi}'(\widetilde{a})$, $\widetilde{a} \in \widetilde{N}_\epsilon$. But from the definition of $\widetilde{f}$ we have $\widetilde{\phi}'' = \widetilde{\phi}'\widetilde{f}$; hence the lemma is proven.

Since t changes the shortest normal to $d(M)$ in $\widetilde{M}$ with respect to $d\widetilde{s}'^2$ to the shortest normal to $d(M)$ with respect to $d\widetilde{s}'^2$, but leaves its foot fixed, we know from the definition of $\widetilde{f}$ that $\widetilde{f}t = t\widetilde{f}$. Hence $\widetilde{f}$ induces the map f in its modular space, $f: N_\epsilon \to N'_\epsilon$. Similarly $\widetilde{\phi}$ and $\widetilde{\phi}'$ commute with t, whence they also induce corresponding maps $\phi: N_\epsilon \to M$ and $\phi': N'_{\epsilon'} \to M$ in the modular space, where ϕ and ϕ' are clearly the projections in $\mathfrak{N}$ and $\mathfrak{N}'$. Similar to the lemma, we see that $\phi \simeq \phi' f$. Hence

$$\phi^* = f^* \phi'^*. \tag{2}$$

Let $U^i = U^i(\widetilde{N}_\epsilon, t)$ and $U'^i = U^i(\widetilde{N}'_{\epsilon'}, t)$. Then from A2° in §1,

$$f^* U'^i = U^i. \tag{3}$$

Let p be an odd prime. Then from (1), according to $(5)'$ in §4, we have the following relations in N_ϵ and $N'_{\epsilon'}$:

$$\sum_{k=0}^{m'q} U^{2qm-4k} \cup \phi^* R_p^{4k}(\mathfrak{T}) = 0, \tag{4}$$

$$\sum_{k=0}^{m'q} U'^{2qm-4k} \cup \phi'^* R_p^{4k}(\mathfrak{T}') = 0. \tag{4$'$}$$

Apply f^* to both sides of $(4)'$. Then from (2) and (3), we have

$$\sum_{k=0}^{m'q} U^{2qm-4k} \cup \phi^* R_p^{4k}(\mathfrak{T}') = 0. \tag{5}$$

Since, by (1), $\mathfrak{N}_\epsilon \sim \mathfrak{X}(\mathfrak{T})$ satisfies the Leray-Hirsch condition, we obtain, on comparison of (4) and (5),

$$R_p^{4k}(\mathfrak{T}) = R_p^{4k}(\mathfrak{T}'). \tag{6}$$

In particular when $p = 3$, (6) becomes

$$P_3^{4k}(\mathfrak{T}) = P_3^{4k}(\mathfrak{T}'). \tag{7}$$

Hence the following theorem holds.

Theorem. *The R characteristic classes $R_p^{4k}(\mathfrak{T})$ (p an odd prime) of a closed manifold with analytic structure, and therefore those of a closed C^∞ manifold* [7], *are independent of the differential structure of the manifold, and hence are topological invariants of it. In particular, the* mod 3 *Pontrjagin classes are topological invariants of this manifold.*

If $p = 2$, then $\widetilde{\mathfrak{N}}_\epsilon \sim \mathfrak{T}$, and $\mathfrak{N}_\epsilon$ is the projection space bundle corresponding to $\mathfrak{T}$. According to the Appendix at the end of §7 in [1], in $\mathfrak{N}_\epsilon$ there is the relation

$$\sum_{i=0}^{m} [W^1(\mathfrak{R})]^{m-i} \cup \pi^* W^i(\mathfrak{T}) = 0,$$

where $\mathfrak{R}$ is the 1-Euclidean space bundle defined naturally in $\mathfrak{N}_\epsilon$. It is easy to prove that $W^1(\mathfrak{R}) = U^1(\widetilde{N}_\epsilon, t)$. Hence the above equality becomes

$$\sum_{i=0}^{m} (U^1)^{m-i} \cup \pi^* W^i(\mathfrak{T}) = 0,$$

where $U^1 = U^1(\widetilde{N}_\epsilon, t)$. With the same method we obtain

Theorem. *The Stiefel-Whitney classes of a closed differentiable manifold are independent of the differential structure of the manifold and hence are topological invariants.*

BIBLIOGRAPHY

[1] Wu Wen-Tsün, *On Pontrjagin classes.* I, Acta Math. Sinica 3 (1953), 291–315; English transl., Amer. Math. Soc. Transl. (2) 92 (1970),

49–62. MR 17, 521.

[2] ———, *On Pontrjagin classes*. II, Acta Math. Sinica 4 (1954), 171–199; English transl., Amer. Math. Soc. Transl. (2) 92 (1970), 63–92. MR 17, 1120.

[3] P. A. Smith, "Fixed points of periodic transforms", in S. Lefschetz, *Algebraic topology*, Amer. Math. Soc. Colloq. Publ., vol. 27, Amer. Math. Soc., Providence, R. I., 1942, Appendix B, pp. 350–373. MR 4, 84.

[4] ———, *Periodic and nearly periodic transformations*, Lectures in Topology, Univ. of Michigan Press, Ann Arbor, Mich., 1941, pp. 159–190. MR 3, 133.

[5] Wu Wen-Tsün, *Sur les puissances de Steenrod*, Colloq. de Topologie de Strasbourg, 1951, no. 9, La Bibliothèque Nationale et Universitaire de Strasbourg, 1952. MR 14, 491.

[6] R. Thom, *Espaces fibrés en sphères et carrés de Steenrod*, Ann. Sci. École Norm. Sup. (3) 69 (1952), 109–182. MR 14, 1004.

[7] H. Whitney, *Differentiable manifolds*, Ann. of Math. (2) 37 (1936), 645–680.

Translated by:
B. B. Murphy

Amer. Math. Soc. Transl.
(2) Vol. 92, 1970

SOME REMARKS ON LIMITING SPACES OF POSTNIKOV SYSTEMS*

UDC 513.836

E. G. SKLJARENKO

Let X be a CW-complex. It is known that there exists a sequence X_1, $X_2, \cdots, X_n, \cdots$ of topological spaces which have the homotopy type of CW-complexes and a sequence of maps $\alpha_n: X_{n+1} \to X_n$ and $i_n: X \to X_n$ such that: 1) the map $\alpha_n: X_{n+1} \to X_n$ is a fibration (satisfying the covering homotopy axiom) whose fiber is an Eilenberg-Mac Lane space $K(\pi_{n+1}(X); n+1)$; 2) $i_n = \alpha_n i_{n+1}$; 3) the map i_n induces an isomorphism $\pi_k(X) \approx \pi_k(X_n)$ for $k \leq n$ and $\pi_k(X_n) = 0$ for $k > n$. Such a sequence $\{X_n, \alpha_n, i_n\}$ is called a Postnikov system of the space X.

We denote by X_∞ the inverse limit $\varprojlim \{X_n, \alpha_n\}$. The maps i_n give rise to a map $i_\infty: X \to X_\infty$. It is known that for any polyhedron Y the map i_∞ induces a one-to-one correspondence between the homotopy classes of maps of Y into X and the homotopy classes of maps of Y into X_∞ (that is, i_∞ is a weak homotopy equivalence). The weak homotopy equivalence, as a rule, replaces the homotopy equivalence. There are, however, problems for which the difference between these two equivalences is essential. This motivates the question put forth by Kahn [5] asking whether the spaces X and X_∞ have the same homotopy type. The following theorem answers this question.

Theorem 1. *The spaces X and X_∞ are of the same homotopy type if and only if X has only a finite number of nontrivial homotopy groups.*

Proof. If X has only a finite number of nontrivial homotopy groups then $X_\infty = X_n$ for some n. In this case X_∞ has the homotopy type of a CW-complex, and since the map $i_\infty = i_n: X \to X_n = X_\infty$ induces an isomorphism of homotopy groups it is a homotopy equivalence by a known theorem of Whitehead.

To prove the converse we shall make use of the following simple concept (cf. [2]): a space Y is called *weakly locally contractible* if every point $y \in Y$ has a neighborhood which is contractible in Y to some point. This property of a space is a homotopy invariant.

---

* Translation of Mat. Sb. 69 (111) (1966), 61–64.

Thus, let X have infinitely many nontrivial homotopy groups. In order to prove that X_∞ is not of the same homotopy type as X it suffices to show that X_∞ is not weakly locally contractible. Let $x_\infty \in X_\infty$ be an arbitrary point and let Ox_∞ be some neighborhood of that point in X_∞. We introduce the following notation. By α_s^r we denote the map $X_r \to X_s$ obtained by composing the corresponding maps $\alpha_n (r > s)$; the natural map $X_\infty \to X_r$ is denoted by α_r^∞; finally, let $x_r = \alpha_r^\infty x_\infty$. By the definition of the topology of X_∞ there is a number k and a neighborhood U_k of the point x_k in X_k such that $x_\infty \in (\alpha_k^\infty)^{-1} U_k = U \subset Ox_\infty$. We shall show that the neighborhood U cannot be contracted in the space X_∞ to a point. Let $U_n = (\alpha_k^n)^{-1} U_k$, and let $N > k$ be a number such that $\pi_N(X) \neq 0$. Consider any map $f\colon S^N \to X$ which represents a nontrivial element of $\pi_N(X)$. Then for any $n \geq N$ the map $i_n f$ is nontrivial. It suffices to show that there is a map $g\colon S^N \to X_\infty$ which is homotopic to $i_\infty f$ and such that $gS^N \subset U$. To this end we first consider the fibration $\alpha_{N-1}\colon X_N \to X_{N-1}$. Since $\pi_N(X_{N-1}) = 0$ the map $i_{N-1} f$ can be contracted to the point x_{N-1}. Let $h_{N,t}$ be the homotopy of the map $i_N f$ which covers this contraction. Then since $x_{N-1} \in U_{N-1}$ and $U_N = \alpha_{N-1}^{-1} U_{N-1}$ we find that $h_{N,1} S^N \subset U_N$. It is clear that for all $n > N$ we can construct homotopies $h_{n,t}$ of maps $i_n f$ which consecutively cover each other and the homotopy $h_{N,t}$. In the limit we obtain a homotopy h_t of the map $i_\infty f$. Since $U = (\alpha_N^\infty)^{-1} U_N$ and h_t covers $h_{N,t}$ we conclude that $h_1 S^N \subset U$. The theorem follows.

The difference between homotopy equivalence and weak homotopy equivalence can be illustrated by the following example. According to Adams [1] there exists a sequence of p-spheres of increasing dimension $X = S_1, S_2, \cdots$ $\cdots, S_n, \cdots$ and maps $\phi_n\colon S_{n+1} \to S_n$ such that for any n the composition $\phi_1 \phi_2 \cdots \phi_{n-1} = \phi_1^n$ is essential.* Let S be the limit space. The compact space S is obviously acyclic in all dimensions (we use the Čech cohomology). By the lemma that we prove below the map $\phi_1^\infty\colon S \to S_1 = X$ is essential, and at the same time, by Theorem 2 (see below), any map of S into the space X_∞ is homotopic to a constant map.

Lemma. *Let the compact space S be the inverse limit of complexes $S_1 \xleftarrow{\varphi_1} S_2 \xleftarrow{\varphi_2} \cdots \xleftarrow{\varphi_{n-1}} S_n \xleftarrow{\varphi_n} \cdots$, where ϕ_n are simplicial maps; if all maps ϕ_n are "onto" and if all compositions $\phi_1^n = \phi_1 \phi_2 \cdots \phi_{N-1}$ are essential, then the map $\phi_1^\infty\colon S \to S_1$ is also essential.*

This lemma is essentially known (cf. [3]), but for completeness we shall

* This result was pointed out to the author by S. P. Novikov.

give a proof.

We can assume that the space S is realized in the subspace Γ_1 of the Hilbert space Γ given by the equation $x_1 = 0$. Since the complexes S_n are nerves of some open coverings of the space S they can be realized in the subspace Γ_1 in such a way that $\sup_{s \in S}\{\rho(S, \phi_n^\infty S)\}$ tends to zero as $n \to \infty$ (here ρ is the metric of Γ). Suppose that the map ϕ_1^∞ can be contracted to a point. This means that there is a map $F: S \times I \to S_1$ such that $F(s, 0) = \phi_1^\infty s$ and $F(s, 1) = x$ for all $x \in S$ (here $I = [0, 1]$ and x is some point in S_1). We clearly can assume that all the spaces $S \times I$, $S_n \times I$, $n = 2, 3, \cdots$ are realized in Γ in such a way that on $S \times 0$ and $S_n \times 0$ this realization concides with the original realization of S and S_n in Γ_1 and such that for any $s \in S$ and any $s_n \in S_n$ the intervals $s \times I$ and $s_n \times I$ are parallel to the x_1-axis. The subset $D \subset \Gamma$ consisting of all $S_n \times I$ and $S \times I$ is, clearly, compact. Let $D_1 \subset D$ be the union of the sets $D \cap \Gamma_1$, $S \times I$ and $S_n \times I$, $n = 2, 3, \cdots$ and let $\Phi_1: D_1 \to S_1$ be the map defined by $\Phi_1(s, t) = F(s, t)$, $\Phi_1(s_n, 0) = \phi_1^n s_n$ and $\Phi_1(s_n \times 1) = x$ (here $s \in S$, $s_n \in S_n$ and $t \in I$). It is easy to check that the map Φ_1 is not trivial. Since S_1 is a finite complex the map Φ_1 can be extended to a map Φ of some neighborhood U of the set D_1 into Γ. It is easy to see that this neighborhood will contain $S_n \times I$ for sufficiently large n. The restriction of Φ to $S_n \times I$ gives a homotopy of ϕ_1^n with the constant map at x, which contradicts the hypothesis. The lemma is proven.

Theorem 2. *Let S be a paracompact space, acyclic in all dimensions (with respect to the Čech cohomology), and let X be any cell complex. Any map of S into X_∞ is contractible to a point.*

Proof. Let $f: S \to X_\infty$ be any continuous map. We must construct a map $F: S \times I \to X_\infty$ which coincides with f on $S \times 0$ and is such that $F(s, 1) = x_\infty$ for all $s \in S$ (x_∞ is some point in X_∞).

We first construct maps $S \times [(k-1)/k, k/(k+1)] \to X_\infty$ which satisfy the following conditions: 1) for every k the maps F_k and F_{k+1} coincide on the intersection of their domains i.e. on the set $S \times k/(k+1)$; 2) the image of the set $S \times k/(k+1)$ under the map $\alpha_k^\infty F_k$ is contained in a fiber of the map $\alpha_k: X_k \to X_{k-1}$ over some point x_{k-1} which belongs to the image of the set $S \times (k-1)/k$ under the map $\alpha_{k-1}^\infty F_{k-1}$. We take the map $f: S \times 0 \to X_\infty$ to be F_0.

Suppose that the maps F_k are constructed for $k < m$. Consider the restriction of the map $\alpha_{m-1}^\infty F_{m-1}$ to the set $S \times (m-1)/m$. Since the fiber of

$\alpha_{m-1}: X_{m-1} \to X_{m-2}$ is an Eilenberg-Mac Lane space $K(\pi_{m-1}(X); m-1)$, and since $H^{m-1}(S) = 0$, the map $\alpha^{\infty}_{m-1} F_{m-1}$ of the set $S \times (m-1)/m$ can be contracted inside this fiber to some point x_{m-1}(cf. [4]). Let us denote the corresponding homotopy of the map $\alpha^{\infty}_{m-1} F_{m-1}$ by $\beta_{m-1}: S \times [(m-1)/m, m/(m+1)] \to X_{m-1}$. Succesively using the covering homotopy property, one can clearly construct homotopies $\beta_m, \beta_{m+1}, \cdots, \beta_n, \cdots$, where $\beta_n: S \times [(m-1)/m, m/(m+1)] \to X_n$, of the maps $\alpha^{\infty}_m F_{m-1}, \alpha^{\infty}_{m+1} F_{m-1}, \cdots, \alpha^{\infty}_n F_{m-1}, \cdots$, which respectively cover the maps $\beta_{m-1}, \beta_m, \cdots, \beta_{n-1}, \cdots$. The maps β_n yield in the limit the map $F_m: S \times [(m-1)/m, m/(m+1)] \to X_\infty$. The first condition is satisfied since F_{m-1} is the limit of $\alpha^{\infty}_n F_{m-1}$. The second condition is satisfied since the maps $\alpha_m F_m$ and β_m coincide on the set $S \times (m-1)/m$.

Having inductively constructed all F_k, we now construct the map F: $S \times I \to X_\infty$. We denote by x_∞ the point in X_∞ for which $\alpha^{\infty}_k x_\infty = x_k$. We set F equal to F_k if the second coordinate of the point (s, t) is smaller than 1 and equal to the constant map at x_∞ for all points $(s, 1)$. The map F is continuous since it is the limit of continuous maps $\alpha^{\infty}_k F$: each $\alpha^{\infty}_k F$ coincides with one of the maps $\alpha^{\infty}_k F_n$, $n \leq k$, on the set $S \times [0, k/(k+1)]$, and takes all other points of $S \times [0, 1]$ into x_k. The theorem follows.

BIBLIOGRAPHY

[1] J. F. Adams, *Cohomological operations*, Summer Inst., Seattle, 1962.

[2] E. Fadell, *On fiber spaces*, Trans. Amer. Math. Soc. 90 (1959), 1–14. MR 21 #330.

[3] H. Freudenthal, *Entwicklungen von Räumen und ihren Gruppen*, Compositio Math. 4 (1937), 145–234.

[4] P. J. Huber, *Homotopical cohomology and Čech cohomology*, Math. Ann. 144 (1961), 73–76. MR 24 #A3646.

[5] D. W. Kahn, *Induced maps for Postnikov systems*, Colloq. Algebraic Topology (Aarhus Univ., 1962) Matematisk Institut, Aarhus Universitet, Aarhus, 1962, pp. 47–51. MR 26 #3565.

Translated by:
S. Feder

Amer. Math. Soc. Transl.
(2) Vol. 92, 1970

ON SIMPLE SYSTEMS OF PATHS ON COMPLETE PRETZELS*

UDC 513.83

H. CIŠANG [HEINER ZIESCHANG]

Let us consider simple closed paths and simple systems of paths (i.e. systems of simple closed paths not intersecting each other) on the surface of a complete pretzel of genus p. Two such systems are called homeomorphic if there exists a homeomorphism of the complete pretzel transforming one system into the other. Now the general problem arises to classify such systems under homeomorphisms of the complete pretzel. In the present work we shall narrow down this problem, and the normal form of a simple system of paths with respect to a cut of the complete pretzel will be constructed (cf. Theorem 1). Further, in §2 we shall give two applications of this theorem:

1) Let K be the element of the fundamental group of the complete pretzel, induced by k. We say that K is of complete rank if K cannot be defined by less that p free generators. Let $\mathfrak{U}_k$ be the subgroup of the fundamental group, defined by paths which lie on the surface and do not pierce k. $\mathfrak{U}_k$ has the rank $2p-1$ if and only if K is of complete rank.

2) Let $\{v_0, \cdots, v_p\}$ be the diagram of Heegaard's decomposition of V and assume that it induces elements $R_1, \cdots, R_p$ of the fundamental group (taken in the sense of free homotopy) such that $R_{i_1}, \cdots, R_{i_q}$ can be expressed by means of q free generators. Then the variety represented by the diagram is either reducible or has Heegaard decomposition of genus less than p.

§1. Normal form of simple systems of paths.

Let V be a complete pretzel of genus p and having the surface F.** A system $Z = \{f_1, \cdots, f_p\}$ consisting of p discs with boundaries $s_1, \cdots, s_p$ is called a cut of the complete pretzel if f_i does not intersect f_j whenever $i \neq j$, and if the intersection of the disk f_i and the surface F is the set s_i and if after the cutting of V by Z one obtains a sphere (cf. [7], p. 232). The set $\{s_1, \cdots, s_p\}$ is called the trace of the cut Z. To each disc f_i we shall assign a symbol S_i. Let $\mathfrak{F}$ be the free group generated by $S_1, \cdots, S_p$.

---

* Translation of Mat. Sb. 66(108) (1965), 230–239.

** In this article all curves, surfaces and three-dimensional varieties are assumed to be simplicial complexes and all homeomorphisms are semilinear mappings.

Assume that a is a point of the complete pretzel which does not lie on Z, and k is an arbitrary path with the origin a. A complete word $\overline{K}$ with respect to the system Z shows by means of the symbols S_i and S_i^{-1} the order and directions in which k pierces the discs of Z. $\overline{K}$ can contain parts of the form $S_iS_i^{-1}$ or also $S_i^{-1}S_i$. $\overline{K}$ defines an element K of the group $\mathfrak{F}$. The mapping $k \to K$ induces an isomorphism between the fundamental groups with the basis a of the complete pretzel and the group $\mathfrak{F}$. By means of this isomorphism we shall identify both groups. Thus, among all isomorphisms between fundamental groups with different origins we have chosen one particular isomorphism and we have gotten rid of the original basic point. Any homeomorphism η: $V \to V$ induces for any point a, for which a and ηa do not lie on Z, an automorphism α of the group $\mathfrak{F}$. For different points one obtains automorphisms which differ only by inner automorphisms; therefore they are all in the same system of automorphisms (cf. [7], pp. 239–240).

Let s' be a path not intersecting the trace of Z. Then s' lies on the surface of the complete pretzel which is cut by the discs of Z; and there exists a disc f' with the boundary s', the rest of the disc f' being inside the sphere. The path s' defines a decomposition of the surface of the sphere into two regions. If, in addition, s' separates the pair f_i, f_i^{-1}, then one can replace f by the disc f' and thus obtain a new cut Z'. Such a procedure will be called *elementary*.

The elements of the free group $\mathfrak{F}$ are the classes of words in symbols S_i, S_i^{-1}. In each class there exists one and only one reduced word, i.e. such a word which does not contain parts of the form $S_iS_i^{-1}$, $S_i^{-1}S_i$. We shall often represent the elements by their reduced words, and we shall not distinguish between elements and words. A reduced word K is called *cyclically reduced* if K does not end by the inverse symbol of the letter which stands at the beginning of the word. Two systems, $\{K_1, \cdots, K_v\}$ and $\{K_1', \cdots, K_v'\}$, are called *(cyclically) equivalent,* if there is an automorphism α of the group $\mathfrak{F}$ such that $\alpha K_i = L_iK_i'L_i^{-1} (L_i \in \mathfrak{F})$. The length of a system is defined as the sum of the lengths of the elements. (The last sum equals the number of symbols S_i, S_i^{-1} in the reduced word.) The cyclic length of a system is the length of the system of its cyclically reduced words. In each class of equivalent systems there are systems of minimal length. If the system $\{K_1, \cdots; K_v\}$ has minimal length, then each word K_i is cyclically reduced.

Theorem 1. *Let $\mathfrak{k} = \{k_1, \cdots, k_v\}$ be a simple system of paths on the surface of a complete pretzel V of arbitrary genus p. Let $\{K_1, \cdots, K_v\}$ be the induced system of elements of $\mathfrak{F}$ and $\{K_1', \cdots, K_v'\}$ the cyclically* *equivalent system of minimal length. Then there exists a homeomorphism* $\eta\colon V \to V$ *such that to the paths* $\eta_1^k, \cdots, \eta_v^k$ *there correspond complete words* $K_1', \cdots, K_v'$.*

Let $\{K_1, \cdots, K_v\}$ be the system of cyclically reduced elements of $\mathfrak{F}$. Denote by α the following automorphism: Let $\binom{1 \cdots p}{n_1 \cdots n_p}$ be a permutation for which $n_1 = i$, let $\epsilon = \pm 1$ and let $1 < v_1 \leq v_2 \leq v_3 \leq p+1$. The automorphism α is then defined by the mapping

$$S_{n_j} \to \begin{cases} S_{n_j} & \text{for } 1 \leqslant j < v_1, \\ S_{n_j} S_i^{\varepsilon} & \text{for } v_1 \leqslant j < v_2, \\ S_i^{-\varepsilon} S_{n_j} & \text{for } v_2 \leqslant j < v_3, \\ S_i^{-\varepsilon} S_{n_j} S_i^{\varepsilon} & \text{for } v_3 \leqslant j \leqslant p. \end{cases}$$

Assume that $\{K_1', \cdots, K_v'\}$ is the system of cyclically reduced words where $\alpha K_j = L_j K_j' L_j^{-1}$ $(L_j \in \mathfrak{F})$. The number l_j indicates how many times the symbols S_j and S_j^{-1} occur in the elements $K_1', \cdots, K_v'$. Let A be a set containing the following elements:

$$S_i^{\varepsilon},\ S_{n_j}\,(v_1 \leqslant j < v_2),\ S_{n_j}^{-1}\,(v_2 \leqslant j < v_3),\quad S_{n_j} \text{ and } S_{n_j}^{-1} \quad (v_3 \leqslant j \leqslant p).$$

By A' we denote the complement of A and by d the number of parts of the form XY^{-1}, or YX^{-1}, $X \in A$, $Y \in A'$, in the words $K_1, \cdots, K_v$ (in the cyclic sense).

Lemma 1.** $l_j' = l_j$ $(j \neq i)$, *and* $l_i' = d$.

Proof. Replace the symbols $S_1, \cdots, S_p$ in the words $K_1, \cdots, K_v$ by the corresponding expressions from (1); thus we get new words $\overline{K}_1, \cdots, \overline{K}_v$. Actually, we have only to put into the words $K_1, \cdots, K_v$ the new symbols S_i and S_i^{-1}. The remaining symbols are the old ones. Since the words $K_1, \cdots, K_v$ are cyclically reduced, only $S_i S_i^{-1}$ and $S_i^{-1} S_i$ can be the places of cancellations (in the cyclic sense). However, at least one symbol of such a part must be a new one. If S_i^{ϵ} is a new letter, then S_i^{ϵ} follows an element from $A \setminus \{S_i^{\epsilon}\}$; since after an element of A' the symbol S_i^{ϵ} cannot follow, in particular, follow the letter $S_i^{-\epsilon}$. The new symbol $S_i^{-\epsilon}$ stands only before the element X^{-1}, where $X \in A \setminus \{S_i^{\epsilon}\}$. Therefore there are no parts of the form $S_i^{-\epsilon} S_i^{\epsilon}$ in the words $\overline{K}_1, \cdots, \overline{K}_v$. In addition, in these words there are no parts such as

---

*If the genus of the complete pretzel equals 2, then a more precise result holds (cf. [8]).

**Cf. [1].

$XS_i^\epsilon S_i^{-\epsilon}X^{-1}$, where X is an arbitrary letter of S_j, S_j^{-1}. Indeed, let S_i^ϵ be a new symbol; S_i^ϵ follows an old element X, $X \in A\backslash\{S_i^\epsilon\}$. If $S_i^{-\epsilon}$ is an old symbol, then it must be the part $XS_i^{-\epsilon}Y$, where Y is some letter. Such a part in the words $\bar{K}_1, \cdots, \bar{K}_v$ gives rise to the part $XS_i^\epsilon S_i^{-\epsilon}Y$ or $XS_i^\epsilon S_i^{-\epsilon} S_i^{-\epsilon}Y$. In the first case $Y^{-1} \in A\backslash\{S_i^\epsilon\}$; hence $Y \neq X^{-1}$, since $X \in A\backslash\{S_i^\epsilon\}$. The second case is also possible because $X \neq S_i^\epsilon$. If $S_i^{-\epsilon}$ is also a new symbol, then one of the words, let us say K_s, contains the part XY^{-1}, $Y \in A\backslash\{S_i^\epsilon\}$. Since the word K_s is cyclically reduced, $X \neq Y$. Hence in the words $\bar{K}_1, \cdots, \bar{K}_v$ there are no parts of the form $XS_i^\epsilon S_i^{-\epsilon}X^{-1}$. From here it follows that we can obtain $K_i', \cdots, K_v'$ (or their cyclic permutations) by removing all parts of the form S_i^ϵ, $S_i^{-\epsilon}$ from the words $\bar{K}_1, \cdots, \bar{K}_v$. Therefore $l_j' = l_j$ for all $j \neq i$.

Let us consider now the parts of form $R(S_i^{\pm\epsilon})^m T$ $(m \geq 0)$ in the words $K_1, \cdots, K_v$; here the letters R and T are different from S_i^ϵ and $S_i^{-\epsilon}$. It is sufficient to investigate the case where the exponent equals ϵ. In $(S_i^\epsilon)^m$ there are $m-1$ parts of the form XY^{-1}, $X \in A$, $Y \in A'$. We can obtain a new part of this form only for $R \in A$ or $T^{-1} \in A'$. The new exponent is therefore $m-1$ if $T^{-1} \in A$ and $R \in A'$; hence, this exponent is equal to the number of parts of the form XY^{-1}, $X \in A$, $Y \in A'$, which are between R and T. However, the new exponent is m if either $R \in A$ and $T^- \notin A'$ or $R \notin A$ and $T^- \in A'$; and it is $m+1$ if $R \in A$ and $T^{-1} \in A'$. Thus we obtain $l_i' = d$.

Automorphisms of form (1) and automorphisms defined by the permutations of the letters S_j, S_j^{-1} are called *Whitehead automorphisms* (cf. [1]). Whitehead [6] and later Rapoport [4] and Higgins and Lyndon [1] proved the following assertion: if a system $\{K_1, \cdots, K_v\}$ does not have minimal length, then there exists an automorphism α of form (1) such that the system $\{\alpha K_1, \cdots, \alpha K_v\}$ has a cyclic length less than the cyclic length of the original system. They also proved that for two equivalent systems of minimal length, $\{K_1, \cdots, K_v\}$ and $\{K_1', \cdots, K_v'\}$, there exists a chain $\tau_1, \cdots, \tau_t$ of Whitehead automorphisms such that $\tau_t \cdot \cdots \cdot \tau_1 K_j = L_j K_j' L_j^{-1}$, where the cyclic lengths of the systems $\{\tau_s \cdot \cdots \cdot \tau_1 K_1, \cdots, \tau_s \cdot \cdots \cdot \tau_1 K_v\}$, $1 \leq s \leq t$, are equal to the length of the system $\{K_1, \cdots, K_v\}$.

Any automorphism α of the free group $\mathfrak{F}$ induces a permutation of free generators: instead of $S_1, \cdots, S_p$ we take new generators $\alpha^{-1}S_1, \cdots, \alpha^{-1}S_p$. In the last group of generators the system $\{K_1, \cdots, K_v\}$ is of the same length as the system $\{\alpha K_1, \cdots, \alpha K_v\}$ in the generators $S_1, \cdots, S_p$. The permutations induced by Whitehead automorphisms are closely related to the elementary cuts of the complete pretzel (cf. [7], p. 238): every elementary process induces the permutation of free generators which is induced by a Whitehead automorphism, and each such permutation of generators is generated by an elementary process.

Proof of Theorem 1. a) Let $\bar{K}_1, \cdots, \bar{K}_v$ be complete words of paths $k_1, \cdots, k_v$. If not all $\bar{K}_i$ are cyclically reduced, then there is a homeomorphism $\eta_1\colon V \to V$ such that to each path $\eta_1 k_i$ corresponds a cyclically reduced complete word of length less than the number of symbols in the word $\bar{K}_i$ (cf. [7], p. 237; [8], Lemma 1).

b) Suppose now that to the system $\mathfrak{k}$ there correspond cyclically reduced complete words $K_1, \cdots, K_v$. If this system is not of minimal length, then there exists a Whitehead automorphism of the form (1) such that the length of the system $\{\alpha K_1, \cdots, \alpha K_v\}$ is less than the length of the system $\{K_1, \cdots \cdots, K_v\}$. Instead of the automorphism α we use the induced permutation of generators. Let A be, as above, a subset of the system of letters $S_1, S_1^{-1}, \cdots \cdots, S_p, S_p^{-1}$, which is attached to the automorphism α. Now, by means of Z we shall cut the surface of the complete pretzel into a two-dimensional sphere with $2p$ holes $f_1^{+1}, f_1^{-1}, \cdots, f_v^{+1}, f_v^{-1}$. Let B denote the set of those f_j^{δ} for which $S_j^{\delta} \in A$.

From our assumptions and Lemma 1 it follows that $d < l_i$. Z cuts the simple system of paths $k_1, \cdots, k_v$ into a system of simple arcs which lie on the sphere with holes and do not intersect each other. The end points of these arcs lie on the boundaries of the holes. Some of the arcs can be closed. We distinguish those arcs which link two holes from B and we cut the sphere with $2p$ holes using these arcs and the boundaries of holes of B. In this way the sphere is decomposed into a number of parts. In one of them lies $f_i^{-\epsilon}$. The boundary of this part, F_0, consists of one or several other curves, and each of these curves decomposes into arcs which lie either on $\mathfrak{k}$ or on the boundaries of holes of B. There exists a chain of parts such that the neighboring parts have a common arc on the boundary. (These arcs lie on $\mathfrak{k}$.) In addition, this chain starts with F_0 and ends by some parts whose boundary contains an arc of the boundary f_i^{ϵ}. One can even find a chain containing any part at least once. On each arc that belongs to two neighbors of the chain we shall choose a point, and we shall link any two neighboring points by a simple curve. Furthermore, we link the first point with the point on the boundary of $f_i^{-\epsilon}$ and the last point with the point on the boundary of f_i^{ϵ}. In this way we obtain on our sphere with holes a path t connecting $f_i^{-\epsilon}$ with f_i^{ϵ} and intersecting the boundary of F_0 only once. We now choose that closed curve s in the boundary of F_0 which pierces t. Then, however, s will decompose the sphere into two parts, to one of which belongs f_i^{ϵ} and to the other $f_i^{-\epsilon}$.

Inside F_0 are some holes that do not belong to the set B. Let E be the set of those arcs which link a hole inside F_0 with s, and let e be the number of arcs of E. We have $e \leq d$. There exists a simple closed path s_i' inside F_0 which intersects each arc of E only once while it does not intersect the other arcs at all. Let s_i' be "parallel" to the path s and close to it (see the figure).

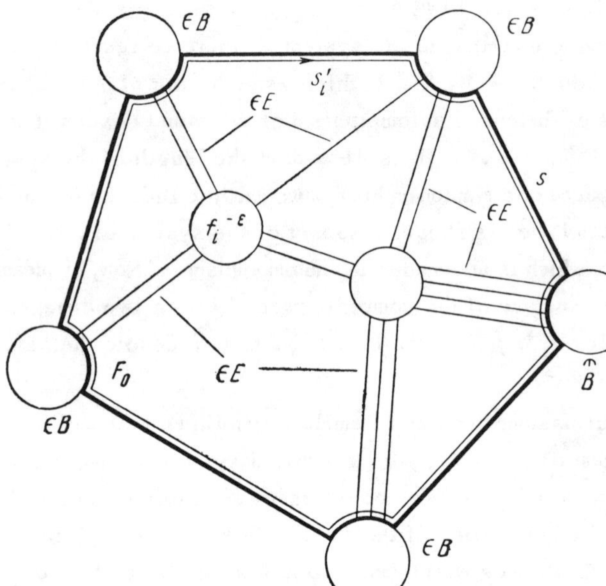

Figure 1

The path s_i' divides the sphere into two parts, and at the same time separates the pair f_i^{ϵ}, $f_i^{-\epsilon}$, since t pierces s_i' exactly once. Hence by means of an elementary process we can replace the disc f_i by the disc f_i' having the boundary s_i'. Thus we obtain a cut Z' with a smaller number of intersections of $\mathfrak{k}$ than in the case of Z. Let η_2 be a homeomorphism $V \rightarrow V$ which takes Z' into Z. Then the number of intersections of $\eta_2 \mathfrak{k}$ and Z is less than the number of intersections of $\mathfrak{k}$ and Z.

Remark. It is clear that, in general, the homeomorphism η_2 does not induce the given Whitehead automorphism α. Only in the case where $e = d$ are the complete words of the system $\eta_2 k$ equal to $L_1 \alpha K_1 L_1^{-1}, \cdots, L_v \alpha K_v L_v^{-1}$. Then the homeomorphism η_2 can induce the automorphism α only "on $\mathfrak{k}$" but not in general.

c) Applying the process described in a) or b) several times, we get a system $\eta_3 \mathfrak{k}$ with complete words $K_1^*, \cdots, K_v^*$. Also, the system $\{K_1^*, \cdots, K_v^*\}$ is of minimal length. Then there exists a chain of Whitehead automorphisms $\tau_1, \cdots, \tau_t$ which brings $\{K_1, \cdots, K_v\}$ and $\{K_1^*, \cdots, K_v^*\}$ into correspondence. Since each of the automorphisms τ_l is of the form (1), then $l_i = d = e$, because each system $\{\tau_s \cdot \cdots \cdot \tau_1 K_1^*, \cdots, \tau_s \cdot \cdots \cdot \tau_1 K_v^*\}$ has the minimal cyclic length. Then the homeomorphism constructed in b) does indeed induce the automorphism τ_l on our system of paths. Finally, we obtain a homeomorphism η: $V \rightarrow V$ such that to the system $\eta \mathfrak{k}$ there correspond complete words $K_1', \cdots, K_v'$. Theorem 1 is proven.

One can deduce from Theorem 1 that in each homotopy class there exists only a finite number of nonhomeomorphic simple closed paths on the surface (cf. [8], § 2).

§ 2. Applications

Let $\mathfrak{F}$ be a free group with p free generators and $\mathfrak{F}_q$ its subgroup generated by the elements $S_1, \cdots, S_q$. We say that the system of elements $K_1, \cdots, K_v \in \mathfrak{F}$ is of rank q, if in $\mathfrak{F}_q$ there exists a system (cyclically) equivalent to it and if there is no such a system in $\mathfrak{F}_l$ for $l < q$. A system of rank p is called a system of *complete rank.*

Lemma 2. *If the system $\{K_1, \cdots, K_v\}$ is of rank q, then each system of minimal length and (cyclically) equivalent to it contains only q generators $S_{i_1}, \cdots, S_{i_q}$.*

Let $L_1, \cdots, L_v \in \mathfrak{F}_q$ be a system equivalent to $\{K_1, \cdots, K_v\}$. If α is a Whitehead automorphism of form (1) with $i \leq q$, then α induces an automorphism of the subgroup $\mathfrak{F}_q$. In this way we obtain all Whitehead automorphisms of the group $\mathfrak{F}_q$. Let the system $M_1, \cdots, M_v \in \mathfrak{F}_q$ be equivalent to the system $L_1, \cdots, L_v$ under the automorphisms of the group $\mathfrak{F}_q$ and let it be of minimal length (with respect to $\mathfrak{F}_q$ and its generators $S_1, \cdots, S_q$). Then $\{K_1, \cdots, K_v\}$ and $\{M_1, \cdots, M_v\}$ are also equivalent. If the system $\{M_1, \cdots, M_v\}$ were not of minimal length with respect to $\mathfrak{F}$ and $S_1, \cdots, S_p$, then there would exist a Whitehead automorphism α of form (1) such that the cyclic length of the system $\{\alpha M_1, \cdots, \alpha M_v\}$ would be smaller that the length of the system $\{M_1, \cdots, M_v\}$. Hence α cannot induce an automorphism of the group $\mathfrak{F}_q$ when $i > q$. From Lemma 1 it follows that in cyclically reduced words which belong to $\alpha M_1, \cdots, \alpha M_v$ the generators $S_1, \cdots, S_q$ occur with the same frequency as in $M_1, \cdots, M_v$. This contradiction proves the lemma.

From Lemma 2 and Theorem 1 there follows the following statement:

Let $k_1, \cdots, k_v$ be a simple system of paths on V such that the induced system $K_1, \cdots, K_v \in \mathfrak{F}$ has the rank q. Then there exists a cut $\{f_1, \cdots, f_p\}$ of the complete pretzel for which $k_1, \cdots, k_v$ intersect only the discs $f_1, \cdots, f_q$.

Let us choose some orientation of the complete pretzel and define the "left side" of one simple closed path on the surface. Let k be such a path not dividing the surface. $\mathfrak{U}_k$ is defined as a subset of the group $\mathfrak{F}$, consisting of those elements which contain at least one path having the following properties:

this path lies on the surface and originates at the same point as k; (2)

its intersection with k consists of this point only; (3)

the path does not pierce k at the beginning point and from there it turns to the left from k (4)

Clearly $\mathfrak{U}_k$ is a subgroup of $\mathfrak{F}$. $\mathfrak{U}_k$ is a "group of paths without intersections".

One can complete k by other paths and obtain a canonical cut $\{k_1, \cdots, k_{2p}\}$ of the surface F by $k_2 = k^{-1}$. (The simple closed paths $k_1, \cdots, k_{2p}$ form a canonical cut if they intersect each other only at the beginning point and cut F into a disc with the boundary $\Pi_{i=1}^{p} k_{2i-1} k_{2i} k_{2i-1}^{-1} k_{2i}^{-1}$.) Let $K_1, \cdots, K_{2p}$ be the induced elements of the group $\mathfrak{F}$. Then $\mathfrak{U}_k$ is a subgroup generated by $K_2, \cdots, K_{2p}$ (cf. [7], p.242)

Theorem 2. *Let k be a simple closed path on the surface F of a complete pretzel V of genus p. Assume that k does not decompose F. The homotopy class K of the path k has the rank p if and only if $\mathfrak{U}_k$ has the rank $2p - 1$.*

Proof. a) Let the rank of K be less than p. In view of Theorem 1 there exists a homeomorphism η: $V \longrightarrow V$ such that k does not intersect the disc $\eta^{-1} f_p$. We link the boundary s_p' of $\eta^{-1} f_p$ with the beginning point of the path k by an arc l which intersects k only at the end point (which is the beginning point of k), and approaches k from the left. Then the path $l^{-1} s_p' l$ can be shrunk on V but not on the surface F. Since $l^{-1} s_p' l$ is a path with properties (2)–(4), it is homotopic on F to some product $\Pi(k_2, \cdots, k_{2p})$. Hence one gets a nontrivial relation between the generators $K_2, \cdots, K_{2p}$ of group $\mathfrak{U}_k$. Thus the rank of $\mathfrak{U}_k$ is less than $2p - 1$.

b) Let the rank of $\mathfrak{U}_k$ be less than $2p - 1$. Then there exists a nontrivial relation among the generators $K_2, \cdots, K_{2p}$ of the group $\mathfrak{U}_k$. This means that there is a path s' on F which does not intersect k and which can be shrunk on V but not on F. Then, however, one can find on F a simple closed path s with the same properties (cf. [3], p. 281). If s does not decompose the surface, then there exists a cut of the complete pretzel which contains the path s in its trace. If s decomposes the surface, there exists a cut of the complete pretzel which does not intersect s (cf. [7], p. 236). Since s decomposes F and does not shrink on F, there is at least one disc of this cut which does not intersect k. In both cases K has a rank less than p.

Remark. Let the rank of K be q, $q < p$, and let $\{f_1, \cdots, f_p\}$ be a cut such that k does not intersect the discs $f_{q+1}, \cdots, f_p$. We shall cut the pretzel V by means of those discs and obtain a complete pretzel V' of genus q. When k does not decompose the surface V' (for instance, when K is not in the commutator of the group $\mathfrak{F}$), then there exists a canonical cut of the surface of V, completing k and inducing the words $K_1, K_2 = K^{-1}, K_3, \cdots, K_{2q}, S_{q+1}, 1, \cdots, S_p, 1$; the words $K_1, \cdots, K_{2q}$ contain only the symbols $S_1, \cdots, S_q$. The group $\mathfrak{U}_k$ has the generators $K_2, \cdots, K_{2q}, S_{q+1}, 1, \cdots$ $\cdots, S_p, 1$. Hence the rank of the group $\mathfrak{U}_k$ equals the rank of $\mathfrak{U}'_k$ plus $(p-q)$; here $\mathfrak{U}'_k$ is the subgroup generated by the elements $K_2, \cdots, K_{2q}$. $\mathfrak{U}'_k$ is the group of paths which lie on V' and do not intersect k. Since K is of rank q, then the rank of $\mathfrak{U}'_k$ is $2q-1$. Hence it follows that the rank of $\mathfrak{U}_k$ is $p+q-1$.

Let M be a three-dimensional bicompact variety decomposed into two complete pretzels V and W of genus p whose intersection consists of the common boundary surface. (V, W) is called a *Heegaard decomposition* of the variety M. Let $\{v_1, \cdots, v_p\}$ be the trace of some cut of the complete pretzel W. We shall call $\{V; v_1, \cdots, v_p\}$ a *diagram of the Heegaard decomposition.* $\{v_1, \cdots, v_p\}$ induces the system $\{R_1, \cdots, R_p\}$ in the fundamental group $\mathfrak{F}$ of the complete pretzel V. The diagram $\{V; v_1, \cdots, v_p\}$ is called *reducible* if there exists a system of generators $S_1, \cdots, S_p$ of the group $\mathfrak{F}$ such that for some one q $(1 \leq q < p)$ the cyclically reduced words $R_{i_1}, \cdots, R_{i_q}$ $(1 \leq i_1 < i_2 < \cdots < i_q \leq p)$ contain only $S_1, \cdots, S_q$.

A simple closed path u on the surface F of a complete pretzel V is called a *strip of degree* q, if u decomposes the surface into two surfaces of genus q and $p-q$ with one hole respectively, and if u shrinks in V. (Instead of "strip of degree 1" we say only "strip".) There exists a disc d which decomposes V in two complete pretzels of genus q and $p-q$.

Theorem 3. *Let $\{V; v_1, \cdots, v_p\}$ be a reducible diagram of Heegaard's decomposition (V, W) of the variety M. Then either M is* *reducible* or M has a Heegaard's decomposition of genus less than p.*

Proof. In view of Lemma 1 and Theorem 1 there exists, for some q, a cut $Z = \{f_1, \cdots, f_p\}$ of the complete pretzel V such that $v_{i_1}, \cdots, v_{i_q}$ intersect only $f_1, \cdots, f_q$. Let $\{s_1, \cdots, s_q\}$ be the trace of the cut Z. Then let $v_{i_1}, \cdots, v_{i_q}, s_{q+1}, \cdots, s_p$ be a simple system of paths on F. We cut F by means of these paths and we obtain in this way several parts $F_1, \cdots, F_m$. Each part is a surface with holes. Let g_i be the genus of F_i. Then at least one of the numbers g_i is zero. Indeed, let us patch together these parts by some of the paths of $v_{i_1}, \cdots, v_{i_q}, s_{q+1}, \cdots, s_p$. As a result, we get a single surface F^* with holes. We need $m - 1$ paths in order to patch all parts. Then there are $p - m + 1$ of pairs of corresponding holes on F^*. The genus of F^* is $\Sigma_{i=1}^m g_i$. Further, we identify the corresponding pairs and get F. Thus

$$p = p - m + 1 + \sum_{i=1}^{m} g_i,$$

so that of the numbers g_i at least one is zero.

The part F_h of genus 0 is a sphere with holes. The holes F_h can be divided into two classes: in the first class are all the holes with boundaries of type $v_{ij}^{\pm 1}$, and in the second class the holes with the boundaries $s_j^{\pm 1}$. We denote by $x_1, \cdots, x_\mu$ the boundaries of the first class and by $x_{\mu+1}, \cdots, x_\gamma$ the boundaries of the second class. Let us choose a point a on F_h. There exist simple paths $l_1, \cdots, l_\gamma$ with the following properties: l_i starts at the point a and ends at x_i; $l_i \cap l_j = \{a\}$ for $i \neq j$; the star of paths at the point a is $l_1, \cdots, l_\gamma$. In other words, the paths $x_1, \cdots, x_\gamma, l_1, \cdots, l_\gamma$ cut F_h into a disc with the boundary $l_1 x_1 l_1^{-1} l_2 x_2 l_2^{-1} \cdots l_\gamma x_\gamma l_\gamma^{-1}$ (here we fixed the appropriate orientations of paths $x_1, \cdots, x_\gamma$). On F_h there exists a simple closed path u such that $u \cap x_i = \phi$ and $u \cap l_i = \{a\}$ $(i = 1, \cdots, \gamma)$, and the new star at the point a is $u, l_1, \cdots, l_\mu, u^{-1}, l_{\mu+1}, \cdots, l_\gamma$. It means, on F_h, that u is homotopic to $\Pi_{i=L}^{\mu} l_i x_i l_i^{-1}$ as well as to $\Pi_{i=\mu+L}^{\gamma} l_i x_i^{-1} l_i^{-1}$. Therefore u can be shrunk in W and also in V. Since $\{v_{i_1}, \cdots, v_{i_q}\}$, $\{s_{q+1}, \cdots, s_p\}$ are the subsets of traces of cuts and none of them decomposes the surface F, we have $0 < \mu < \gamma$. Hence u cannot be shrunk on F.

If u decomposes the surface F, then u is a strip of some particular degree for both complete pretzels V and W. Let us take some discs $d_V \subset V$ and $d_W \subset W$ which have the common boundary u and are such that $d_V \cap F =$

---

* in the sense of Kneser ([2], p. 252)

$d_W \cap F = u$. Then d_V and d_W form a two-dimensional sphere decomposing M in two parts. Therefore either M is the sum of two three-dimensional varieties or one of them is the sphere.

If u does not decompose F we choose such a simple closed path s which pierces u exactly once at the common beginning point and, in addition, does not intersect u. The paths u and s cut F into a surface of genus $p - 1$ with one hole, the boundary of which is $usu^{-1}s^{-1}$. Let u^* be a simple closed path parallel to the boundary of the hole. Let u^* be homotopic on F to the product $usu^{-1}s^{-1}$. Since u can be shrunk in both V and W, u^* is also shrinkable in both pretzels. Hence u^* is the strip for both pretzels.

The theorem is proved.

BIBLIOGRAPHY

[1] P. J. Higgins and R. C. Lyndon, *Equivalence of elements under automorphisms of a free group*, Queen Mary College, London. (preprint)

[2] H. Kneser, *Geschlossene Flächen in dreidimensionalen Mannigfaltigkeiten*, Jber. Deutsch. Math. Verein. 38 (1929), 248–260.

[3] C. D. Papakyriakopoulos, *On solid tori*, Proc. London Math. Soc. (3) 7 (1957), 281–299. MR 19,441.

[4] E. S. Rapaport, *On free groups and their automorphisms*, Acta Math. 99 (1958), 139–163. MR 24 #A1302.

[5] K. Reidemeister, *Über Heegaard-Diagramme*, Abh. Math. Sem. Univ. Hamburg 25 (1961/62), 140–145. MR 25 #4527.

[6] J. H. C. Whitehead, *On equivalent sets of elements in a free group*, Ann. of Math. (2) 37 (1936), 782–800.

[7] H. Zieschang, *Über einfache Kurven auf Vollbrezeln*, Abh. Math. Sem. Univ. Hamburg 25 (1961/62), 231–250. MR 26 #6957.

[8] ———, *Über einfache Kurvensysteme auf einer Vollbrezel vom Geschlecht* 2, Abh. Math. Sem. Univ. Hamburg 26 (1963/64), 237–247. MR 28 #4524.

Translated by:
M. A. Dostal

Amer. Math. Soc. Transl.
(2) Vol. 92, (1970)

PARALLEL AND NORMAL CORRESPONDENCE OF TWO-DIMENSIONAL SURFACES IN THE FOUR-DIMENSIONAL EUCLIDEAN SPACE E_4 *

L. N. KRIVONOSOV

§1. The fundamental equations of a two-dimensional surface in E_4

Let $M(u, v)$ be a given two-dimensional surface in E_4. We associate with each point of the surface four orthonormal vectors $\bar{e}_i$, of which $\bar{e}_1$ and $\bar{e}_2$ lie in the tangent plane and $\bar{e}_3$ and $\bar{e}_4$ lie in the normal plane of the surface. We write the equations of infinitesimal displacements of the frame so obtained as

$$dM = \omega^1\bar{e}_1 + \omega^2\bar{e}_2, \quad d\bar{e}_i = \omega_i^k\bar{e}_k, \tag{1}$$
$$\omega_i^k + \omega_k^i = 0 \quad (i, k = 1, 2, 3, 4).$$

The integrability condition of equations (1) have the form

$$[\omega^1\omega_1^3] + [\omega^2\omega_2^3] = 0, \quad [\omega^1\omega_1^4] + [\omega^2\omega_2^4] = 0, \tag{2}$$
$$D\omega^1 = [\omega_1^2\omega^2], \quad D\omega^2 = [\omega^1\omega_1^2], \quad D\omega_1^2 = [\omega_2^3\,\omega_1^3] + [\omega_2^4\,\omega_1^4],$$
$$D\omega_1^3 = [\omega_1^2\,\omega_2^3] + [\omega_3^4\,\omega_1^4], \quad D\omega_2^3 = [\omega_1^3\,\omega_1^2] + [\omega_3^4\,\omega_2^4], \quad D\omega_1^4 = [\omega_1^2\,\omega_2^4] + [\omega_1^3\,\omega_3^4],$$
$$D\omega_2^4 = [\omega_1^4\,\omega_1^2] + [\omega_2^3\,\omega_3^4], \quad D\omega_3^4 = [\omega_1^4\,\omega_1^3] + [\omega_2^4\,\omega_2^3]. \tag{3}$$

The elements of the frame being considered are determined with accuracy up to the transformations

$$\bar{e}_1' = \cos\alpha\bar{e}_1 - \sin\alpha\bar{e}_2, \quad \bar{e}_2' = \varepsilon(\sin\alpha\bar{e}_1 + \cos\alpha\bar{e}_2), \tag{4}$$
$$\bar{e}_3' = \cos\beta\bar{e}_3 - \sin\beta\bar{e}_4, \quad \bar{e}_4' = \varepsilon_1(\sin\beta\bar{e}_3 + \cos\beta\bar{e}_4), \quad \varepsilon = \pm 1, \quad \varepsilon_1 = \pm 1. \tag{5}$$

Under these transformations the Pfaffian forms ω_1^2 and ω_3^1 are transformed as follows:

$$\bar{\omega}_1^2 = \varepsilon(\omega_1^2 - d\alpha), \quad \bar{\omega}_3^4 = \varepsilon_1(\omega_3^4 - d\beta). \tag{6}$$

Solving equations (2) algebraically by Cartan's lemma, we obtain

$$\omega_1^3 = a\omega^1 + b\omega^2, \quad \omega_2^3 = b\omega^1 + c\omega^2, \quad \omega_1^4 = A\omega^1 + B\omega^2, \quad \omega_2^4 = B\omega^1 + C\omega^2. \tag{7}$$

By using these expansions we can write the third and the last of equations

* Translation of Izv. Vysš. Učebn. Zaved. Matematika 1966, No. 5 (54), 78–87.

(3) in the form

$$D\omega_1^2 = (b^2 - ac + B^2 - AC)[\omega^1\omega^2], \quad D\omega_3^4 = (Ab - aB + Bc - bC)[\omega^1\omega^2]. \tag{8}$$

Following Cartan ([1], Russian p. 254), we call the invariants

$$K_r = b^2 - ac + B^2 - AC = \frac{D\omega_1^2}{[\omega^1\omega^2]} \text{ and } K_H = Ab - aB + Bc - bC = \frac{D\omega_3^4}{[\omega^1\omega^2]}$$

the Gaussian curvature (torsion) of $M(u, v)$.

In view of the transformation law (6) the Pfaffian forms ω_1^2 and ω_3^4 generate two linear connections on the surface $M(u, v)$. With the help of the first connection we define a parallel transport of the tangent vectors of surface, and with the help of the second connection we define a parallel transport of the normal vectors of the surface. The analytical condition for the field of normal vectors $\bar{\eta} = \xi^3\bar{e}_3 + \xi^4\bar{e}_4$ to be parallel along the curve $u = u(s)$, $v = v(s)$ on the surface can be written as

$$\begin{aligned} \frac{d\xi^3}{ds} - \xi^4\omega_3^4\left(u(s), v(s), \frac{du}{ds}, \frac{dv}{ds}\right) &= 0, \\ \frac{d\xi^4}{ds} + \xi^3\omega_3^4\left(u(s), v(s), \frac{du}{ds}, \frac{dv}{ds}\right) &= 0. \end{aligned} \tag{9}$$

We can define the parallel transport of an arbitrary space vector $\bar{p}$ along a curve on the surface by requiring that the projections of $\bar{p}$ onto the tangent plane and onto the normal plane of the surface are respectively transported in a parallel way. Similarly to the way in which a system of geodesic lines is defined on the surface by using the first connection, with the help of the second connection we can also introduce a system of curves on the surface. The curves of this system can be characterized by the property that the projection of the normal curvature vector of the curve onto the normal plane of the surface is transported along this curve in a parallel way. Suppose we take arc length as the parameter s of the curve $u = u(s)$, $v = v(s)$. Then the normal curvature vector $\bar{m}$ is simply the second derivative with respect to arc length of the radius vector of the curve:

$$m = \frac{d^2M}{ds^2}$$

$$= \frac{(d\omega^1 - \omega^2\omega_1^2)\bar{e}_1 + (d\omega^2 + \omega^1\omega_1^2)\bar{e}_2 + (\omega^1\omega_1^3 + \omega^2\omega_2^3)\bar{e}_3 + (\omega^1\omega_1^4 + \omega^2\omega_2^4)\bar{e}_4}{ds^2}.$$

From here the projection $\bar{m}_H$ of the vector $\bar{m}$ onto the normal plane is expressed as $\bar{m}_H = \xi^3 \bar{e}_3 + \xi^4 \bar{e}_4$, where

$$\xi^3 = \frac{\omega^1\omega_1^3 + \omega^2\omega_2^3}{ds^2}, \quad \xi^4 = \frac{\omega^1\omega_1^4 + \omega^2\omega_2^4}{ds^2}.$$

The expressions for ξ^3 and ξ^4 are equivalent forms in du/ds and dv/ds. For convenience we temporarily write $u = u^1$, $v = u^2$. Then

$$\xi^3 = a_{ij}\frac{du^i}{ds}\frac{du^j}{ds}, \quad \xi^4 = b_{ij}\frac{du^i}{ds}\frac{du^j}{ds} \quad (i, j = 1, 2).$$

Next, by putting $\omega_3^4 = l_i\, du^i/ds$ and by writing out the conditions (9) for the parallel transport of the vector $\bar{m}_H$, we obtain the following system of equations to determine the curves under consideration:

$$\begin{aligned} a_{ij}\frac{d^2u^i}{ds^2}\frac{du^j}{ds} + \frac{1}{2}\left(\frac{\partial a_{ij}}{\partial u^k} - b_{ij}l_k\right)\frac{du^i}{ds}\frac{du^j}{ds}\frac{du^k}{ds} &= 0, \\ b_{ij}\frac{d^2u^i}{ds^2}\frac{du^j}{ds} + \frac{1}{2}\left(\frac{\partial b_{ij}}{\partial u^k} + a_{ij}l_k\right)\frac{du^i}{ds}\frac{du^j}{ds}\frac{du^k}{ds} &= 0. \end{aligned} \tag{10}$$

The determinant of the coefficients of the highest terms in the system is

$$(a_{1j}b_{2i} - a_{2j}b_{1i})\frac{du^i}{ds}\frac{du^j}{ds}.$$

The net of lines on the surface defined by

$$(a_{1j}b_{2l} - a_{2j}b_{1i})\, du^i du^j = \omega_1^3\,\omega_2^4 - \omega_2^3\,\omega_1^4 = 0 \tag{11}$$

is a conjugate set. If the vector $(du_0^1/ds, du_0^2/ds)$ does not touch a line of the conjugate net at (u_0^1, u_0^2), then in a neighborhood of the point $(u_0^1, u_0^2, du_0^1/ds, du_0^2/ds)$ the system (10) can be solved uniquely for the highest derivatives; that is, it can be reduced to normal form, and we can conclude that one and only one curve of the form (10), in a direction not tangential to a line of the conjugate net, passes through each point of the surface.

In the normal plane to the surface of any of its points we consider the locus of intersection of this normal plane with all infinitely near normal planes. Let the point $\tilde{M} = M + \alpha\bar{e}_3 + \beta\bar{e}_4$ belong to this locus. Then, for some direction of a displacement on the surface $M(u, v)$, the point $\tilde{M}_1 = \tilde{M} + d\tilde{M}$ must remain in the normal plane which contains the point $\tilde{M}$; that is, the vector $\overline{\tilde{M}\tilde{M}_1} = d\tilde{M}$ must be perpendicular to $\bar{e}_1$ and $\bar{e}_2$:

$$(d\tilde{M}\bar{e}_1) = (d\tilde{M}\bar{e}_2) = 0. \tag{12}$$

Since

$$\begin{aligned} d\tilde{M} &= dM + d\alpha\bar{e}_3 + d\beta\bar{e}_4 + \alpha d\bar{e}_3 + \beta d\bar{e}_4 \\ &= (\omega^1 - \alpha\omega_1^3 - \beta\omega_1^4)\,\bar{e}_1 + (\omega^2 - \alpha\omega_2^3 - \beta\omega_2^4)\,\bar{e}_2 \\ &\quad + (d\alpha - \beta\omega_3^4)\,\bar{e}_3 + (d\beta + \alpha\omega_3^4)\,\bar{e}_4, \end{aligned} \tag{13}$$

we can rewrite (12) as

$$\omega^1 - \alpha\omega_1^3 - \beta\omega_1^4 = 0, \quad \omega^2 - \alpha\omega_2^3 - \beta\omega_2^4 = 0. \tag{12'}$$

The Pfaffian form on the left-hand sides of the last equations are not zero identically, but only for a displacement in some direction. Hence the exterior derivative of these forms is zero:

$$\begin{aligned} [\omega^1 - \alpha\omega_1^3 - \beta\omega_1^4,\ \omega^2 - \alpha\omega_2^3 - \beta\omega_2^4] &= \alpha^2[\omega_1^3\,\omega_2^3] + \beta^2[\omega_1^4\,\omega_2^4] \\ + \alpha\beta([\omega_1^3\,\omega_2^4] + [\omega_1^4\,\omega_2^3]) &- \alpha([\omega_1^3\,\omega^2] + [\omega^1\omega_2^3]) \\ - \beta([\omega_1^4\,\omega^2] &+ [\omega^1\omega_2^4]) + [\omega^1\omega^2] = 0. \end{aligned} \tag{14}$$

This is the equation of the required locus. We rewrite this equation in a more convenient form by using (7) and by cancelling by $[\omega^1\omega^2]$:

$$\begin{aligned} \alpha^2(ac - b^2) + \beta^2(AC - B^2) + \alpha\beta(aC + Ac - 2bB) - \alpha(a + c) \\ - \beta(A + C) + 1 = 0. \end{aligned} \tag{14'}$$

The curve of second order defined in the normal plane by (14′) will be said to be a focal curve. [1)]

Similarly we can define a curve in the tangent plane as the locus of intersection of a given tangent plane with all infinitely near tangent planes. The equations that are the analogs of (12′) have the form

$$\begin{aligned} \lambda\omega_1^3 + \mu\omega_2^3 &= \lambda(a\omega^1 + b\omega^2) + \mu(b\omega^1 + c\omega^2) = 0, \\ \lambda\omega_1^4 + \mu\omega_2^4 &= \lambda(A\omega^1 + B\omega^2) + \mu(B\omega^1 + C\omega^2) = 0. \end{aligned} \tag{15}$$

By taking the exterior product of the Pfaffian forms on the left-hand sides of these equations we obtain the equation of a reducible curve of the second order with center at the point M:

$$\lambda^2[\omega_1^3\,\omega_1^4] + \mu^2[\omega_2^3\,\omega_2^4] + \lambda\mu([\omega_1^3\,\omega_2^4] + [\omega_2^3\,\omega_1^4]) = 0, \tag{16}$$

or, after cancelling by $[\omega^1\omega^2]$,

$$\lambda^2(aB - Ab) + \mu^2(bC - Bc) + \lambda\mu(aC - Ac) = 0. \tag{16'}$$

The net of curves on the surface for which the two lines defined by (16′) are

1) These curves are considered in [3].

tangents at the point M is a conjugate net. The equation of this net is obtained by dropping the parameters λ, μ from (15):

$$\omega_1^3\,\omega_2^4-\omega_1^4\,\omega_2^3=(aB-Ab)\,(\omega^1)^2+(bC-Bc)\,(\omega^2)^2+(aC-Ac)\,\omega^1\omega^2=0. \tag{17}$$

This equation is the same as (11). The curves (14′) and (16′) permit us to introduce an invariant frame on the surface in many ways. For example, we can direct the vector $\bar{e}_1$ along the bisector of the angle between the lines into which the curve (16′) degenerates and choose the vector $\bar{e}_3$ to be parallel to the polar of M with respect to the curve (14′). This invariant frame is characterized analytically by the equations $a = c = 0$. It is impossible by this method to choose a frame only for minimal surfaces for which, in the general frame (1),

$$a+c=A+C=0.$$

It is interesting to note the following link between the focal curve (14′) in the normal plane and the conjugate net (17): 1. if the conjugate net is real nonorthogonal, then the curve (14′) is a hyperbola; 2. if the conjugate net is imaginary nonorthogonal, then the curve (14′) is an ellipse; 3. if the conjugate net is orthogonal (real or imaginary), then the curve (14′) degenerates into a pair of straight lines (respectively real or imaginary), not parallel to one another [2]; 4. if the conjugate net degenerates to a single family of lines, then the curve (14′) is a parabola; 5. if the conjugate net is isotropic then the curve (14′) is an ellipse with center at M; 6. if, finally, the conjugate net is indeterminate, that is, (17) is satisfied identically, then the curve (14′) is a pair of distinct or coincident parallel straight lines, but they are necessarily real.

It is easy to verify all these facts (except 5) if we write (14′) and (17) in the invariant frame indicated above.

§2. Parallel correspondence of surfaces

We will say that the surfaces $M(u, v)$ and $\overset{*}{M}(u, v)$ correspond in a parallel way if by a motion of one of the surfaces we can bring it into a position in which the tangent planes at corresponding points are parallel to the tangent planes of the second surface. As is well known, in the euclidean space E_3, this correspondence between surfaces is called a Peterson correspondence. When considering the properties of parallel correspondence in what is to follow, we will always assume that the motion referred to in the definition has taken place, and two surfaces with correspondingly parallel tangent planes will simply be said to be parallel.

We are going to find all the surfaces parallel to a given surface $M(u, v)$, taken in the invariant frame given at the end of §1. Let the surface $\overset{*}{M}(u, v)$ be parallel to the surface $M(u, v)$. Then the tangent vectors $\bar{e}_1$, $\bar{e}_2$ of $M(u, v)$ are parallel to the tangent plane of $\overset{*}{M}(u, v)$ at the corresponding point. Hence we can write

$$d\overset{*}{M} = \sigma^1\bar{e}_1 + \sigma^2\bar{e}_2. \tag{18}$$

The integrability condition of this equation is

$$D\sigma^1 = [\omega_1^2\sigma^2],\ D\sigma^2 = [\sigma^1\omega_1^2],\ [\sigma^1\omega_1^3] + [\sigma^2\omega_2^3] = 0,\ [\sigma^1\omega_1^4] + [\sigma^2\omega_2^4] = 0. \tag{19}$$

We restrict our discussion to the case when the equation of the conjugate net (17) is not satisfied identically. In the invariant frame this restriction reduces to the following: $b \neq 0$, and at least one of the functions $A, C \neq 0$. In this case, on solving equations (19) by Cartan's lemma we obtain

$$\sigma^1 = x\omega^1 + Cy\omega^2,\ \sigma^2 = Ay\omega^1 + x\omega^2. \tag{20}$$

After substituting the expressions (20) into the first two equations at (19) we arrive at a system of equations in x, y:

$$\begin{gathered}[dx - y(A + C)\,\omega_1^2,\ \omega^1] + [d(Cy),\ \omega^2] = 0,\\ [d(Ay),\ \omega^1] + [dx + y(A + C)\,\omega_1^2,\ \omega^2] = 0.\end{gathered} \tag{21}$$

This system imposes two independent linear relations on the unknown functions x, y and their partial derivatives. The solution of this system exists with arbitrariness in two functions of one argument. Since the system (21) is linear and homogeneous in the unknown functions x, y and their partial derivatives, its solutions form an infinite-dimensional linear space over the field of real numbers. Correspondingly, the set of all surfaces parallel to a given surface $M(u, v)$ forms an infinite-dimensional linear space which we denote by D. We can think of the geometrical operation of adding two surfaces from D as the addition of the corresponding radius vectors of the surfaces; the multiplication of a surface from D by a number must be represented as the multiplication of each radius vector of this surface by the number. The geometrical object that plays the role of the zero of D is the collection of planes passing through a constant point and respectively parallel to the tangent planes of $M(u, v)$. Next we consider the properties of the parallel correspondence.

Theorem 1. *The conjugate nets on two parallel surfaces are in parallel correspondence; that is, the tangents to the lines of the conjugate net on one surface are respectively parallel to the tangents to the lines of the conjugate net at corresponding points on the second surface.*

Corollary. *The angle between the lines of a conjugate net is unchanged under parallel correspondence, although, in general, the angle between curves is not preserved.*

Proof. We can take the vectors of a local frame for the surface $\overset{*}{M}(u, v)$ parallel to $M(u, v)$ to be the vectors $\bar{e}_i$ of a local frame of $M(u, v)$. In this case all the forms $\overset{*}{\omega}{}_i^k$, except the parametric forms σ^1 and σ^2, on the surface $\overset{*}{M}(u, v)$ will be equal to the corresponding forms ω_i^k on $M(u, v)$, because $\omega_i^k = (d\bar{e}_i \bar{e}_k) = (d\overset{*}{\bar{e}}_i \overset{*}{\bar{e}}_k) = \overset{*}{\omega}{}_i^k$. Hence equation (17) of the conjugate net on the surface $M(u, v)$ is at the same time the equation of the conjugate net on $\overset{*}{M}(u, v)$. Equation (16) for the tangential directions of the lines of the conjugate net on $M(u, v)$ defines, simultaneously, the directions of the lines of the conjugate net on $\overset{*}{M}(u, v)$. The assertion of the theorem follows from this result.

Theorem 2. *If M_0 and $\overset{*}{M}_0$ are corresponding points of two parallel surfaces $M(u, v)$ and $\overset{*}{M}(u, v)$, then the asymptotes of the focal curve (14′), constructed for the point M_0 of the surface $M(u, v)$ will be parallel to the asymptotes of the focal curve (14′), constructed for the point $\overset{*}{M}_0$ on the surface $\overset{*}{M}(u, v)$.*

Proof. Suppose that local frames at corresponding points of two parallel surfaces $M(u, v)$ and $\overset{*}{M}(u, v)$ are chosen with the condition $\bar{e}_i = \overset{*}{\bar{e}}_i$. Then, by (14), the equation of the focal curve in the normal plane of $\overset{*}{M}(u, v)$ can be written as

$$\overset{*}{\alpha}{}^2 [\omega_1^3 \omega_2^3] + \overset{*}{\beta}{}^2 [\omega_1^4 \omega_2^4] + \overset{*}{\alpha} \overset{*}{\beta} ([\omega_1^3 \omega_2^4] + [\omega_1^4 \omega_2^3])$$

$$- \overset{*}{\alpha} ([\omega_1^3 \sigma^2] + [\sigma^1 \omega_2^3]) - \overset{*}{\beta} ([\omega_1^4 \sigma^2] + [\sigma^1 \omega_2^4]) + [\sigma^1 \sigma^2] = 0. \tag{22}$$

The directions of the asymptotes depend only on the highest coefficients of a curve of the second order. In our case, (14) and (22) have the same highest coefficients and, because the systems of coordinates in which (14) and (22) have been expressed have respectively parallel axes, the assertion of the theorem is obvious.

Theorem 3. *The ratio of the Gaussian curvature to the Gaussian torsion is invariant under the parallel correspondence.*

Theorem 4. *If the vector* $\bar{p}$ *is transported in a parallel way along a curve on the surface* $M(u, v)$, *then this same vector will be transported in a parallel way along the corresponding curve on the parallel surface* $\overset{*}{M}(u, v)$.

Proof of Theorems 3 and 4. If the local frames on $M(u, v)$ and $\overset{*}{M}(u, v)$ are coordinated by the condition $\overset{*}{\bar{e}} = \bar{e}_i$, then $\overset{*}{\omega}{}_i^k = \omega_i^k$. Hence

$$\frac{K_\Gamma}{K_H} = \frac{D\omega_1^2}{D\omega_3^4} = \frac{D\overset{*}{\omega}{}_1^2}{D\overset{*}{\omega}{}_3^4} = \frac{\overset{*}{K}_\Gamma}{\overset{*}{K}_H},$$

which proves Theorem 3. To prove Theorem 4 it is sufficient to note that the conditions (9) for the parallel transport of the normal component $\bar{p}_N$ of $\bar{p}$ do not change under a transfer to $\overset{*}{M}(u, v)$, and that the conditions for the parallel transport of the tangential component $\bar{p}_T$ of $\bar{p}$ do not vary under a transfer to the surface $M(u, v)$. Equations (20) and (21), which are used to investigate surfaces parallel to a given surface, can be given another geometrical interpretation. We associate with (20) the linear transformation of differential forms

$$\alpha_1\omega^1 + \alpha_2\omega^2 \to \alpha_1\sigma^1 + \alpha_2\sigma^2. \tag{23}$$

The algebraic structure of the expressions (20) is characterized by the fact that the transformation (23) sends the families of lines of the conjugate net into themselves. The differential relations placed on the functions x, y by equations (21) are characterized by the result that the quadratic forms $(\omega^1)^2 + (\omega^2)^2$ and $(\sigma^1)^2 + (\sigma^2)^2$ induce the same affine connection; this is defined by the linear form ω_1^2. Thus the transformation (23) leaves both families of lines of the conjugate net (17) invariant and, while changing the metric, leaves the connection in $M(u, v)$ invariant. The set of all transformations of the form (23) forms an infinite continuous group. Each surface parallel to a given surface $M(u, v)$ uniquely determines a transformation (23) on $M(u, v)$. Hence we can regard the collection of surfaces parallel to a given surface as a collection of frames of an infinite group of transformations (23).

§3. Normal correspondence of surfaces

Two surfaces $M(u, v)$ and $S(u, v)$ will be said to correspond normally (or simply, to be normal) if at corresponding points the tangent planes of one surface are parallel to the normal planes of the second surface. It is clear that if the surface $S(u, v)$ is in normal correspondence with the surface $M(u, v)$, then any surface $\overset{*}{S}(u, v)$ parallel to $S(u, v)$ is also in normal correspondence with $M(u, v)$.

Conversely, if each of $S(u, v)$ and $\overset{*}{S}(u, v)$ is in normal correspondence with the surface $M(u, v)$, then they are in parallel correspondence. Thus, if there exists at least one surface $S(u, v)$ that is in normal correspondence with the surface $M(u, v)$, then all the remaining surfaces normal to $M(u, v)$ are determined from the condition of being parallel to $S(u, v)$. It follows from here that the collection of all surfaces normal to $M(u, v)$ forms a linear space over the field of real numbers. We denote this space by D'.

Let the surface $S(u, v)$ be normal to the surface $M(u, v)$. We denote the vectors of a local frame of $S(u, v)$ by $\overline{f}_i$, and the components of infinitesimal displacements of this frame by θ_i^j. By the definition of a normal surface we can assume that

$$\overline{f}_1 = \overline{e}_3, \ \overline{f}_2 = \overline{e}_4, \ \overline{f}_3 = -\overline{e}_1, \ \overline{f}_4 = -\overline{e}_2. \tag{24}$$

With this choice of the vectors $\overline{f}_i$ the Pfaffian forms θ_i^j have the following expression:

$$\theta_1^2 = \omega_3^4, \ \theta_1^3 = \omega_1^3, \ \theta_2^3 = \omega_1^4, \ \theta_1^4 = \omega_2^3, \ \theta_2^4 = \omega_2^4, \ \theta_3^4 = \omega_1^2. \tag{25}$$

We write out the integrability conditions (3) for the forms θ_i^j; then, after substituting from (25), each of these integrability equations, apart from the first two, coincides with a similar equation of the system (3). Only for the parametric forms θ^1, θ^2 do we obtain new integrability conditions:

$$D\theta^1 = [\omega_3^4\theta^2], \ D\theta^2 = [\theta^1\omega_3^4], \quad [\theta^1\omega_1^3] + [\theta^2\omega_1^4] = 0, \quad [\theta^1\omega_2^3] + [\theta^2\omega_2^4] = 0. \tag{26}$$

The system (26) is used to investigate the forms θ^1, θ^2. If, on $M(u, v)$, the equation of the conjugate net does not degenerate into an identity, then the system (26) defines the forms θ^1, θ^2 with arbitrariness in two functions of a single argument.

We consider properties of the normal correspondence.

Theorem 1. *The conjugate nets of two mutually normal surfaces correspond.*

Proof. Suppose that, on the surface $S(u, v)$ normal to $M(u, v)$ we have selected a frame in accordance with (24). Then, by (17) and (25), the conjugate net on $S(u, v)$ will be defined by

$$\theta_1^3\theta_2^4 - \theta_2^3\theta_1^4 = \omega_1^3\omega_2^4 - \omega_1^4\omega_2^3 = 0;$$

that is, the conjugate net on both mutually normal surfaces is given by one and the same equation; this proves our theorem.

Theorem 2. *If the surfaces $M(u, v)$ and $S(u, v)$ are in normal correspondence, then the directions of the lines of the conjugate net of each of these surfaces are, at corresponding points, parallel to the asymptotic directions of the focal curve* (14′) *of the second surface.*

Corollary. *The angle between lines of the conjugate net on one surface is equal to the angle between the asymptotes of the focal curve at the corresponding point on a second surface normal to the first.*

Proof. The directions of the lines of the conjugate net on $M(u, v)$ are determined from equations (16), whilst the asymptotic directions of the focal curve (14) are found from

$$\alpha^2 [\omega_1^3 \omega_2^3] + \beta^2 [\omega_1^4 \omega_2^4] + \alpha\beta ([\omega_1^3 \omega_2^4] + [\omega_1^4 \omega_2^3]) = 0. \tag{27}$$

Analogously, on the surface $S(u, v)$ the equation

$$\overline{\lambda}^2 [\theta_1^3 \theta_1^4] + \overline{\mu}^2 [\theta_2^3 \theta_2^4] + \overline{\lambda}\, \overline{\mu} ([\theta_1^3 \theta_2^4] + [\theta_2^3 \theta_1^4]) = 0 \tag{28}$$

defines the directions of lines of the conjugate net, and the equation

$$\overline{\alpha}^2 [\theta_1^3 \theta_2^3] + \overline{\beta}^2 [\theta_1^4 \theta_2^4] + \overline{\alpha}\, \overline{\beta} ([\theta_1^3 \theta_2^4] + [\theta_1^4 \theta_2^3]) = 0 \tag{29}$$

defines the directions of the asymptotes of the focal curve.

If next we choose the vectors of a local frame on $S(u, v)$ in accordance with (24) then, by (25), equation (28) coincides with (27) and (29) with (16); this proves the theorem.

Theorem 3. *The ratios of the Gaussian curvature to the Gaussian torsion on mutually normal surfaces are reciprocals of one another.*

Proof. Coordinating these frames on mutually normal surfaces by (24), we have

$$\frac{K_\Gamma}{K_H} = \frac{D\omega_1^2}{D\omega_3^4} = \frac{D\theta_3^4}{D\theta_1^2} = \frac{\overline{K}_H}{\overline{K}_\Gamma},$$

where $\overline{K}_\Gamma (\overline{K}_H)$ is the Gaussian curvature (torsion) of $S(u, v)$.

Theorem 4. *If the space vector $\overline{p}$ is transported in a parallel way along a curve on the surface $M(u, v)$, then this vector will be transported in a parallel way along the corresponding curve on the normal surface $S(u, v)$.*

The proof is obtained directly if, along with the conditions (9) for the parallel transport of the normal component of $\overline{p}$ we write out the conditions for the parallel transport of the tangential component of $\overline{p}$, and note that, in view of (24) and (25), the roles of these conditions are switched on $S(u, v)$.

§ 4. Parallel and normal correspondence of special classes of surfaces

1. Let the surface $M(u, v)$ be characterized by the property that the parallel transport of normal vectors of this surface according to (9) does not depend on the path of the displacement. From here we can obtain, by a well-known method, that the Pfaffian form ω_3^4, which is used to define the parallel transport of normal vectors, must be an exact differential: $\omega_3^4 = df$; that is equivalent to the vanishing of the exterior differential of ω_3^4: $D\omega_3^4 = 0$. It follows from (8) that the surfaces of the class being considered have zero Gaussian torsion. In view of Theorem 4 of § 2, on all surfaces parallel to $M(u, v)$ the parallel transport of normal vectors does not depend on the path of the displacement; that is, the Gaussian torsion is zero on each of the parallel surfaces. Let the surface $S(u, v)$ be in normal correspondence with the surface $M(u, v)$. By choosing a frame on $S(u, v)$ in accordance with (24), we obtain from (25) that $D\theta_1^2 = D\omega_3^4 = 0$. This means that the Gaussian torsion of $S(u, v)$ is zero, or, equivalently, that the intrinsic metric is euclidean.

2. If the Gaussian curvature and torsion of $M(u, v)$ are zero, then any surface parallel to $M(u, v)$ and any surface normal to $M(u, v)$ has this property.

3. Suppose that the ratio of the Gaussian curvature and torsion is constant on $M(u, v)$. Then it follows from Theorems 3 of § 2 and 3 of § 3 that any surface parallel to $M(u, v)$ and any surface normal to $M(u, v)$ has this property.

4. We consider a surface $M(u, v)$ for which the conjugate net degenerates into a family of lines. In view of Theorem 1 (§ 2), the conjugate net on each of the parallel surfaces also degenerates into a family of lines; moreover, in view of Theorem 1 of § 3, each surface normal to $M(u, v)$ has this property.

5. In a general frame minimal surfaces are characterized by the equation $a + c = A + C = 0$ ([1], Russian p. 275). It is clear from (14′) and (16′) that these equations are equivalent to the requirement that the directions of the conjugate net be isotropic, and are also equivalent to the requirement that the point M coincides with the center of the focal curve (14′). It follows from Theorem 1 of § 2 that the conjugate net on each surface parallel to $M(u, v)$ has isotropic directions; that is, that each parallel sur-

face is minimal. It follows from Theorem 2 of §3 that on each surface normal to $M(u, v)$ the asymptotic directions of the focal curve are isotropic; that is, that the focal curve is a circle.

BIBLIOGRAPHY

[1] É. Cartan, *Riemannian geometry in an orthogonal coordinate system*, Lectures (Sorbonne, 1926/27); Russian transl., Izdat. Moskov. Univ., Moscow, 1960. MR 23 #A1316.

[2] A. V. Čakmazjan, *On surfaces D_m in the space K_n*, Akad. Nauk Armjan. SSR Dokl. 36 (1963), 71–75. (Russian) MR 27 #4162.

[3] K. Š. Ramazanova, *On the theory of two-dimensional surfaces in E_4*, Volž. Mat. Sb. No. 3 (1965), 296–311. (Russian) MR 34 #693.

Translated by:
L. W. Longdon

Amer. Math. Soc. Transl.
(2) Vol. 92, 1970

ON DEMOULIN TRANSFORMS OF PROJECTIVE MINIMAL SURFACES. I*†

SU BU-CHIN

I. Introduction

It is well known since the time of the discoverer, G. Thomsen, that a projective minimal surface and any one of its Demoulin transforms are in asymptotic correspondence. Conversely, only projective minimal surfaces and so called Q-surfaces are of this behavior [1]. In the investigations of many authors, for instance, O. Mayer [2], a projective minimal surface was often characterized by means of its Demoulin transforms or simply D-transforms. About two decades ago the present author already defined the associate quadric covariantly connected with a generic point of a surface [3] and then obtained certain different characterizations of a projective minimal surface by utilizing the associate quadric [4]. Any one D-transform $\hat{S}$ of a projective minimal surface S is also projective minimal and S is in turn a D-transform of $\hat{S}$, as was shown by Mayer. It is natural to inquire: How are the other three D-transforms of the surface $\hat{S}$? In §IV we find the parametric representations of these transforms in precise form and thence demonstrate that two of them are W-transforms of S (cf. §V). Moreover, the four Demoulin quadrilaterals at the corresponding points of the D-transforms have two by two a side in common, and these four sides thus obtained form the rectilinear congruences W, each of which has S and one of the second D-transforms of S for its focal sheets. In consequence, there are nine second D-transforms of S, namely, the original surface S, the four focal sheets and four other surfaces (§VI).

It is of interest to consider the Godeaux sequences of S and $\hat{S}$ (§VII). Let us arrange the corresponding points of these sequences in the matrix of three rows:

$$\left\| \begin{matrix} * & * & \mathfrak{U}_{(4)} & \mathfrak{U}_{(3)} & \mathfrak{U}_{(2)} & \mathfrak{U}_{(1)} & \mathfrak{U} & \mathfrak{V} & \mathfrak{V}_{(1)} & \mathfrak{V}_{(2)} \\ \hat{\mathfrak{U}}_{(4)} & \hat{\mathfrak{U}}_{(3)} & \hat{\mathfrak{U}}_{(2)} & \hat{\mathfrak{U}}_{(1)} & \hat{\mathfrak{U}} & \hat{\mathfrak{V}} & \hat{\mathfrak{V}}_{(1)} & \hat{\mathfrak{V}}_{(2)} & \hat{\mathfrak{V}}_{(3)} & \hat{\mathfrak{V}}_{(4)} \\ \mathfrak{U}_{(2)} & \mathfrak{U}_{(1)} & \mathfrak{U} & \mathfrak{V} & \mathfrak{V}_{(1)} & \mathfrak{V}_{(2)} & \mathfrak{V}_{(3)} & \mathfrak{V}_{(4)} & * & * \end{matrix} \right\|,$$

* First published in Chinese in *Acta Math. Sinica* 7 (1957), 28–50.
† A facsimile of the translation in Scientia Sinica 6 (1957), 941–965.

the asterisk denoting the lack of the corresponding point, we can show that in the 5-space S_5 the join of two consecutive points in the middle row must intersect the join of the corresponding points standing on the same columns of the upper or lower row. Thus the Godeaux sequences of S and $\hat{S}$ at the corresponding points constitute two broken lines, which intersect each other at an interval of three sides. This result has an intimate relation with the one due to L. Godeaux [5].

In §II we make use of the theory of surfaces in Wilczynski system and the semi-invariant differentiation of G. Bol [6] as an analytic basis. It is easily shown by means of these formulae that the associate quadric just coincides with the second quadric of Godeaux system (§III).

II. Fundamental Formulae

Let S be an analytic non-parabolically curved surface $\mathfrak{x}(u, v)$ referred to its asymptotic parameters u and v; if the surface is elliptically curved, then we can take these parameters such that they assume conjugate complex values for a real surface. Furthermore, we can suitably normalize the vector $\mathfrak{x}$, so that the following determinant equals to a constant:

$$(\mathfrak{x}, \mathfrak{x}_u, \mathfrak{x}_v, \mathfrak{x}_{uv}) = \kappa; \tag{2.1}$$

by hypothesis, this determinant is not zero, and is supposed to assume real values on an elliptically curved surface.

If a point $\mathfrak{y}$ is suitably chosen and the semi-invariant differentiation of Bol is utilized to the case of a non-ruled surface, then we obtain the derived equations in the projective theory:

$$\begin{aligned} \mathfrak{x}_{11} &= \beta\mathfrak{x}_2 + L\mathfrak{x}, & \mathfrak{x}_{12} &= \mathfrak{y} + P\mathfrak{x}, \\ \mathfrak{x}_{21} &= \mathfrak{y} + Q\mathfrak{x}, & \mathfrak{x}_{22} &= \gamma\mathfrak{x}_1 + M\mathfrak{x}, \\ \mathfrak{y}_1 &= Q\mathfrak{x}_1 + L\mathfrak{x}_2 + \beta M\mathfrak{x}, & \mathfrak{y}_2 &= M\mathfrak{x}_1 + P\mathfrak{x}_2 + \gamma L\mathfrak{x}, \end{aligned} \tag{2.2}$$

where the invariants obey the following laws:

$$\begin{gathered} P = \frac{1}{2}\beta\gamma - \sigma, \qquad Q = \frac{1}{2}\beta\gamma - \tau, \\ L_2 = P_1, \qquad M_1 = Q_2, \\ \gamma L_1 = \beta M_2 \end{gathered} \tag{2.3}$$

and

$$\sigma = \frac{1}{2}(\log \gamma)_{uv}, \qquad \tau = \frac{1}{2}(\log \beta)_{uv} \tag{2.4}$$

with $\beta\gamma \neq 0$. The semi-invariant derivatives of a semi-invariant $A(u, v)$ of weights (m, n) are given by

$$A_1 = A_u + m\frac{\gamma_u}{\gamma}A, \qquad A_2 = A_v + n\frac{\beta_v}{\beta}A. \tag{2.5}$$

Here we have also the integrability condition:

$$A_{12} - A_{21} = 2(m\sigma - n\tau)A. \tag{2.6}$$

Moreover, it is easily shown that β and γ are respectively of weights $(2, -1)$ and $(-1, 2)$ and accordingly

$$\beta_2 = 0, \qquad \gamma_1 = 0. \tag{2.7}$$

As $\beta\gamma \neq 0$, the last equation of (2.3) may be replaced by

$$L_1 = 2\beta N, \qquad M_2 = 2\gamma N. \tag{2.8}$$

If we merely consider the case where $P \neq 0$, $Q \neq 0$, then follow the equations

$$L_1 = 2\beta N, \quad M_2 = 2\gamma N, \quad L_2 = P_1 = 2aP, \quad M_1 = Q_2 = 2bQ; \tag{2.9}$$

$$N_1 - \beta M = Qd, \qquad N_2 - \gamma L = Pc \tag{2.10}$$

with integrability conditions

$$\begin{aligned} a_1 &= \beta c + 2(L - a^2), & b_2 &= \gamma d + 2(M - b^2), \\ c_1 &= Qg + 2(N - ac), & d_2 &= Pg + 2(N - bd), \end{aligned} \tag{2.11}$$

where g denotes a new function.

Let us now take in the 5-space S_5 the six points defined by the vectors

$$\begin{aligned} \mathfrak{U} &= (\mathfrak{x}, \mathfrak{x}_1), & \mathfrak{V} &= (\mathfrak{x}, \mathfrak{x}_2), \\ \mathfrak{A} &= \frac{1}{2}[(\mathfrak{x}, \mathfrak{y}) - (\mathfrak{x}_1, \mathfrak{x}_2)], & \mathfrak{B} &= \frac{1}{2}[(\mathfrak{x}, \mathfrak{y}) + (\mathfrak{x}_1, \mathfrak{x}_2)], \\ \mathfrak{P} &= (\mathfrak{x}_2, \mathfrak{y}), & \mathfrak{Q} &= (\mathfrak{x}_1, \mathfrak{y}), \end{aligned} \tag{2.12}$$

and consider them as the vertices of the moving simplex at a generic point of the surface S. Denoting by the symbol $\circ$ the polar mapping with respect to the absolute hyperquadric Q_4 in S_5, so that

$$\mathfrak{G} \circ \mathfrak{G} = 0$$

denotes the condition in order that $\mathfrak{G}$ should be a straight line, we have from (2.1)

$$\begin{gathered} \mathfrak{A} \circ \mathfrak{A} = -\frac{\kappa}{2}, \quad \mathfrak{B} \circ \mathfrak{B} = \frac{\kappa}{2}, \quad \mathfrak{P} \circ \mathfrak{U} = \mathfrak{U} \circ \mathfrak{P} = \kappa, \\ \mathfrak{Q} \circ \mathfrak{V} = \mathfrak{V} \circ \mathfrak{Q} = -\kappa, \end{gathered} \tag{2.13}$$

the other products of the vectors (2.12) being all zero.

Derivation of (2.12) and reduction by means of (2.2) immediately lead to the fundamental equations in S_5. For convenience' sake we write down the corresponding weights of all the above quantities as follows:

$\mathfrak{x}$	$\mathfrak{y}$	β	γ	P	Q	L	M	N
$-\frac{1}{2}$	$\frac{1}{2}$	2	-1	1	1	2	0	1
$-\frac{1}{2}$	$\frac{1}{2}$	-1	2	1	1	0	2	1

(2.14)

$\mathfrak{U}$	$\mathfrak{V}$	$\mathfrak{A}$	$\mathfrak{B}$	$\mathfrak{P}$	$\mathfrak{Q}$
0	-1	0	0	0	1
-1	0	0	0	1	0

(2.15)

The fundamental equations are found to be

$$\begin{aligned}
&\mathfrak{U}_1 = \beta\mathfrak{V}, && \mathfrak{U}_2 = 2\mathfrak{A},\\
&\mathfrak{V}_1 = 2\mathfrak{B}, && \mathfrak{V}_2 = \gamma\mathfrak{U},\\
&\mathfrak{A}_1 = Q\mathfrak{U}, && \mathfrak{A}_2 = \mathfrak{P} + M\mathfrak{U},\\
&\mathfrak{B}_1 = \mathfrak{Q} + L\mathfrak{V}, && \mathfrak{B}_2 = P\mathfrak{V},\\
&\mathfrak{P}_1 = 2Q\mathfrak{A} - \beta M\mathfrak{V}, && \mathfrak{P}_2 = \gamma\mathfrak{Q} + 2M\mathfrak{A} - \gamma L\mathfrak{V},\\
&\mathfrak{Q}_1 = \beta\mathfrak{P} + 2L\mathfrak{B} - \beta M\mathfrak{U}, && \mathfrak{Q}_2 = 2P\mathfrak{B} - \gamma L\mathfrak{U}.
\end{aligned} \tag{2.16}$$

Let the Godeaux sequence of the surface $\mathfrak{x}(u, v)$ be

$$\cdots, \mathfrak{V}_{(3)}, \mathfrak{V}_{(2)}, \mathfrak{V}_{(1)}, \mathfrak{V}, \mathfrak{U}, \mathfrak{U}_{(1)}, \mathfrak{U}_{(2)}, \mathfrak{U}_{(3)}, \cdots,$$

the direction u being from right to left and v from left to right; we obtain

$$\begin{aligned}
&\mathfrak{U}_{(1)} = \mathfrak{A}, && \mathfrak{U}_{(2)} = \mathfrak{P} + M\mathfrak{U} - 2b\mathfrak{A},\\
&\mathfrak{V}_{(1)} = \mathfrak{B}, && \mathfrak{V}_{(2)} = \mathfrak{Q} + L\mathfrak{V} - 2a\mathfrak{B},
\end{aligned} \tag{2.17}$$

$$\begin{aligned}
\mathfrak{U}_{(3)} &= \mathfrak{Q} + 2N\mathfrak{U} - L\mathfrak{V} - 2d\mathfrak{A} + E\mathfrak{U}_{(2)},\\
\mathfrak{V}_{(3)} &= \mathfrak{P} + 2N\mathfrak{V} - M\mathfrak{U} - 2c\mathfrak{B} + F\mathfrak{V}_{(2)},
\end{aligned} \tag{2.18}$$

where E and F are defined by

$$E(b_1 - Q) = \beta b - d_1, \qquad F(a_2 - P) = \gamma a - c_2. \tag{2.19}$$

In virtue of (2.13) and (2.16) we can easily derive two formulae for determining L and M, namely,

$$\mathfrak{B}_1 \bigcirc \mathfrak{B}_1 = -2L\kappa, \qquad \mathfrak{A}_2 \bigcirc \mathfrak{A}_2 = 2M\kappa. \tag{2.20}$$

III. The Associate Quadric

The points

$$\mathfrak{P} \pm 2K\mathfrak{A} + M\mathfrak{U}, \qquad \mathfrak{Q} \pm 2H\mathfrak{B} + L\mathfrak{B} \tag{3.1}$$

in S_5 represent the four edges of the Demoulin quadrilateral of the surface $\mathfrak{x}(u, v)$, where

$$H^2 = L, \qquad K^2 = M. \tag{3.2}$$

From (2.9) we may derive the equations:

$$\begin{aligned} H_1 &= \frac{\beta N}{H}, & H_2 &= \frac{aP}{H}, \\ K_1 &= \frac{bQ}{K}, & K_2 &= \frac{\gamma N}{K}. \end{aligned} \tag{3.3}$$

Now consider one edge of the Demoulin quadrilateral,

$$\mathfrak{R}^{(\epsilon)} = \mathfrak{P} + M\mathfrak{U} + 2\epsilon K\mathfrak{A} \qquad (\epsilon = \pm 1) \tag{3.4}$$

as well as its intersection with the tangent plane of the surface at $\mathfrak{x}$:

$$\mathfrak{F}^{(\epsilon)} = \epsilon K\mathfrak{x} + \mathfrak{x}_2. \tag{3.5}$$

$\mathfrak{F}^{(1)}$ and $\mathfrak{F}^{(2)}$ are evidently the two flecnodes on the asymptotic tangent $(\mathfrak{x}, \mathfrak{x}_2)$. On account of (2.2) and (3.3) we have

$$\mathfrak{F}_2^{(\epsilon)} = \left(M + \epsilon\frac{\gamma N}{K}\right)\mathfrak{x} + \gamma\mathfrak{x}_1 + \epsilon K\mathfrak{x}_2.$$

Using $\mathfrak{G}^{(\epsilon)} = (\mathfrak{F}^{(\epsilon)}, \mathfrak{F}_2^{(\epsilon)})$ to denote the tangent of the flecnode curve $u = \text{const.}$ at the point $\mathfrak{F}^{(\epsilon)}$, we easily obtain

$$\mathfrak{G}^{(\epsilon)} = \mathfrak{A} - \mathfrak{B} + \epsilon K\mathfrak{U} - \frac{\epsilon N}{K}\mathfrak{B}. \tag{3.6}$$

From (3.4), (3.6), (2.17) and (2.18) it follows that the matrix of the coefficients of $\mathfrak{P}, \mathfrak{Q}, \mathfrak{A}, \mathfrak{B}, \mathfrak{U}, \mathfrak{B}$ in the expressions of five lines $\mathfrak{R}^{(\epsilon)}, \mathfrak{G}^{(\epsilon)}, \mathfrak{B}_{(1)}, \mathfrak{B}_{(2)}, \mathfrak{B}_{(3)}$ is

$$\left\|\begin{matrix} 1 & 0 & 2\epsilon K & 0 & M & 0 \\ 0 & 0 & 1 & -1 & \epsilon K & -\dfrac{\epsilon N}{K} \\ 0 & 0 & 0 & 1 & 0 & 0 \\ 0 & 1 & 0 & * & 0 & L \\ 1 & 0 & 0 & * & -M & 2N \end{matrix}\right\|.$$

Since the latter is easily shown to be of rank 4, we may infer that the line $\mathfrak{R}^{(\epsilon)}$ at the flecnode $\mathfrak{F}^{(\epsilon)}$ is tangent to the curve of intersection of the second Godeaux quadric Φ_1 and the tangent plane $\{\mathfrak{x}, \mathfrak{x}_1, \mathfrak{x}_2\}$, namely, the conic K_2.

In a similar way we can prove that when any flecnode $\overline{\mathfrak{F}}^{(\epsilon')}$ $(\epsilon' = \pm 1)$ on another asymptotic tangent $(\mathfrak{x}, \mathfrak{x}_1)$ describes the curve $v =$ const., the tangent at this point is also tangent to the conic K_2.

Hence K_2 is the associate conic [3] of the surface $\mathfrak{x}(u, v)$ at the point $\mathfrak{x}$. Consequently, *the quadric* Φ_1 *coincides with the associate quadric.* Thus we have proved the theorem due to L. Godeaux [7].

IV. D-Transforms

The surface described by one vertex of the Demoulin quadrilateral of $\mathfrak{x}(u, v)$ is, as usual, called *the first D-transform* or simply *D-transform.* Suppose that the surface $\mathfrak{x}(u, v)$ has four distinct non-ruled D-transforms. We shall demonstrate

Theorem 4.1. *Suppose that a non-ruled non-degenerate surface S have four distinct non-ruled D-transforms, and S and one D-transform $\hat{S}$ be in asymptotic correspondence. Then S and any D-transform must be in asymptotic correspondence and S is projective minimal, and vice versa.*

Consider one D-transform $\hat{S}$ of the surface S; the corresponding point is given by

$$\mathfrak{z} = \mathfrak{y} + K\mathfrak{x}_1 + H\mathfrak{x}_2 + HK\mathfrak{x}\,, \tag{4.1}$$

where H and K are determined by (3.2).

In virtue of (2.2) and (3.3) we have

$$\begin{aligned} \mathfrak{z}_1 &= H\mathfrak{z} + Q\left(1 + \frac{b}{K}\right)\mathfrak{x}_1 + \beta\left(K + \frac{N}{H}\right)\mathfrak{x}_2 \\ &\quad + \left\{\beta K\left(K + \frac{N}{H}\right) + HQ\left(1 + \frac{b}{K}\right)\right\}\mathfrak{x}, \\ \mathfrak{z}_2 &= K\mathfrak{z} + \gamma\left(H + \frac{N}{K}\right)\mathfrak{x}_1 + P\left(1 + \frac{a}{H}\right)\mathfrak{x}_2 \\ &\quad + \left\{\gamma H\left(H + \frac{N}{K}\right) + KP\left(1 + \frac{a}{H}\right)\right\}\mathfrak{x}. \end{aligned} \tag{4.2}$$

Upon calculating $\mathfrak{z}_{11}$ from the first equation,

$$\mathfrak{z}_{11} = (*)\,\mathfrak{z} + H\mathfrak{z}_1 + \left\{\left(Q\left(1 + \frac{b}{K}\right)\right)_1 + HQ\left(1 + \frac{b}{K}\right)\right\}\mathfrak{x}_1 +$$

$$+\left\{\beta Q\left(1+\frac{b}{K}\right)-\beta H\left(K+\frac{N}{H}\right)+\left(\beta\left(K+\frac{N}{H}\right)\right)_1\right\}\mathfrak{x}_2$$

$$+\left[LQ\left(1+\frac{b}{K}\right)-\beta HK\left(K+\frac{N}{H}\right)+\beta Q\left(K+\frac{N}{H}\right)\right.$$
$$\left.+\left\{\beta K\left(K+\frac{N}{H}\right)+HQ\left(1+\frac{b}{K}\right)\right\}_1\right]\mathfrak{x}$$

and putting

$$\nabla\equiv PQ\left(1+\frac{a}{H}\right)\left(1+\frac{b}{K}\right)-\beta\gamma\left(H+\frac{N}{K}\right)\left(K+\frac{N}{H}\right), \quad (4.3)$$

we easily compute the determinant

$$(\vartheta,\vartheta_1,\vartheta_2,\vartheta_{11})=2\beta\frac{\nabla}{H}\left(1+\frac{b}{K}\right)QN\kappa. \quad (4.4)$$

Similarly, we find

$$(\vartheta,\vartheta_1,\vartheta_2,\vartheta_{22})=2\gamma\frac{\nabla}{K}\left(1+\frac{a}{H}\right)PN\kappa. \quad (4.5)$$

$\beta\gamma\neq 0$, since S is non-ruled. Furthermore, $\nabla\neq 0$, as otherwise the rank of the matrix $(\vartheta,\ \vartheta_1,\ \vartheta_2)$ would be less than 2 and S would become a plane, which contradicts the hypothesis.

In order that the surfaces S and $\hat{S}$ should be in asymptotic correspondence, it is therefore necessary and sufficient that

$$\left(1+\frac{b}{K}\right)QN=0, \qquad \left(1+\frac{a}{H}\right)PN=0. \quad (4.6)$$

These last conditions are satisfied when $N=0$. In this case, the surface S is projective minimal and the remaining three D-transforms and S are also in asymptotic correspondence; the theorem is thus proved.

It remains for us to consider the case $N\neq 0$. If

$$P=0 \qquad Q=0, \quad (4.7)$$

then (4.6) are evidently satisfied. In this case the asymptotic curves of two systems of S all belong to linear complexes and accordingly its D-transforms become one and the same quadric [3], which is to be excluded.

Thus the conditions (4.6) hold good when and only when

$$K=-b, \qquad P=0; \quad (4.8)$$

$$H=-a, \qquad Q=0; \quad (4.9)$$

or

$$K=-b, \qquad H=-a. \quad (4.10)$$

Let us first suppose that (4.8) are valid; we have

$$\vartheta = \mathfrak{y} - b\mathfrak{x}_1 + H\mathfrak{x}_2 - bH\mathfrak{x},$$

$$\vartheta_1 = H\vartheta + \left(b - \frac{N}{H}\right)(\beta b\mathfrak{x} - \beta\mathfrak{x}_2),$$

where $b - N/H \neq 0$, as otherwise $\vartheta_1 = H\vartheta$ and $\hat{S}$ would degenerate into a curve. Hence

$$\frac{\vartheta_1 - H\vartheta}{\beta\left(b - \frac{N}{H}\right)} = -\mathfrak{x}_2 + b\mathfrak{x}.$$

Differentiation of the last equation with respect to u gives

$$\left[\frac{\vartheta_1 - H\vartheta}{\beta\left(b - \frac{N}{H}\right)}\right]_1 + H\frac{\vartheta_1 - H\vartheta}{\beta\left(b - \frac{N}{H}\right)} + \vartheta = 0.$$

That is, the asymptotic curves u of the surface $\hat{S}$ are straight lines, which contradicts the hypothesis.

Similar conclusion is also applicable to the case (4.9).

Therefore we have merely to discuss the case (4.10).

Noticing that $PQ \neq 0$, we may utilize the formulae (2.9)–(2.11). Since

$$M = b^2, \qquad L = a^2, \tag{4.11}$$

(2.11) and (2.9) respectively lead to the relations:

$$a_1 = \beta c, \qquad b_2 = \gamma d; \tag{4.12}$$

$$N = ac = bd, \tag{4.13}$$

and therefore $abcd \neq 0$ according to $N \neq 0$.

It is easily seen that the surface S in question is determined by the following *system of differential equations in involution*:

$$\begin{aligned} a_1 &= \beta c, & a_2 &= P, \\ b_1 &= Q, & b_2 &= \gamma d, \\ c_1 &= Qg, & c_2 &= \gamma a, \\ d_1 &= \beta b, & d_2 &= Pg, \\ g_1 &= 2(d - ag), & g_2 &= 2(c - bg), \end{aligned} \tag{4.14}$$

the last two equations being necessary consequences of the integrability conditions of the remaining equations in (4.14). As to the integrability condition

$$g_{12} - g_{21} = 0, \tag{4.15}$$

we have to substitute

$$g_{12} = -4a(c - bg), \qquad g_{21} = -4b(d - ag),$$

so that (4.15) reduces to (4.13).

Now, we come to prove that the D-transforms of the surfaces of this class must necessarily be ruled surfaces.

In fact, we have

$$\vartheta = \mathfrak{y} - b\mathfrak{z}_1 - a\mathfrak{z}_2 + ab\mathfrak{z},$$

$$\vartheta_1 = -a\vartheta + \beta(b+c)(-\mathfrak{z}_2 + b\mathfrak{z}).$$

There is no loss of generality to assume that $b + c \neq 0$, as, otherwise, the surface $\hat{S}$ would degenerate into a curve.

Thus we find

$$(b+c)[\vartheta_{11} + \{\beta(b+2c) - a^2\}\vartheta] = \left\{\frac{\beta_1}{\beta}(b+c) + Q(1+g)\right\}(\vartheta_1 + a\vartheta).$$

That is to say, the asymptotic curves u of the surface $\hat{S}$ are straight lines.

Therefore, the proof of the theorem is complete.

Hereafter, we shall assume that the surface S is projective minimal, so that

$$N = 0. \tag{4.16}$$

Fundamental equations (2.8)–(2.11) now take a simple form [8]:

$$L_1 = 0, \qquad M_2 = 0; \tag{4.17}$$

$$L_2 = P_1 = 2aP, \qquad M_1 = Q_2 = 2bQ; \tag{4.18}$$

$$\gamma L = -Pc, \qquad \beta M = -Qd; \tag{4.19}$$

$$a_1 = \beta c + 2(L - a^2), \qquad b_2 = \gamma d + 2(M - b^2),$$

$$c_1 = -2ac, \qquad d_2 = -2bd. \tag{4.20}$$

It shall be mentioned that by differentiating (4.19) and using (2.11) we obtain $g = 0$ in this case.

The derived equations (3.3) can also be written more simply as follows:

$$H_1 = 0, \qquad H_2 = \frac{aP}{H},$$

$$K_1 = \frac{bQ}{K}, \qquad K_2 = 0. \tag{4.21}$$

The derived equations (4.2) of the D-transform (4.1) of the surface S now become

$$\begin{aligned} \vartheta_1 - H\vartheta &= Q\left(1+\frac{b}{K}\right)(\mathfrak{x}_1 + H\mathfrak{x}) + \beta K(\mathfrak{x}_2 + K\mathfrak{x}), \\ \vartheta_2 - K\vartheta &= \gamma H(\mathfrak{x}_1 + H\mathfrak{x}) + P\left(1+\frac{a}{H}\right)(\mathfrak{x}_2 + K\mathfrak{x}), \end{aligned} \tag{4.22}$$

whence

$$\begin{aligned} \vartheta_{11} = \beta K\vartheta + H\vartheta_1 &- \left[\beta HK - \left\{\beta_1 K + 2\beta Q\frac{b}{K} + \beta Q\right\}\right](\mathfrak{x}_2 + K\mathfrak{x}) \\ &+ \left\{HQ\left(1+\frac{b}{K}\right) + \left(Q\left(1+\frac{b}{K}\right)\right)_1\right\}(\mathfrak{x}_1 + H\mathfrak{x}), \end{aligned} \tag{4.23}$$

$$\begin{aligned} \vartheta_{12} = (*)\,\vartheta + (*)\,\vartheta_2 &+ \left\{\gamma\left(\beta K + \frac{dQ}{K}\right) + QK\left(1+\frac{b}{K}\right)\right\}\mathfrak{x}_1 \\ &+ \left\{PQ\left(1+\frac{a}{H}\right)\left(1+\frac{b}{K}\right) + HKQ\left(1+\frac{b}{K}\right) + \beta KM \right. \\ &\left. + \gamma H\frac{dQ}{K}\right\}\mathfrak{x}. \end{aligned} \tag{4.24}$$

In calculating the last equation we have made use of a relation, namely, the second of the following relations, which may be deduced from (4.18), (4.20) and (4.21):

$$\begin{aligned} \left(P\left(1+\frac{a}{H}\right)\right)_1 &= 2PH\left(1+\frac{a}{H}\right) - \beta\gamma H, \\ \left(Q\left(1+\frac{b}{K}\right)\right)_2 &= 2QK\left(1+\frac{b}{K}\right) - \beta\gamma K. \end{aligned} \tag{4.25}$$

We solve the equations (4.22) for $\mathfrak{x}_1 + H\mathfrak{x}$ and $\mathfrak{x}_2 + K\mathfrak{x}$:

$$\begin{aligned} \mathfrak{x}_1 + H\mathfrak{x} &= \frac{1}{\nabla}\left\{P\left(1+\frac{a}{H}\right)\vartheta_1 - \beta K\vartheta_2 + (*)\,\vartheta\right\}, \\ \mathfrak{x}_2 + K\mathfrak{x} &= \frac{1}{\nabla}\left\{-\gamma H\vartheta_1 + Q\left(1+\frac{b}{K}\right)\vartheta_2 + (*)\,\vartheta\right\}, \end{aligned}$$

where ∇ is given by (4.3):

$$\nabla = PQ\left(1+\frac{a}{H}\right)\left(1+\frac{b}{K}\right) - \beta\gamma HK. \tag{4.26}$$

If these expressions are substituted into (4.23), then follows one fundamental equation for ϑ, in which the coefficient $\hat{\beta}$ of ϑ_2 just corresponds to β in the equation (2.2) and is independent of the normalization of coordinates. Thus we obtain the fundamental quantity $\hat{\beta}$ and its analogous $\hat{\gamma}$ as follows:

$$\hat{\beta} = -\frac{1}{\nabla}\Big[2\beta HKQ\Big(1+\frac{b}{K}\Big)+\beta K\Big\{Q\Big(1+\frac{b}{K}\Big)\Big\}_1 - Q\Big(1+\frac{b}{K}\Big)\Big(\beta_1 K + 2\beta\frac{bQ}{K}+\beta Q\Big)\Big],$$
$$\hat{\gamma} = -\frac{1}{\nabla}\Big[2\gamma HKP\Big(1+\frac{a}{H}\Big)+\gamma H\Big\{P\Big(1+\frac{a}{H}\Big)\Big\}_2 - P\Big(1+\frac{a}{H}\Big)\Big(\gamma_2 H + 2\gamma\frac{aP}{H}+\gamma P\Big)\Big]. \tag{4.27}$$

In order to normalize the coordinates $\vartheta(u, v)$ of the D-transform $\hat{S}$, we have from (4.22) and (4.24) to calculate the determinant

$$(\vartheta, \vartheta_1, \vartheta_2, \vartheta_{12}) = -\nabla^2 (\mathfrak{x}, \mathfrak{x}_1, \mathfrak{x}_2, \mathfrak{y}),$$

whence follow the normalized coordinates

$$\hat{\mathfrak{x}} = \frac{1}{\sqrt{\nabla}}(HK\mathfrak{x} + K\mathfrak{x}_1 + H\mathfrak{x}_2 + \mathfrak{y}) \tag{4.28}$$

with the relation

$$(\hat{\mathfrak{x}}, \hat{\mathfrak{x}}_u, \hat{\mathfrak{x}}_v, \hat{\mathfrak{x}}_{uv}) = -\kappa. \tag{4.29}$$

We shall adopt the same notion but covered with Λ to denote the corresponding fundamental quantities for the surface $\hat{S}$ as in (4.27).

In the attempt of finding out the expressions for the quantities $\hat{Z}$ and $\hat{M}$ we put

$$\hat{\mathfrak{U}} = (\hat{\mathfrak{x}}, \hat{\mathfrak{x}}_{\hat{1}}), \qquad \hat{\mathfrak{B}} = (\hat{\mathfrak{x}}, \hat{\mathfrak{x}}_{\hat{2}}),$$

where the subscripts $\hat{1}$ and $\hat{2}$ denote the semi-invariant derivatives with respect to $\hat{\beta}$ and $\hat{\gamma}$ respectively. From the definition it can be shown that

$$\hat{\mathfrak{U}} = (\hat{\mathfrak{x}}, \hat{\mathfrak{x}}_1), \qquad \hat{\mathfrak{B}} = (\hat{\mathfrak{x}}, \hat{\mathfrak{x}}_2),$$

so that in virtue of (4.28) and (4.22) the coordinates of the asymptotic tangents of the surface $\hat{S}$ are found to be

$$\hat{\mathfrak{U}} = -\frac{1}{\nabla}\Big[\beta MK\mathfrak{U} + LQ\Big(1+\frac{b}{K}\Big)\mathfrak{B} + 2\beta M\mathfrak{A} + 2HQ\Big(1+\frac{b}{K}\Big)\mathfrak{B} + Q\Big(1+\frac{b}{K}\Big)\mathfrak{Q} + \beta K\mathfrak{P}\Big], \tag{4.30}$$

$$\hat{\mathfrak{B}} = -\frac{1}{\nabla}\Big[MP\Big(1+\frac{a}{H}\Big)\mathfrak{U} + \gamma LH\mathfrak{B} + 2KP\Big(1+\frac{a}{H}\Big)\mathfrak{A} + 2\gamma L\mathfrak{B} + \gamma H\mathfrak{Q} + P\Big(1+\frac{a}{H}\Big)\mathfrak{P}\Big]. \tag{4.31}$$

It is convenient to point out here another method of finding $\hat{\beta}$ and $\hat{\gamma}$. As $\hat{\mathfrak{U}}$ and $\hat{\mathfrak{V}}$ are respectively of the weights $(0, -1)$ and $(-1, 0)$, we have

$$\hat{\mathfrak{U}}_{\hat{1}} = \hat{\mathfrak{U}}_1, \qquad \hat{\mathfrak{V}}_{\hat{2}} = \hat{\mathfrak{V}}_2.$$

Utilizing further the following relations:

$$\nabla_1 = 2HPQ\left(1+\frac{a}{H}\right)\left(1+\frac{b}{K}\right) + P\left(1+\frac{a}{H}\right)\left\{Q\left(1+\frac{b}{K}\right)\right\}_1 - \gamma H\left(\beta_1 K + 2\beta\frac{bQ}{K} + \beta Q\right), \tag{4.32}$$

$$\nabla_2 = 2KPQ\left(1+\frac{a}{H}\right)\left(1+\frac{b}{K}\right) + Q\left(1+\frac{b}{K}\right)\left\{P\left(1+\frac{a}{H}\right)\right\}_2 - \beta K\left(\gamma_2 H + 2\gamma\frac{aP}{H} + \gamma P\right), \tag{4.33}$$

$$\hat{\mathfrak{U}}_{\hat{1}} = \hat{\beta}\hat{\mathfrak{V}}, \qquad \hat{\mathfrak{V}}_{\hat{2}} = \hat{\gamma}\hat{\mathfrak{U}}, \tag{4.34}$$

we are newly led to (4.27).

Let us now attend to the calculation of certain expressions. Evidently, $P(1+a/H)$ is a semi-invariant of the weights $(1, 1)$ and

$$\sigma - \tau = Q - P, \tag{4.35}$$

whence follows the relation

$$\left(P\left(1+\frac{a}{H}\right)\right)_{12} - \left(P\left(1+\frac{a}{H}\right)\right)_{21} = 2(Q-P)\,P\left(1+\frac{a}{H}\right).$$

But from (4.25)

$$\left(P\left(1+\frac{a}{H}\right)\right)_{12} = 2H\left(P\left(1+\frac{a}{H}\right)\right)_2 + 2\frac{aP^2}{H}\left(1+\frac{a}{H}\right) - \beta\left(\gamma_2 H + \gamma\frac{aP}{H}\right);$$

therefore

$$\left(P\left(1+\frac{a}{H}\right)\right)_{21} = 2H\left(P\left(1+\frac{a}{H}\right)\right)_2 + 2P^2\left(1+\frac{a}{H}\right)^2 - 2PQ\left(1+\frac{a}{H}\right) - \beta\left(\gamma_2 H + \gamma\frac{aP}{H}\right), \tag{4.36}$$

and similarly

$$\left(Q\left(1+\frac{b}{K}\right)\right)_{12} = 2K\left(Q\left(1+\frac{b}{K}\right)\right)_1 + 2Q^2\left(1+\frac{b}{K}\right)^2 - 2PQ\left(1+\frac{b}{K}\right) - \gamma\left(\beta_1 K + \beta\frac{bQ}{K}\right). \tag{4.37}$$

Write down, for brevity, the equations (4.27), (4.30), (4.31) in the form

$$\hat{\beta} = -\frac{1}{\nabla}[\hat{\beta}], \qquad \hat{\gamma} = -\frac{1}{\nabla}[\hat{\gamma}], \tag{4.38}$$

$$\hat{\mathfrak{U}} = -\frac{1}{\nabla}[\hat{\mathfrak{U}}], \qquad \hat{\mathfrak{V}} = -\frac{1}{\nabla}[\hat{\mathfrak{V}}], \tag{4.39}$$

where $[\hat{\mathfrak{V}}]$ denotes, for instance, the expression in the brackets of the right-hand side of (4.31).

For the quantity $\hat{\mathfrak{V}}$ of the weights $(-1, 0)$ we have

$$\begin{aligned} \hat{\mathfrak{V}}_{\hat{1}} &= \hat{\mathfrak{V}}_1 - \left(\frac{\hat{\gamma}_u}{\hat{\gamma}} - \frac{\gamma_u}{\gamma}\right)\hat{\mathfrak{V}} = \hat{\mathfrak{V}}_1 - \frac{\hat{\gamma}_1}{\hat{\gamma}}\hat{\mathfrak{V}} \\ &= -\frac{1}{\nabla[\hat{\gamma}]}\left([\hat{\gamma}][\hat{\mathfrak{V}}]_1 - [\hat{\gamma}]_1[\hat{\mathfrak{V}}]\right). \end{aligned} \tag{4.40}$$

On the other hand, it is easily shown in virtue of (2.7), (3.3), (4.18), (4.20) and (4.36) that

$$[\hat{\mathfrak{V}}]_1 = 2H[\hat{\mathfrak{V}}] + 2\nabla(K\mathfrak{U} + \mathfrak{A}), \tag{4.41}$$

$$[\hat{\gamma}]_1 = 2H[\hat{\gamma}] + 2\nabla\gamma H. \tag{4.42}$$

In computing these derivatives we have utilized the relation

$$\gamma_{21} = 2(\sigma + 2\tau)\gamma \tag{4.43}$$

derived from (2.7).

Substituting (4.41) and (4.42) onto (4.40) and noticing that

$$\hat{\mathfrak{V}}_{\hat{1}} = 2\hat{\mathfrak{V}},$$

we obtain

$$\mathfrak{S} = -\frac{1}{[\hat{\gamma}]}\left[[\hat{\gamma}](K\mathfrak{U} + \mathfrak{A}) - \gamma H[\hat{\mathfrak{V}}]\right], \tag{4.44}$$

a semi-invariant of the weights (0, 0). On account of (4.41) and (4.42) we have

$$\mathfrak{S}_{\hat{1}} = -\frac{2\nabla\gamma H}{[\hat{\gamma}]}\mathfrak{S} - \left\{\varrho\left(1 + \frac{b}{K}\right)\mathfrak{U} + \beta K\mathfrak{V}\right\}, \tag{4.45}$$

whence

$$\begin{aligned} \mathfrak{S}_{\hat{1}} \circ \mathfrak{S}_{\hat{1}} = \frac{4\gamma H\nabla}{[\hat{\gamma}]^2}\Big[&\gamma H\nabla(\mathfrak{S} \circ \mathfrak{S}) - \varrho\left(1 + \frac{b}{K}\right)([\mathfrak{S}] \circ \mathfrak{U}) \\ &- \beta K([\mathfrak{S}] \circ \mathfrak{V})\Big]. \end{aligned}$$

A reference to (2.13) and (4.29) gives the final expression of $\hat{L}$:

$$\hat{L} = \left(\frac{\gamma}{\hat{\gamma}}\right)^2 L. \tag{4.46}$$

Similarly,

$$\hat{M} = \left(\frac{\beta}{\hat{\beta}}\right)^2 M. \tag{4.47}$$

From the table (2.14) it is seen that $\hat{\gamma}^2 \hat{L}$ is of the weights (0, 4) and therefore that

$$(\hat{\gamma}^2 \hat{L})_{\hat{1}} = (\hat{\gamma}^2 \hat{L})_1 .$$

But (4.46) shows that

$$\hat{\gamma}^2 \hat{L} = \gamma^2 L$$

and $\gamma_1 = 0$, $L_1 = 0$. Hence we may conclude

$$\hat{N} = 0, \tag{4.48}$$

which proves

Theorem 4.2. *The D-transforms of a projective minimal surface are also projective minimal.*

V. The Second D-Transforms

According to the result in the last paragraph, any D-transform of a projective minimal surface S is projective minimal; the corresponding point of one of surfaces, $\hat{S}$, is given by

$$\hat{\mathfrak{x}} = \frac{1}{\sqrt{\nabla}} (HK\mathfrak{x} + K\mathfrak{x}_1 + H\mathfrak{x}_2 + \mathfrak{y}) \tag{5.1}$$

with

$$H = \varepsilon \sqrt{L}, \qquad K = \varepsilon' \sqrt{M} \quad (\varepsilon, \varepsilon' = +1 \text{ or } -1). \tag{5.2}$$

For the purpose of determining the D-transforms, $\hat{\hat{S}}$, of $\hat{S}$, namely, the second D-transforms of the original S, we first mention the relations

$$\hat{H} = \omega \frac{\gamma}{\hat{\gamma}} H, \qquad \hat{K} = \omega' \frac{\beta}{\hat{\beta}} K \quad (\omega, \omega' = \pm 1). \tag{5.3}$$

For given values of H and K, four D-transforms of $\hat{S}$ can be found according to different signs ω and ω'.

We have concretely expressed certain fundamental quantities and allied fundamental vectors of $\hat{S}$ in (4.27), (4.30), (4.31), (4.44), (4.46) and (4.47); especially from (4.44) we may write down the expression for $\hat{\mathfrak{A}}$. The result of computation is as follows:

$$\hat{\mathfrak{A}} = -\frac{1}{\nabla}\left\{\frac{\beta^2 M^2}{\hat{\beta}}\mathfrak{U} + H\left(\nabla + \frac{\beta}{\hat{\beta}}HKQ\left(1+\frac{b}{K}\right)\right)\mathfrak{B}\right.$$

$$+\left(\nabla + \frac{2\beta}{\hat{\beta}}HKQ\left(1+\frac{b}{K}\right)\right)\mathfrak{B}$$

$$\left.+\frac{\beta}{\hat{\beta}}K\left(2\beta M\mathfrak{A} + \beta K\mathfrak{P} + Q\left(1+\frac{b}{K}\right)\mathfrak{Q}\right)\right\}, \tag{5.4}$$

$$\hat{\mathfrak{B}} = -\frac{1}{\nabla}\left\{K\left(\nabla + \frac{\gamma}{\hat{\gamma}}HKP\left(1+\frac{a}{H}\right)\right)\mathfrak{U} + \frac{\gamma^2 L^2}{\hat{\gamma}}\mathfrak{V}\right.$$

$$+\left(\nabla + \frac{2\gamma}{\hat{\gamma}}HKP\left(1+\frac{a}{H}\right)\right)\mathfrak{A}$$

$$\left.+\frac{\gamma}{\hat{\gamma}}H\left(2\gamma L\mathfrak{B} + P\left(1+\frac{a}{H}\right)\mathfrak{P} + \gamma H\mathfrak{Q}\right)\right\}. \tag{5.5}$$

In virtue of (4.45) and an analogous formula, we easily obtain

$$\hat{\mathfrak{P}} = \frac{2\beta K}{\hat{\beta}}\hat{\mathfrak{A}} - \frac{\beta^2 M^2}{\hat{\beta}^2}\hat{\mathfrak{U}} - \left\{P\left(1+\frac{a}{H}\right)\mathfrak{B} + \gamma H\mathfrak{U}\right\}, \tag{5.6}$$

$$\hat{\mathfrak{Q}} = \frac{2\gamma H}{\hat{\gamma}}\hat{\mathfrak{B}} - \frac{\gamma^2 L^2}{\hat{\gamma}^2}\hat{\mathfrak{V}} - \left\{Q\left(1+\frac{b}{K}\right)\mathfrak{U} + \beta K\mathfrak{B}\right\}. \tag{5.7}$$

By definition

$$\hat{\mathfrak{x}}_{\hat{1}} = \hat{\mathfrak{x}}_1 - \frac{1}{2}\frac{\hat{\gamma}_1}{\hat{\gamma}}\hat{\mathfrak{x}},$$

or, rewritten in virtue of (4.32) and (4.42),

$$\hat{\mathfrak{x}}_{\hat{1}} = \frac{1}{\sqrt{\nabla}}\left\{Q\left(1+\frac{b}{K}\right)\mathfrak{x}_1 + \beta K\mathfrak{x}_2 + \left(\beta M + HQ\left(1+\frac{b}{K}\right)\right)\mathfrak{x}\right\} + \omega\hat{H}\hat{\mathfrak{x}}. \tag{5.8}$$

Similarly, we have

$$\hat{\mathfrak{x}}_{\hat{2}} = \frac{1}{\sqrt{\nabla}}\left\{\gamma H\mathfrak{x}_1 + P\left(1+\frac{a}{H}\right)\mathfrak{x}_2 + \left(\gamma L + KP\left(1+\frac{a}{H}\right)\right)\mathfrak{x}\right\} + \omega'\hat{K}\hat{\mathfrak{x}}. \tag{5.9}$$

To determine

$$\hat{\mathfrak{y}} = x\mathfrak{x} + y\mathfrak{x}_1 + z\mathfrak{x}_2 + t\mathfrak{y}, \tag{5.10}$$

we have to utilize the equation

$$(\hat{\mathfrak{x}}, \hat{\mathfrak{y}}) = \hat{\mathfrak{A}} + \hat{\mathfrak{B}}. \tag{5.11}$$

On account of (5.1) and (5.10) the left-hand side of (5.11) is equal to

$$(\hat{\mathfrak{x}}, \hat{\mathfrak{y}}) = \frac{1}{\sqrt{\nabla}}\left(K(-x + Hy)\mathfrak{U} + H(-x + Kz)\mathfrak{V} + \right.$$

$$+ (-x+Hy-Kz+HKt)\,\mathfrak{A}+(-Hy-x+Kz+HKt)\,\mathfrak{B}$$
$$+ (Ht-z)\,\mathfrak{P}+(Kt-y)\,\mathfrak{Q}\}, \tag{5.12}$$

while the right-hand side of (5.11) is to be calculated by means of (5.4) and (5.5). The result of comparison of the corresponding coefficients in both sides runs:

$$Ht - z = -\frac{1}{\sqrt{\nabla}}\left\{\omega\hat{H}P\left(1+\frac{a}{H}\right)+\omega'\hat{K}\beta K\right\},$$
$$Kt - y = -\frac{1}{\sqrt{\nabla}}\left\{\omega'\hat{K}Q\left(1+\frac{b}{K}\right)+\omega'\hat{H}\gamma H\right\},$$
$$Hy - x = -\frac{1}{\sqrt{\nabla}}\left\{\nabla+\omega\hat{H}KP\left(1+\frac{a}{H}\right)+\omega'\hat{K}\beta M\right\},$$

namely,

$$x = HKt+\frac{1}{\sqrt{\nabla}}\left\{\nabla+\omega\gamma L\hat{H}+\omega'\beta M\hat{K}\right.$$
$$\left.+\omega K\hat{H}P\left(1+\frac{a}{H}\right)+\omega' H\hat{K}Q\left(1+\frac{b}{K}\right)\right\},$$

$$y = Kt+\frac{1}{\sqrt{\nabla}}\left\{\omega'\hat{K}Q\left(1+\frac{b}{K}\right)+\omega\hat{H}\gamma H\right\},$$
$$z = Ht+\frac{1}{\sqrt{\nabla}}\left\{\omega\hat{H}P\left(1+\frac{a}{H}\right)+\omega'\hat{K}\beta K\right\}. \tag{5.13}$$

In determining t, another equation is necessary; take, for example,

$$\hat{\mathfrak{P}} = (\hat{\mathfrak{x}}_{\mathfrak{z}}, \mathfrak{h}) \tag{5.14}$$

and compare one pair of corresponding coefficients in both sides; we are led to the required

$$\beta Ky - Q\left(1+\frac{b}{K}\right)z+\left\{\frac{\hat{\beta}}{\beta K}\nabla+HQ\left(1+\frac{b}{K}\right)-\beta M\right\}t = 0.$$

Substitution of y and z given by (5.13) onto the last equation gives t:

$$t = \frac{1}{\sqrt{\nabla}}\omega\omega'\hat{H}\hat{K}, \tag{5.15}$$

and consequently,

$$x = \frac{1}{\sqrt{\nabla}}\Big\{\omega\omega' HK\hat{H}\hat{K} + \omega\gamma L\hat{H} + \omega'\beta M\hat{K} + \nabla$$
$$+ \omega\hat{H}KP\Big(1 + \frac{a}{H}\Big) + \omega'\hat{K}HQ\Big(1 + \frac{b}{K}\Big)\Big\},$$
$$y = \frac{1}{\sqrt{\nabla}}\Big\{\omega\omega' K\hat{H}\hat{K} + \omega'\hat{K}Q\Big(1 + \frac{b}{K}\Big) + \omega\gamma H\hat{H}\Big\},$$
$$z = \frac{1}{\sqrt{\nabla}}\Big\{\omega\omega' H\hat{H}\hat{K} + \omega\hat{H}P\Big(1 + \frac{a}{H}\Big) + \omega'\beta K\hat{K}\Big\}. \tag{5.16}$$

Upon substituting these expressions onto (5.10) so as to determine $\hat{\mathfrak{H}}$ and, in consequence,

$$\hat{\mathfrak{x}} = \hat{H}\hat{K}\mathfrak{x} + \hat{K}\mathfrak{x}_1 + \hat{H}\mathfrak{x}_2 + \mathfrak{y}, \tag{5.17}$$

we obtain the *second D-transform* $S_{\omega,\omega'}$ of the surface S. By Theorem 4.2 this surface $\hat{\hat{S}}$ is also projective minimal. An easy calculation gives the final formula:

$$\hat{\hat{\mathfrak{x}}} = \Big\{\hat{H}\hat{K}K\,(1+\omega)(1+\omega') + (1+\omega')\,\hat{K}Q\Big(1+\frac{b}{K}\Big) + \gamma H\hat{H}\,(1+\omega)\Big\}\mathfrak{x}_1$$
$$+ \Big\{H\hat{H}\hat{K}\,(1+\omega)(1+\omega') + (1+\omega)\,\hat{H}P\Big(1+\frac{a}{H}\Big) + \beta K\hat{K}\,(1+\omega')\Big\}\mathfrak{x}_2$$
$$+ \Big\{\nabla + (1+\omega)(1+\omega')\,HK\hat{H}\hat{K} + (1+\omega')\,\beta M\hat{K} + (1+\omega)\,\gamma L\hat{H}$$
$$+ (1+\omega)\,K\hat{H}P\Big(1+\frac{a}{H}\Big) + (1+\omega')\,H\hat{K}Q\Big(1 + \frac{b}{K}\Big)\Big\}\mathfrak{x}$$
$$+ (1+\omega)(1+\omega')\,\hat{H}\hat{K}\mathfrak{y}. \tag{5.18}$$

This transformation $S \to S_{\omega,\omega'}$ we call the *double D-transformation*.

In particular, when $\omega = \omega' = -1$, the corresponding $\hat{\hat{\mathfrak{x}}} = \nabla\mathfrak{x}$, whence follows a theorem of O. Mayer [2]:

Theorem 5.1. *A projective minimal surface is one of its own second D-transforms.*

The other three surfaces $\hat{\hat{S}}$ are also given by (5.18) with $(\omega, \omega') = (1, -1), (-1, 1)$ or $(1, 1)$. The relation between them will be shown in the next section.

VI. Demoulin Quadrilaterals of D-Transforms

First of all, put $\omega = 1$, $\omega' = -1$ or $\omega = -1$, $\omega' = 1$ in (5.18); it is seen that $\hat{\hat{\mathfrak{x}}}$ contains no term in $\mathfrak{y}$. That is to say, the surface $\hat{S}$ has such two D-transforms that the corresponding points lie on the tangent plane of the original surface S. The surface $S_{1,-1}$ is, in fact, given by

$$\mathfrak{z} = \gamma H \mathfrak{x}_1 + P\left(1+\frac{a}{H}\right)\mathfrak{x}_2 + \left\{\frac{\nabla\hat{\gamma}}{2\gamma H} + \gamma L + KP\left(1+\frac{a}{H}\right)\right\}\mathfrak{x}, \quad (6.1)$$

where we have omitted the common factor 2.

If the value of $\hat{\gamma}$ given by (4.27) is substituted onto (6.1), then the coefficient of $\mathfrak{x}$ is found to be independent of K and equal to

$$\gamma L - \frac{1}{2}\left(P\left(1+\frac{a}{H}\right)\right)_2 + \frac{1}{2}P\left(1+\frac{a}{H}\right)\left(\frac{\gamma_2}{\gamma} + 2\frac{aP}{L} + \frac{P}{H}\right).$$

Therefore we adopt the notation $\mathfrak{x}_H(1,-1)$ in the place of $\hat{\mathfrak{x}}$, where $1,-1$ in the parentheses denote the values of ω, ω' and the subscript H shows the only dependence of the surface $\hat{S}$ on H.

In a similar way, we have another second D-transform $\mathfrak{x}_K(-1,1)$, which corresponds to the values $\omega=-1$, $\omega'=1$ and depends merely upon K.

Thus we have found two second D-transforms:

$$\begin{aligned}\mathfrak{x}_H(1,-1) &= \gamma H\mathfrak{x}_1 + P\left(1+\frac{a}{H}\right)\mathfrak{x}_2 \\ &+ \left\{\gamma L - \frac{1}{2}\left(P\left(1+\frac{a}{H}\right)\right)_2 + \frac{1}{2}P\left(1+\frac{a}{H}\right)\left(\frac{\gamma_2}{\gamma}+2\frac{aP}{L}+\frac{P}{H}\right)\right\}\mathfrak{x},\end{aligned} \quad (6.2)$$

$$\begin{aligned}\mathfrak{x}_K(-1,1) &= Q\left(1+\frac{b}{K}\right)\mathfrak{x}_1 + \beta K\mathfrak{x}_2 + \\ &+ \left\{\beta M - \frac{1}{2}\left(Q\left(1+\frac{b}{K}\right)\right)_1 + \frac{1}{2}Q\left(1+\frac{b}{K}\right)\left(\frac{\beta_1}{\beta}+2\frac{bQ}{M}+\frac{Q}{K}\right)\right\}\mathfrak{x}.\end{aligned} \quad (6.3)$$

The original surface S has four D-transforms, and the points (5.1) of them differ from one another only in the signs of H and K. According to Theorem 5.1 each of these D-transforms possesses S as its D-transform. That is, S stands four-fold for its own second D-transform. To distinguish these D-transforms more clearly, we shall use the notations $\hat{\mathfrak{x}}_{H,K}$, $\hat{\mathfrak{x}}_{H,-K}$, $\hat{\mathfrak{x}}_{-H,K}$, $\hat{\mathfrak{x}}_{-H,-K}$ to denote them.

Summing up what was stated above we may conclude the following result:

The surfaces $\hat{\mathfrak{x}}_{H,K}$ and $\hat{\mathfrak{x}}_{H,-K}$ have two D-transforms $\mathfrak{x}$ and $\mathfrak{x}_H(1,-1)$ in common; the surfaces $\hat{\mathfrak{x}}_{H,-K}$ and $\hat{\mathfrak{x}}_{-H,-K}$ have two D-transforms $\mathfrak{x}$ and $\mathfrak{x}_{-K}(-1,1)$ in common, etc. These D-transforms are the second D-transforms of the original surface S, among them the corresponding points of four surfaces $\mathfrak{x}_H(1,-1)$, $\mathfrak{x}_K(-1,1)$, $\mathfrak{x}_{-H}(1,-1)$, $\mathfrak{x}_{-K}(-1,1)$ are on the tangent plane of S at the point $\mathfrak{x}$, and the join of each point with $\mathfrak{x}$ generates a rectilinear congruence W. Thus we have proved

Theorem 6.1. *A projective minimal surfaces has, in general, nine second D-transforms including the original surface, and four of them are W-transforms of S.*

Beside these surfaces there are four second D-transforms, the corresponding point of each of them being not on the tangent plane of S at the point $\mathfrak{x}$. Putting $\omega = \omega' = 1$ in (5.18) we get one of them as follows:

$$\begin{aligned}\mathfrak{x}_{H,K}(1,1) = &\left\{2K\hat{H}\hat{K} + \hat{K}Q\left(1+\frac{b}{K}\right) + \gamma H\hat{H}\right\}\mathfrak{x}_1 \\ &+\left\{2H\hat{H}\hat{K} + \hat{H}P\left(1+\frac{a}{H}\right) + \beta K\hat{K}\right\}\mathfrak{x}_2 \\ &+\left\{\frac{1}{2}\nabla + 2HK\hat{H}\hat{K} + \beta M\hat{K} + \gamma L\hat{H}\right. \\ &\left. + H\hat{K}Q\left(1+\frac{b}{K}\right) + K\hat{H}P\left(1+\frac{a}{H}\right)\right\}\mathfrak{x} + 2\hat{H}\hat{K}\mathfrak{y},\end{aligned} \tag{6.4}$$

where

$$\hat{H} = \frac{\gamma}{\hat{\gamma}}H, \qquad \hat{K} = \frac{\beta}{\hat{\beta}}K. \tag{6.5}$$

The other three surfaces turn out by changing the signs of H and K, namely,

$$\mathfrak{x}_{H,-K}(1,1), \quad \mathfrak{x}_{-H,K}(1,1) \quad \text{and} \quad \mathfrak{x}_{-H,-K}(1,1).$$

It shall be mentioned that $\mathfrak{x}_{H,K}(1,1)$, $\mathfrak{x}_H(1,-1)$, $\mathfrak{x}_K(-1,1)$ and $\mathfrak{x}$ are related to each other by the equation:

$$\mathfrak{x}_{H,K}(1,1) = 2\frac{\beta\gamma}{\hat{\beta}\hat{\gamma}}HK\hat{\mathfrak{x}}_{H,K} + \frac{\gamma}{\hat{\gamma}}H\mathfrak{x}_H(1,-1) + \frac{\beta}{\hat{\beta}}K\mathfrak{x}_K(-1,1) - \frac{\nabla}{2}\mathfrak{x}, \tag{6.6}$$

where $\hat{\mathfrak{x}}_{H,K}$ denotes one D-transform of S:

$$\hat{\mathfrak{x}}_{H,K} = \mathfrak{y} + K\mathfrak{x}_1 + H\mathfrak{x}_2 + HK\mathfrak{x}. \tag{6.7}$$

Let us now consider again the join of the points $\mathfrak{x}$ and $\mathfrak{x}_H(1,-1)$; it is the common side of Demoulin quadrilaterals of the two D-transforms $\hat{\mathfrak{x}}_{H,K}$ and $\hat{\mathfrak{x}}_{H,-K}$, while the join of the corresponding points of the latter two is in turn one side of Demoulin quadrilateral of the original surface S. This fact is also valid for the two D-transforms on another side. Thus we get

Theorem 6.2. *If we consider the two points of D-transforms of a projective minimal surface S on any one side of Demoulin quadrilateral of S, then the Demoulin quadrilaterals of these D-transforms at these two points have*

one common side, which is a tangent of S at the corresponding point. These four common sides thus obtained generate four rectilinear congruences W with S and one of the projective minimal surfaces $\mathfrak{x}_H(1,-1)$, $\mathfrak{x}_K(-1,1)$, $\mathfrak{x}_{-H}(1,-1)$ *and* $\mathfrak{x}_{-K}(-1,1)$ *for focal sheets.*

On account of the relation (6.6) we can further find the two Wilczynski directrices of the D-transform at the point $\hat{\mathfrak{x}}_{H,K}$. In fact, the points $\mathfrak{x}$, $\mathfrak{x}_H(1,-1)$, $\mathfrak{x}_K(-1,1)$ and $\mathfrak{x}_{H,K}(1,1)$ are the vertices of Demoulin quadrilateral of the D-transform $\hat{S}$, so that $(\mathfrak{x}, \mathfrak{x}_{H,K}(1,1))$ and $(\mathfrak{x}_H(1,-1), \mathfrak{x}_K(-1,1))$ are its diagonals. Using the classical theorems due to Bompiani and Godeaux we easily deduce from (6.6) the equations to the directrices:

$$\mathfrak{x}^{(1)}_{H,K} = \rho\mathfrak{x}_{H,K} + \frac{\gamma}{\hat{\gamma}} H\mathfrak{x}_H(1,-1) + \frac{\beta}{\hat{\beta}} K\mathfrak{x}_K(-1,1)\,, \tag{6.8}$$

$$\begin{aligned}\mathfrak{x}^{(2)}_{H,K} = 2\frac{\beta\gamma}{\hat{\beta}\hat{\gamma}} HK\mathfrak{x}_{H,K} &+ \frac{\gamma}{\hat{\gamma}} H(1+\sigma)\,\mathfrak{x}_H(1,-1)\\ &+ \frac{\beta}{\hat{\beta}} K(1-\sigma)\,\mathfrak{x}_K(-1,1)\,,\end{aligned} \tag{6.9}$$

ρ and σ being parameters.

VII. Godeaux Sequences

We are now in a position to investigate the relation between the Godeaux sequences of a projective minimal surface S and its D-transform $\hat{S}$.

In the first place, rewrite (4.30) and (4.31).

$$\hat{\mathfrak{U}} = -\frac{1}{\nabla}\left\{\beta K(\mathfrak{P}+M\mathfrak{U}+2K\mathfrak{A}) + Q\left(1+\frac{b}{K}\right)(\mathfrak{Q}+L\mathfrak{V}+2H\mathfrak{B})\right\}, \tag{7.1}$$

$$\hat{\mathfrak{V}} = -\frac{1}{\nabla}\left\{P\left(1+\frac{a}{H}\right)(\mathfrak{P}+M\mathfrak{U}+2K\mathfrak{A}) + \gamma H(\mathfrak{Q}+L\mathfrak{V}+2H\mathfrak{B})\right\}. \tag{7.2}$$

As (2.17) gives

$$\mathfrak{P}+M\mathfrak{U}+2K\mathfrak{A} = 2K\left(1+\frac{b}{K}\right)\mathfrak{U}_{(1)} + \mathfrak{U}_{(2)}\,,$$

$$\mathfrak{Q}+L\mathfrak{V}+2H\mathfrak{B} = 2H\left(1+\frac{a}{H}\right)\mathfrak{V}_{(1)} + \mathfrak{V}_{(2)}\,,$$

so we get

$$2K\left(1+\frac{b}{K}\right)\mathfrak{U}_{(1)} + \mathfrak{U}_{(2)} = -\gamma H\hat{\mathfrak{U}} + Q\left(1+\frac{b}{K}\right)\hat{\mathfrak{V}}\,, \tag{7.3}$$

$$2H\left(1+\frac{a}{H}\right)\mathfrak{V}_{(1)} + \mathfrak{V}_{(2)} = P\left(1+\frac{a}{H}\right)\hat{\mathfrak{U}} - \beta K\hat{\mathfrak{V}}\,. \tag{7.4}$$

From these equations we may infer that in the five-dimensional space S_5 the join $(\hat{\mathfrak{U}}, \hat{\mathfrak{V}})$ must intersect the joins $(\mathfrak{U}_{(1)}, \mathfrak{U}_{(2)})$, and $(\mathfrak{V}_{(1)}, \mathfrak{V}_{(2)})$, and the points of intersection are different from these six points, since the coefficients in (7.3) and (7.4) are not zero.

In the second place, from (4.44) and

$$\hat{\mathfrak{V}}_{(1)} = \hat{\mathfrak{V}}$$

follows the relation:

$$\hat{\mathfrak{V}}_{(1)} - \frac{\gamma H}{\hat{\gamma}} \hat{\mathfrak{V}} = -(K\mathfrak{U} + \mathfrak{U}_{(1)}) \,. \tag{7.5}$$

Rewriting (4.31) in the form

$$\begin{aligned} -\nabla\hat{\mathfrak{V}} &= 2KP\left(1 + \frac{a}{H}\right)(K\mathfrak{U} + \mathfrak{A}) \\ &+ \gamma H(\mathfrak{Q} + L\mathfrak{V} + 2H\mathfrak{V}) + P\left(1 + \frac{a}{H}\right)(\mathfrak{P} - M\mathfrak{U}) \end{aligned} \tag{7.6}$$

and deducing from (2.17) and (2.18) the relations

$$\mathfrak{Q} + L\mathfrak{V} + 2H\mathfrak{V} = \mathfrak{V}_{(2)} + 2H\left(1 + \frac{a}{H}\right)\mathfrak{V},$$

$$\mathfrak{P} - M\mathfrak{U} = \mathfrak{V}_{(3)} + 2c\mathfrak{V} - F\mathfrak{V}_{(2)},$$

we have that the coefficient of $\mathfrak{V}$ in the right-hand side of (7.6) after substitution is equal to

$$2\left(1 + \frac{a}{H}\right)(Pc + \gamma L) = 0$$

on account of (4.19), and

$$\begin{aligned} 2KP\left(1 + \frac{a}{H}\right)\hat{\mathfrak{V}}_{(1)} - \left(\nabla + 2\frac{\gamma H}{\hat{\gamma}}KP\left(1 + \frac{a}{H}\right)\right)\hat{\mathfrak{V}} \\ = \left(\gamma H - FP\left(1 + \frac{a}{H}\right)\right)\mathfrak{V}_{(2)} + P\left(1 + \frac{a}{H}\right)\mathfrak{V}_{(3)}. \end{aligned} \tag{7.7}$$

(7.5) and (7.7) suffice to demonstrate that in the space S_5 the join $(\hat{\mathfrak{V}}, \mathfrak{V}_{(1)})$ must intersect both the joins $(\mathfrak{U}_{(1)}, \mathfrak{U})$ and $(\mathfrak{V}_{(2)}, \mathfrak{V}_{(3)})$, and the points of intersection are different from these points.

Similarly, the join $(\hat{\mathfrak{U}}, \hat{\mathfrak{U}}_{(1)})$ must intersect both the joins $(\mathfrak{V}_{(1)}, \mathfrak{V})$ and $(\mathfrak{U}_{(2)}, \mathfrak{U}_{(3)})$, and the points of intersection are different from the end points.

In summary, we get

Theorem 7.1. *Suppose that S be a projective minimal surface with four distinct D-transforms; then in the arrangement of the points in the space S_5*

$$\left\| \begin{matrix} \mathfrak{U}_{(3)} & \mathfrak{U}_{(2)} & \mathfrak{U}_{(1)} & \mathfrak{U} \\ \hat{\mathfrak{U}}_{(1)} & \hat{\mathfrak{U}} & \hat{\mathfrak{V}} & \hat{\mathfrak{V}}_{(1)} \\ \mathfrak{V} & \mathfrak{V}_{(1)} & \mathfrak{V}_{(2)} & \mathfrak{V}_{(3)} \end{matrix} \right\|, \tag{7.8}$$

the join of any two consecutive points in the middle row must intersect both the joins of the points of the same columns in the upper and lower rows, the points of intersection of each pair being different from the joining points.

We can prove that this property is characteristic of a projective minimal surface and ist D-transform. The proof will be given in another paper [9].

It is natural to inquire whether or not the arrangement (7.8) can be extended to infinity in both direction. First of all, the original surface and its D-transform are involutive in the sense of O. Mayer [2], and therefore the intersection property can be generalized to a more extensive arrangement

$$\left\| \begin{matrix} * & * & \mathfrak{U}_{(3)} & \mathfrak{U}_{(2)} & \mathfrak{U}_{(1)} & \mathfrak{U} & \mathfrak{V} & \mathfrak{V}_{(1)} \\ \hat{\mathfrak{U}}_{(3)} & \hat{\mathfrak{U}}_{(2)} & \hat{\mathfrak{U}}_{(1)} & \hat{\mathfrak{U}} & \hat{\mathfrak{V}} & \hat{\mathfrak{V}}_{(1)} & \hat{\mathfrak{V}}_{(2)} & \hat{\mathfrak{V}}_{(3)} \\ \mathfrak{U}_{(1)} & \mathfrak{U} & \mathfrak{V} & \mathfrak{V}_{(1)} & \mathfrak{V}_{(2)} & \mathfrak{V}_{(3)} & * & * \end{matrix} \right\|, \tag{7.9}$$

the asterisk denoting the lack of the corresponding point.

More generally, we proceed to demonstrate

Theorem 7.2. *In the arrangement*

$$\left\| \begin{matrix} * & * & \mathfrak{U}_{(4)} & \mathfrak{U}_{(3)} & \mathfrak{U}_{(2)} & \mathfrak{U}_{(1)} & \mathfrak{U} & \mathfrak{V} & \mathfrak{V}_{(1)} & \mathfrak{V}_{(2)} \\ \hat{\mathfrak{U}}_{(4)} & \hat{\mathfrak{U}}_{(3)} & \hat{\mathfrak{U}}_{(2)} & \hat{\mathfrak{U}}_{(1)} & \hat{\mathfrak{U}} & \hat{\mathfrak{V}} & \hat{\mathfrak{V}}_{(1)} & \hat{\mathfrak{V}}_{(2)} & \hat{\mathfrak{V}}_{(2)} & \mathfrak{V}_{(3)} \\ \mathfrak{U}_{(2)} & \mathfrak{U}_{(1)} & \mathfrak{U} & \mathfrak{V} & \mathfrak{V}_{(1)} & \mathfrak{V}_{(2)} & \mathfrak{V}_{(3)} & \mathfrak{V}_{(4)} & * & * \end{matrix} \right\|, \tag{7.10}$$

where the asterisk denotes the lack of the corresponding point, the join of any two consecutive points in the middle row must intersect both the joins of the points of the same columns in the upper and lower rows.

To this end we first calculate $\mathfrak{V}_{(4)}$ and $\hat{\mathfrak{V}}_{(2)}$. From (2.18) we have

$$\mathfrak{V}_{(3)} = F\mathfrak{V}_{(2)} + \mathfrak{P} - M\mathfrak{U} - 2c\mathfrak{V}, \tag{7.11}$$

$$\mathfrak{V}_{(3)1} = \mathfrak{V}_{(4)} + (F\beta - \rho)\, \mathfrak{V}_{(3)}, \tag{7.12}$$

where

$$\mathfrak{V}_{(4)} = 2Q\,(-b\mathfrak{U} + d\mathfrak{V} + \mathfrak{A}) + \{F_1 - F\,(\beta F + 2a) - 2c\}\, \mathfrak{V}_{(2)} + \rho\mathfrak{V}_{(3)} \tag{7.13}$$

and ρ denotes a function to be determined.

Here, it is necessary to deduce an identity:

$$(P - a_2)\,\{F_1 - F\,(\beta F + 2a) - 2c\} + 2cQ = 0\,. \tag{7.14}$$

Differentiating semi-invariantly the second equation of (2.19) and noticing that a and c are respectively of the weights (1, 0) and (0, 1), we get

$$F_1 (P - a_2) + F (2aP - a_{21}) = c_{21} - \gamma \{\beta c + 2 (L - a^2)\},$$

$$2aP - a_{21} = a\beta\gamma - \beta c_2 - 4a (P - a_2),$$

$$c_{21} - \gamma \{\beta c + 2 (L - a^2)\} = - 2 (a_2 c + a c_2) - 2Qc - 2\gamma (L - a^2),$$

whence, after applying the first equation of (4.19), follows (7.14).

There is no difficulty in computing $\mathfrak{B}_{(4)2}$:

$$\mathfrak{B}_{(4)2} = (\rho_2 + 2Q) \mathfrak{B}_{(3)} + \{F_{12} - 2F_2 (\beta F + a) - 2Fa_2 - 2c_2 - 2QF + \rho (\gamma + F_2)\} \mathfrak{B}_{(2)}.$$

Since $\mathfrak{B}_{(2)}$, $\mathfrak{B}_{(3)}$ and $\mathfrak{B}_{(4)}$ are linearly independent, ρ is therefore determined by the following equation:

$$\rho (\gamma + F_2) + F_{12} - 2F_2 (\beta F + a) - 2F - (a_2 + Q) - 2c_2 = 0. \quad (7.15)$$

Consequently,

$$\mathfrak{B}_{(4)2} = (\rho_2 + 2Q) \mathfrak{B}_{(3)}. \quad (7.16)$$

From (4.45) we have

$$\hat{\mathfrak{Q}} = - 2\hat{H} (K\mathfrak{U} + \mathfrak{A}) + \hat{L}\hat{\mathfrak{B}} - \left\{Q\left(1 + \frac{b}{K}\right)\mathfrak{U} + \beta K\mathfrak{B}\right\}. \quad (7.17)$$

On account of an equation analogous to (2.17) we get

$$\hat{\mathfrak{B}}_{(2)} = - 2 (\hat{a} - \hat{H}) \hat{\mathfrak{B}}_{(1)} - \left\{Q\left(1 + \frac{b}{K}\right)\mathfrak{U} + \beta K\mathfrak{B}\right\}. \quad (7.18)$$

It remains merely to show that there is a linear relation among $\hat{\mathfrak{B}}_{(1)}$, $\hat{\mathfrak{B}}_{(2)}$, $\mathfrak{B}_{(3)}$, $\mathfrak{B}_{(4)}$; in other words, there exist certain functions λ, μ and ν such that holds good the identity:

$$2KP\left(1 + \frac{a}{H}\right)\lambda\hat{\mathfrak{B}}_{(1)} - \mu\left\{Q\left(1 + \frac{b}{K}\right)\mathfrak{U} + \beta K\mathfrak{B}\right\} - \nu\mathfrak{B}_{(3)}$$
$$= \{F_1 - F (\beta F + 2a) - 2c\} \mathfrak{B}_{(2)} + 2Q (- b\mathfrak{U} + d\mathfrak{B} + \mathfrak{A}). \quad (7.19)$$

In virtue of (7.7) the last equation can be written in the equivalent one:

$$\lambda\left[\left\{\nabla + 2\hat{H}KP\left(1 + \frac{a}{H}\right)\right\}\hat{\mathfrak{B}} + \left\{\gamma H - FP\left(1 + \frac{a}{H}\right)\right\}\mathfrak{B}_{(2)}\right] - T\mathfrak{B}_{(3)}$$
$$= \{F_1 - F (\beta F + 2a) - 2c\} \mathfrak{B}_{(2)} - Q \left\{2b - \mu\left(1 + \frac{b}{K}\right)\right\}\mathfrak{U}$$
$$+ (2dQ + \mu\beta K) \mathfrak{B} + 2Q\mathfrak{A},$$

or, when $\hat{\mathfrak{B}}$ is substituted by (4.31),

$$-\lambda\left\{1+2\frac{\hat{H}K}{\nabla}P\left(1+\frac{a}{H}\right)\right\}\left\{2MP\left(1+\frac{a}{H}\right)\mathfrak{U}+\gamma H\,(\mathfrak{Q}+L\mathfrak{V})\right.$$

$$\left.+P\left(1+\frac{a}{H}\right)(\mathfrak{P}-M\mathfrak{U})+2KP\left(1+\frac{a}{H}\right)\mathfrak{A}+2\gamma L\mathfrak{B}\right\}$$

$$=\left[\lambda\left\{FP\left(1+\frac{a}{H}\right)-\gamma H\right\}+TF+F_1-F\,(\beta F+2a)-2c\right]\,(\mathfrak{Q}+L\mathfrak{V}-2a\mathfrak{V})$$

$$-Q\left\{2b-\mu\left(1+\frac{b}{K}\right)\right\}\mathfrak{U}+(2dQ+\mu\beta K)\,\mathfrak{V}+2Q\mathfrak{A}$$

$$+T\,(\mathfrak{P}-M\mathfrak{U})-2cT\mathfrak{V}\,.$$

Comparison of the coefficients of $\mathfrak{U}$, $\mathfrak{B}$ and $\mathfrak{P}-M\mathfrak{U}$ in both sides of the last equation gives μ, T and λ:

$$\mu=2K\,, \tag{7.20}$$

$$T=-\lambda P\left(1+\frac{a}{H}\right)\left\{1+2\frac{\hat{H}K}{\nabla}P\left(1+\frac{a}{H}\right)\right\}=\frac{Q}{K}. \tag{7.21}$$

From (7.20) it is easy to show that the corresponding coefficients of $\mathfrak{A}$ there appeared are equal. Further application of the first equation of (4.19) suffices to verify that the coefficients of $\mathfrak{B}$ are also equal. It remains therefore to consider the coefficients of $\mathfrak{Q}+L\mathfrak{B}$ in both sides of the equation, namely, to show that

$$\lambda\left\{FP\left(1+\frac{a}{H}\right)-\gamma H\right\}+TF+F_1-F\,(\beta F+2a)-2c=\frac{\gamma HQ}{PK\left(1+\frac{a}{H}\right)}. \tag{7.22}$$

For this purpose we make use of (7.14) in rewriting (7.22),

$$\left\{\lambda+\frac{Q}{P\left(1+\frac{a}{H}\right)K}\right\}\left\{FP\left(1+\frac{a}{H}\right)-\gamma H\right\}=-\frac{2cQ}{a_2-P}, \tag{7.23}$$

so that the left-hand side of (7.23) is simplified by means of (7.21), (5.3) and (4.27). After some calculations we arrive at the first equation of (4.19).

Thus we have proved that the joins $(\hat{\mathfrak{B}}_{(1)}, \hat{\mathfrak{B}}_{(2)})$ and $(\mathfrak{B}_{(3)}, \mathfrak{B}_{(4)})$ must intersect each other.

In a similar way the joins $(\hat{\mathfrak{U}}_{(1)}, \hat{\mathfrak{U}}_{(2)})$ and $(\mathfrak{U}_{(3)}, \mathfrak{U}_{(4)})$ are shown to intersect each other. Again applying the theorem of O. Mayer and proceeding as in the proof of (7.9), we finally obtain Theorem 7.2.

Finally, we have to point out a question: May we extend the arrangement (7.10) to infinity in both directions? As was stated above, the property of intersection in the original arrangement (7.8) is characteristic of a projective minimal surface, while Theorem 7.2 shows the possibility of extending (7.8) to (7.10); it seems valid that the question has an affirmative answer. We shall, however, consider this question on another occasion.

BIBLIOGRAPHY

[1] G. Bol, *Projektive Differentialgeometrie*. Teil 2, Vandenhoeck & Ruprecht, Göttingen, 1954. MR 16, 1150.

[2] O. Mayer, *Contribution à l'étude des surfaces minima projectives*, Bull. Sci. Math. (2) 56 (1932), 146–168, 188–200.

[3] Su Buchin, *On the surfaces whose asymptotic curves belong to linear complexes*. I, Tôhoku Math. J. 40 (1935), 408–420.

[4] ——, *Some characteristic properties of projective minimal surfaces*, Sci. Rep. Tôhoku Imp. Univ. Ser. I 24 (1936), 595–600.

[5] L. Godeaux, *Note sur l'enveloppe des quadriques de Lie d'une surface*, Acad. Roy. Belg. Bull. Cl. Sci. (5) 39 (1953), 156–164. MR 14, 897.

[6] G. Bol. *Über die Flächen, deren Godeaux-Kette sich schliesst und die Periode 8 hat*, Arch. Math. 4 (1953), 61–74. MR 15, 252; 1139.

[7] L. Godeaux, *Remarques sur les quadriques associées aux points d'une surface*, J. Chinese Math. Soc. 2 (1937), 1–5.

[8] G. Bol, *Über die Flächen, bei denen die Diagonalen der Vierseite von Demoulin linearen Kongruenzen angehören*, Math. Z. 59 (1953), 97–150. MR 15, 553.

[9] Su Buchin, *On Demoulin transforms of projective minimal surfaces*. II, Futan Univ. J. Nat. Sci. 1 (1956), 111–119. (Chinese)

Amer. Math. Soc. Transl.
(2) Vol. 92, 1970

ON DEMOULIN TRANSFORMS OF PROJECTIVE MINIMAL SURFACES. III*

SU BU-CHIN

Abstract: As a generalization of the main theorem given in the first paper we give here an almost geometrical demonstration of the following

Theorem. *If the corresponding points of the Godeaux sequences of a projective minimal surface S and one of its Demoulin transforms $\hat{S}$ in the line space S_5 be arranged in three rows*:

$$\begin{array}{l} \cdots U_{n+1}\, U_n \;\cdots U_3\; U_2\; U_1\; U \;\cdots V_m \;\; V_{m+1} \cdots, \\ \cdots \hat{U}_{n-1}\, \hat{U}_{n-2} \cdots \hat{U}_1\; \hat{U}\;\; \hat{V}\;\; \hat{V}_1 \;\cdots \hat{V}_{m+2}\, \hat{V}_{m+3} \cdots, \\ \cdots U_{n-3}\, U_{n-4} \cdots V\;\; V_1\; V_2\; V_3 \;\cdots V_{m+4}\, V_{m+5} \cdots, \end{array}$$

then the join of any two consecutive points in the middle row must intersect the join of the two consecutive points standing in the same columns of the first or third row, and therefore the corresponding points of the Godeaux sequences constitute two broken lines, generally extended in both directions, which intersect each other at intervals of three sides.

Suppose that a projective minimal surface S and one of its Demoulin transforms $\hat{S}$ are referred to the same asymptotic curvilinear parameters, u, v, and let

$$\cdots,\ U_m,\ \cdots,\ U_1,\ U,\ V,\ V_1,\ \cdots,\ V_n,\ \cdots \tag{L}$$

and

$$\cdots,\ \hat{U}_m,\ \cdots,\ \hat{U}_1,\ \hat{U},\ \hat{V},\ \hat{V}_1,\ \cdots,\ \hat{V}_n,\ \cdots \tag{$\hat{L}$}$$

be the Godeaux sequences of S and $\hat{S}$ respectively, where every point in each sequence is the Demoulin transform of the point above it along the direction u.

In part I of this paper [1] we proved a fundamental fact, namely in the

* Translation of Acta Math. Sinica 7 (1957), 123–126.

three rows of the array

$$\begin{array}{cccc} U_3 & U_2 & U_1 & U \\ \hat{U}_1 & \hat{U} & \hat{U} & \hat{V}_1 \\ V & V_1 & V_2 & V_3 \end{array} \tag{1}$$

the join of any two consecutive points in the second row intersects the join of the two consecutive points in the first or third row and in the same columns as the former points at a point different from these points. In part II [2] we showed that this property is a characterization of projective minimal surfaces and any of their Demoulin transforms. If we apply the theorem of O. Mayer [3] to the array (1), then it is easy to expand the array (1) a little bit so that the above property still holds in the following array:

$$\begin{array}{cccccccc} & & U_3 & U_2 & U_1 & U & V & V_1 \\ \hat{U}_3 & \hat{U}_2 & \hat{U}_1 & \hat{U} & \hat{V} & \hat{V}_1 & \hat{V}_2 & \hat{V}_3 \\ U_1 & U & V & V_1 & V_2 & V_3 & & \end{array} \tag{2}$$

except that there lack two points in the upper left and lower right corners. In part I we explained that the array (2) can be further expanded in both left and right directions to the array

$$\begin{array}{cccccccccc} & & U_4 & U_3 & U_2 & U_1 & U & V & V_1 & V_2 \\ \hat{U}_4 & \hat{U}_3 & \hat{U}_2 & \hat{U}_1 & \hat{U} & \hat{V} & \hat{V}_1 & \hat{V}_2 & \hat{V}_3 & \hat{V}_4 \\ U_2 & U_1 & U & V & V_1 & V_2 & V_3 & V_4 & & \end{array} \tag{3}$$

Problem. Can this array be expanded indefinitely in both left and right directions? In this paper we shall use the fundamental property of the Godeaux sequences to show that the answer to this problem is affirmative.

The proof uses mathematical induction; that is, by assuming the above property holds for the array (n), we shall prove that the same property also holds in the next array $(n+1)$. For the convenience of discussion, we shall give the demonstration in details only for the case $n = 3$, since the same method is also good for a general n. When $n = 3$, from the array (3) we can deduce the following fact: In the line space S_5 the two lines (U_4U_5) and $(\hat{U}_2\hat{U}_3)$ must intersect, and therefore according to the symmetric property the two lines (V_4V_5) and $(\hat{V}_2\hat{V}_3)$ must also intersect. By applying Mayer's theorem the array (3) can be expanded to the array (4).

In fact, it is known [4] that the Godeaux sequence (L) is selfconjugate with respect to the Klein hyperquadric Q in S_5. In particular, the conjugate hyperplane with respect to Q of U_n is $V_{n-2}V_{n-1}V_nV_{n+1}V_{n+2}$, and that of V_n is $U_{n-2}U_{n-1}U_nU_{n+1}U_{n+2}$. For the sequence $(\hat{L})$ there are similar properties.

According to these facts we can conclude that the lines (V_3V_4) and $(\hat{V}_1\hat{V}_2)$ are respectively conjugate to the three-dimensional spaces $\Sigma_3 \equiv (U_2U_3U_4U_5)$ and $\hat{\Sigma}_3 \equiv (\hat{U}\hat{U}_1\hat{U}_2\hat{U}_3)$ with respect to Q. Since we have known that the two lines (V_3V_4) and $(\hat{V}_1\hat{V}_2)$ intersect (suppose the point of intersection to be P), the conjugate hyperplane S_4 of P with respect to the hyperquadric Q must contain Σ_3 and $\hat{\Sigma}_3$. That is to say, the eight points U_2, U_3, U_4, U_5; $\hat{U}, \hat{U}_1, \hat{U}_2, \hat{U}_3$ are in a four-dimensional space S_4.

On the other hand, we have known that the two lines (U_2U_3) and $(\hat{U}\hat{U}_1)$ intersect, and therefore determine the plane $\Sigma_2 \equiv (U_2U_3\hat{U}\hat{U}_1)$. Similarly, the two lines (U_3U_4) and $(\hat{U}_1\hat{U}_2)$ also determine a plane $\Sigma_2' \equiv (U_3U_4\hat{U}_1\hat{U}_2)$.

We first investigate the case where the two planes Σ_2 and Σ_2' coincide. In this case, U_2, U_3, U_4 and $\hat{U}, \hat{U}_1, \hat{U}_2$ are on a plane. Since any three consecutive points of a Godeaux sequence cannot be collinear, this plane is determined by the three points U_2, U_3, U_4 and also by the three points $\hat{U}, \hat{U}_1, \hat{U}_2$. But, according to formulas of Godeaux [4],

$$\begin{aligned} U_n^{01} &= U_{n+1} + U_n (\log bh_1h_2 \cdots h_n)^{01}, \\ \hat{U}_n^{01} &= \hat{U}_{n+1} + \hat{U}_n (\log \hat{b}\hat{h}_1\hat{h}_2 \cdots \hat{h}_n)^{01}, \end{aligned} \tag{4}$$

we can deduce that the points U_5 and $\hat{U}_3$ are also on this plane. In fact, since the point of intersection of the two lines (U_2U_3) and $(\hat{U}\hat{U}_1)$ is not $\hat{U}$, we have $\hat{U}_2 = A\hat{U} + BU_2 + CU_3$. Taking the partial derivatives of both sides of this equation with respect to v, we obtain

$$\hat{U}_3 = A'\hat{U} + B'\hat{U}_1 + C'U_2 + D'U_3 + E'U_4$$

which expresses that the point $\hat{U}_3$ belongs to the plane under discussion. Similarly, the same is true for the point U_5. In this case the two lines (U_4U_5) and $(\hat{U}_2\hat{U}_3)$ are coplanar, and therefore must intersect each other.

Next, let us discuss the case where the two planes do not coincide. In this case the points $\hat{U}$ and U_2 do not belong to the plane Σ_2'. If the two lines (U_4U_5) and $(\hat{U}_2\hat{U}_3)$ do not intersect, the four points $U_4, U_5, \hat{U}_2, \hat{U}_3$

determine a three-dimensional space S_3. We claim that this S_3 must contain the plane Σ_2'.

In fact, if it were not true, then this S_3 does not contain the plane Σ_2', and Σ_2' and S_3 determine a four-dimensional space, which must be the conjugate hyperplane S_4 of the above-mentioned point P. Since the point $\hat{U}$ does not belong to Σ_2', and the point $\hat{U}_3$ belongs to S_3, $\hat{U}_3$ is a linear combination of $\hat{U}, \hat{U}_2, U_4, U_5$:

$$\hat{U}_3 = A\hat{U} + C\hat{U}_2 + DU_4 + EU_5. \tag{5}$$

Taking partial derivatives of both sides of (5) with respect to u and applying the formulas [4] of Godeaux:

$$\begin{aligned} U_n^{10} &= h_n U_{n-1}, \\ \hat{U}_n^{10} &= \hat{h}_n \hat{U}_{n-1}, \end{aligned} \tag{6}$$

we then obtain

$$\hat{h}_3\hat{U}_2 = -2bA\hat{V} + B'\hat{U}_1 + C'\hat{U}_2 + D'U_3 + E'U_4 + F'U_5. \tag{7}$$

If $A \neq 0$, since $b\hat{h}_3 \neq 0$, then equation (7) expresses that the point $\hat{V}$ belongs to the above-mentioned four-dimensional space S_4. Since the two lines $(\hat{U}\hat{V})$ and $(\overline{U_1 U_2})$ intersect, the point U_1 also belongs to S_4. Thus the four-dimensional space $\overline{S}_4$ determined by the eight points $U_1, U_2, U_3, U_4; \hat{V}, \hat{U}, \hat{U}_1, \hat{U}_2$ must coincide with S_4. Since the two lines (V_2V_3) and $(\hat{V}\hat{V}_1)$ intersect at a point $\overline{P}$ and this space $\overline{S}_4$ is the conjugate hyperplane of $\overline{P}$ with respect to the hyperquardic Q, in general the points V_2, V_3, V_4 and $\hat{V}, \hat{V}_1, \hat{V}_2$ are respectively not collinear, and therefore the point P cannot coincide with $\overline{P}$. Thus S_4 cannot coincide with $\overline{S}_4$.

From this it follows that $A = 0$, and (5) becomes

$$\hat{U}_3 = C\hat{U}_2 + DU_4 + EU_5,$$

which implies that the two lines (U_4U_5) and $(\hat{U}_2\hat{U}_3)$ intersect, and contradicts the original assumption. Thus we have proved that the S_3 determined by the four points $U_4, U_5, \hat{U}_2, \hat{U}_3$ must contain the plane $\Sigma_2' \equiv (U_3U_4\hat{U}_1\hat{U}_2)$.

Now we can write

$$\hat{U}_3 = B\hat{U}_1 + C\hat{U}_2 + DU_4 + EU_5. \tag{8}$$

Partially differentiating this equation and applying (6), we then obtain

$$\hat{h}_3\hat{U}_2 = \hat{h}_1 B\hat{U} + B'\hat{U}_1 + C'\hat{U}_2 + D'U_3 + E'U_4 + F'U_5.$$

If $B \neq 0$, since $\hat{h}_1 \neq 0$, by solving the last two equations we have

$$\hat{U} = \bar{A}\hat{U}_1 + \bar{B}\hat{U}_2 + \bar{C}U_4 + \bar{D}U_5.$$

This expresses that the point $\hat{U}$, and therefore the point U_2, belong to the same $S_3 \equiv (\hat{U}_1\hat{U}_2U_4U_5)$. In other words, the two three-dimensional spaces Σ_3 and Σ_3' coincide with one S_3. Thus their conjugate lines (V_3V_4) and $(\hat{V}_1\hat{V}_2)$ must coincide, and we have a contradiction.

Therefore $B = 0$. From (8) it is seen that the two lines $(\hat{U}_2\hat{U}_3)$ and (U_4U_5) intersect. This deduction contradicts the original assumption again.

Summing the above results, we obtain

Theorem. *If the Godeaux sequences of a projective minimal surface and any one of its Demoulin transforms*

$$\begin{gathered} \cdots, U_n, \cdots, U_1, U, V, V_1, \cdots, V_m, \cdots, \\ \cdots, \hat{U}_n, \cdots, \hat{U}_1, \hat{U}, \hat{V}, \hat{V}_1, \cdots, \hat{V}_m, \cdots \end{gathered}$$

are extended to infinity in both directions, then in the three rows of the array

$$\begin{array}{cccccccccc} \cdots U_{n+1} & U_n & \cdots U_3 & U_2 & U_1 & U & \cdots V_m & V_{m+1} \cdots \\ \cdots \hat{U}_{n-1} & \hat{U}_{n-2} & \cdots \hat{U}_1 & \hat{U} & \hat{V} & \hat{V}_1 & \cdots \hat{V}_{m+2} & \hat{V}_{m+3} \cdots \\ \cdots U_{n-3} & U_{n-4} & \cdots V & V_1 & V_2 & V_3 & \cdots V_{m+4} & V_{m+5} \cdots \end{array}$$

the join of any two consecutive points in the second row intersects the join of the two consecutive points in the first or third row and in the same columns of the former points, and the corresponding points of the two Godeaux sequences form two infinite crossing broken lines, of which every side intersects the next fourth side.

Hence we have extended the main theorem of part I.

BIBLIOGRAPHY

[1] Su Buchin, *On Demoulin transforms of projective minimal surfaces*. I, Acta Math. Sinica 7 (1957), 28–50; English transl., Amer. Math. Soc. Transl. (2) 92 (1970), 151–175. MR 21 #5979.

[2] ———, *On Demoulin transforms of projective minimal surfaces*. II, Futan Univ. J. Nat. Sci. 1 (1956), 111–119. (Chinese)

[3] O. Mayer, *Contribution à l'étude des surfaces minima projectives*, Bull. Sci. Math. (2) 56 (1932), 146–168, 188–200.

[4] L. Godeaux, *La théorie des surfaces et l'espace réglé*, Paris, 1934.

Translated by:
C. C. Hsiung

Amer. Math. Soc. Transl.
(2) Vol. 92, 1970

A METHOD FOR DETERMINING A QUARTERNARY DIFFERENTIAL QUADRATIC FORM WITH PRESCRIBED CHARACTERISTICS*

P. I. PETROV

1. The curvature tensor R_{hijk} for a form in four variables

$$ds^2 = g_{ij}(u)\, du^i\, du^j, \qquad g \neq 0 \tag{1}$$

is completely reducible (É. Cartan). The corresponding classical result in tensor analysis says that this tensor decomposes into four irreducible parts with 1, 9, 5 and 5 parameters respectively. The first of these irreducible tensors, with only one component, is the scalar curvature of the Riemannian space. The second of them is defined by the formula $B_{ij} \equiv R_{ij} - (R/4)\, g_{ij}$, where R_{ij} is the Ricci tensor. In view of the condition $B^{\sigma}_{\sigma} = 0$ it has 9 essential components. The quantity B_{ij} can also be expressed in terms of

$$\bar{R}_{hijk} = R_{\alpha\beta\gamma\delta}\, \underset{h}{\lambda^{\alpha}}\, \underset{i}{\lambda^{\beta}}\, \underset{j}{\lambda^{\gamma}}\, \underset{k}{\lambda^{\delta}},$$

where $\underset{s}{\lambda^{\tau}}$ are the components of the sth unit vector in an orthonormal frame induced by the metric form (1) of a space V_4. Below, by a realization we will mean just such a representation of B_{ij}, without hinting at this by using any special symbol. The two tensors (in the wide sense of the word) with five parameters each were interpreted by me in [1] with respect to an orthogonal frame (linear substitutions preserving the quadratic $\Sigma_{i=1}^{4} \epsilon_i u_i^2$, where $\epsilon_i = \pm 1$) as Euclidean symmetric tensors with zero trace and of the second valency $T_{\alpha\beta} \overline{T}_{\bar{\alpha}\bar{\beta}}$ in three-dimensional Euclidean space. On the basis of what was said in the above-mentioned article there are associated with the form (1) three arithmetic invariants, which I called the *characteristics* of this form ds^2:

1) The Weierstrass characteristic $[e_1 e_2 e_3 e_4]$ of the λ-matrix

$$\|B_{ij} - \lambda g_{ij}\|, \text{ where } g_{ij} = \begin{cases} \pm 1, & \text{if} \quad i = j \\ 0, & \text{if} \quad i \neq j \end{cases};$$

$$i, j = 1, 2, 3, 4;$$

---

*Translation of Bull. Acad. Polon. Sci. Sér. Sci. Math. Astronom. Phys. 12 (1964), 311–316.

2) The characteristics $[e_1 e_2 e_3]$ and $[\bar{e}_1 \bar{e}_2 \bar{e}_3]$ of the λ-matrices

$$\| T_{\alpha\beta} - \lambda\delta_{\alpha\beta} \|, \qquad \| \bar{T}_{\bar{\alpha}\bar{\beta}} - \lambda\delta_{\bar{\alpha}\bar{\beta}} \|,$$

where $\delta_{\alpha\beta} = 1$ if $\alpha = \beta$ and is zero if $\alpha \neq \beta$; $\alpha, \beta = 1, 2, 3$.

It is possible to carry out a classification of quarternary differential quadratic forms which depends on the value of the signature s and the characteristics. It turns out that when $s = 4$ there exists only one type of form with characteristic $[1\,1\,1\,1]$, $[1\,1\,1]$, $[\bar{1}\,\bar{1}\,\bar{1}]$. When $s = -2$ there are in all 9 types of form, the characteristics of which are obtained from the scheme

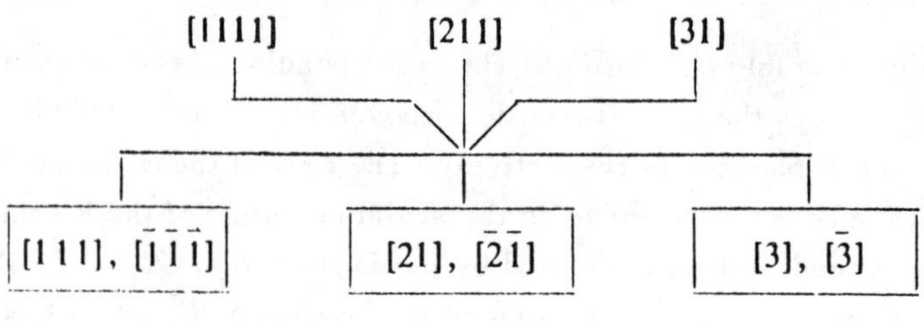

Finally, when $s = 0$ there are 45 types of forms like (1). Their characteristics are all possible combinations of the symbols $[e_1 e_2 e_3 e_4]$, $[e_1 e_2 e_3]$ and $[\bar{e}_1 \bar{e}_2 \bar{e}_3]$.

The problem of the present communication is to establish a differential system for finding the coefficients $g_{ij}(u_1 u_2 u_3 u_4)$ of a quarternary differential quadratic form with prescribed characteristics.

2. The problem reduces to determining what the elements of the linear vector functions of the first kind A (or the matrix corresponding to them) must be in the space of three or four dimensions in order that they have the prescribed Weierstrass characteristics. Because of the novelty of this difficult algebraic problem, which goes far beyond the scope of the specialized theme of the given paper, I consider it necessary to describe in detail an auxiliary technique which is essential to formulate the conclusions of the proposed investigation. The proofs of the assertions which are utilized for this form the contents of another paper.*

Editor's note.* P. I. Petrov, *Characterization of types of conformally flat four-dimensional Riemann spaces*, Ukrain. Mat. Ž. **18 (1966), no. 2, 42–49; English transl., Amer. Math. Soc. Transl. (2) 73 (1968), 251–260.

3. In place of B_{ij} we take its mixed components, putting $B^{\alpha}_{\beta} = g^{\alpha\sigma} B_{\sigma\beta}$ and denoting by B the matrix

$$\begin{pmatrix} B^1_1 & B^1_2 & B^1_3 & B^1_4 \\ B^2_1 & B^2_2 & B^2_3 & B^2_4 \\ \cdots & \cdots & \cdots & \cdots \\ B^4_1 & B^4_2 & B^4_3 & B^4_4 \end{pmatrix}. \tag{2}$$

The invariants of the characteristic polynomial $\phi(\lambda) = |\lambda E - B|$ of the vector function of the first kind B can be put in matrix form:

$$i = \frac{1}{3.2^3} (\varphi''^2(B) - 2\varphi'(B)\varphi'''(B)),$$

$$j = \frac{1}{3^2.2^5} (3\varphi'(B)\varphi''(B)\varphi'''(B) - \varphi''^3(B) - 108\,\varphi'^2(B).$$

Hence we obtain an expression for the discriminant – the matrix

$$R = j^2 - \frac{1}{6} i^3.$$

We now introduce three nonnegative integers ρ_1, ρ_2, ρ_3 defined in the following way:

ρ_3 – the rank of the matrix $\phi'(B)$;

ρ_2 – the maximum of the ranks of the matrices $\phi'(B)$, $\phi''(B)$, i, j;

ρ_1 – the maximum of the ranks of the matrices $\phi'(B)$, $\phi''(B)$, $\phi'''(B)$, i, j.

We distinguish 5 types of matrices B, having characteristics

I. [1111] II. [211] III. [31] IV. [22] V. [4]

respectively. By enclosing in parentheses the powers of the elementary divisors corresponding to a single linear factor it is possible to distinguish among one type several different species. To each of these species of given type we associate a choice of $(\rho_1\rho_2\rho_3)$ and a certain number of algebraic equalities and inequalities. Repeating this process for all of the types enumerated above, we obtain the arithmetic and algebraic characteristics of the linear vector functions B:

(I, k)

1. (444), $R \neq 0$, if [1111].
2. (442), $R = 0$, $i \neq 0$, $j \neq 0$, if [(11) 11].
3. (440), $R = 0$, $i \neq 0$, $j \neq 0$, $\varphi'(B) = 0$, if [(11) (11)].
4. (411), $i = 0$, $j = 0$, if [(111) 1].
5. (000), $\varphi''(B) = 0$, $\varphi'''(B) = 0$, if [(1111)].

(II, l)
1. (443), $R = 0,\ i \neq 0,\ j \neq 0,$ if [211].
2. (441), $R = 0,\ i \neq 0,\ j \neq 0,\ \varphi'^2(B) = 0,$ if [2 (11)].
3. (421), $i = 0,\ j = 0,$ if [(21) 1].
4. (100), $\varphi''(B) = 0,\ \varphi'''^2(B) = 0,\ \varphi'''(B) \neq 0,$ if [(211)].

(III, m)
1. (432), $i = 0,\ j = 0,$ if [31].
2. (210), $\varphi''^2(B) = 0,\ \varphi''(B) \neq 0,\ \varphi'''^3(B) = 0,\ \phi'''^2(B) \neq 0,$ if [(31)].

(IV, p)
1. (442), $R = 0,\ i \neq 0,\ j \neq 0,\ [\varphi'(B)]^2 = 0,$ if [22].
2. (200), $\varphi''(B) = 0,\ \varphi'''^2(B) = 0,\ \varphi'''(B) \neq 0,$ if [(22)].

(V)
1. (321), $\varphi'^2(B) = 0,\ \varphi'(B) \neq 0,\ \varphi''^2(B) = 0,\ \phi''(B) \neq 0,\ \phi'''^4(B) = 0,$ $\varphi'''^3(B) \neq 0,$ if [4].

4. For $n = 3$ there exist three types of linear vector functions $T = (T^\alpha_\beta)$ $(\alpha, \beta = 1, 2, 3)$ with Weierstrass characteristics

$$\text{I}^*.\ [111] \quad \text{II}^*.\ [21] \quad \text{III}^*.\ [3].$$

Let

$$\varphi(\lambda) = |\lambda E - T|, \quad H = \varphi'^2(T) - \frac{1}{16}\varphi''^2(T)\varphi'(T),$$

let ρ_2^* be the rank of the matrix $\phi'(T)$, and let ρ_1^* be the maximum of the ranks of the matrices $\phi'(T)$ and $\phi''(T)$.

Repeating the method of relating a species of this type with a choice of $(\rho_1^*\ \rho_2^*)$ and algebraic equalities and inequalities, we get the arithmetic and algebraic characteristics of the linear vector functions T:

(I*, a)
1. (33), $H\varphi'(T) \neq 0,$ if [111].
2. (31), $H = 0,\ \varphi'^2(T) \neq 0,$ if [(11) 1].
3. (00), $\varphi''(T) = 0,$ if [(111)].

(II*, в)
1. (32), $H\varphi'(T) = 0,\ H \neq 0,$ if [21].
2. (10), $\varphi''^2(T) = 0,\ \varphi''(T) \neq 0,$ if [(21)].

(III*, c)
1. (21), $\varphi'^2(T) = 0,\ \varphi'(T) \neq 0,$ if [3].

5. The problem of determining the forms (1) with prescribed characteristics reduces to the problem of integrating a differential system (S) with 10 unknown functions $g_{ij}(u)$ in four independent variables u_1, u_2, u_3, u_4. First of all the criteria indicating the signature s of the form ds^2 gives a certain number of inequalities (S') in its coefficients. Further, the requirements that the characteristics of this form are expressed by means of the symbols

$[e_1e_2e_3e_4]$, $[e_1e_2e_3]$ and $[\bar{e}_1\bar{e}_2\bar{e}_3]$ may, on the basis of the auxiliary propositions in §§3 and 4, be registered by how the differential system S'' is composed from three subsystems taken one from each line:

$$(\mathrm{I}, k),\ (\mathrm{II}, l),\ (\mathrm{III}, m),\ (\mathrm{IV}, p),\ (\mathrm{V}), \tag{3}$$

$$(\mathrm{I}^*, a),\ (\mathrm{II}^*, b),\ (\mathrm{III}^*, c), \tag{4}$$

$$(\overline{\mathrm{I}}^*, a),\ (\overline{\mathrm{II}}^*, b),\ (\overline{\mathrm{III}}^*, c). \tag{5}$$

The system $S = S' + S''$ represents necessary and sufficient conditions for the coefficients $g_{ij}(u)$ of the required form ds^2.

In the case $s = 4$ the system S'' contains three subsystems, each of which is taken from the systems (I, k), (I^*, a), ($\overline{\mathrm{I}}^*$, a). The existence of a solution of the differential system (S) constructed in such a way, at least for certain values of $k = 1, \cdots, 5$ and $a = 1, 2, 3$, is obvious: B, T and $\overline{T}$ are matrices of simple type for any form $ds^2 > 0$.

When $s = -2$ the matrices T and $\overline{T}$ are complex conjugates, and therefore their characteristics coincide. Therefore the coefficients of the 9 types of forms ds^2 with signature $s = -2$ may be defined by integral systems which are obtained according to the scheme below:

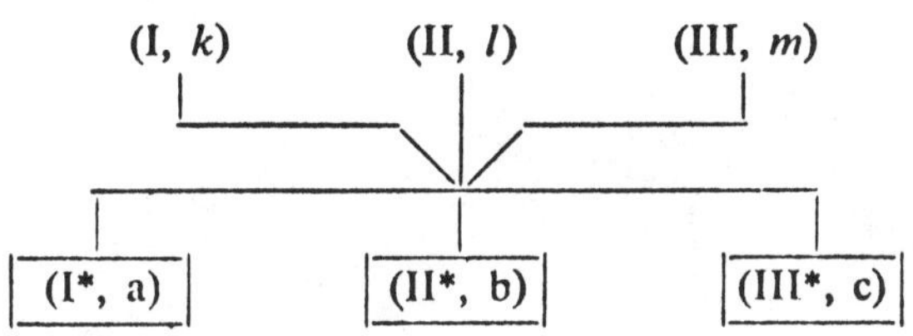

The compatibility of the differential systems arising from this (the question of the existence of a form ds^2, $s = -2$, of the chosen type) at least for certain values of k, l, m, a, b and c has been proved by the author by constructing particular examples of ds^2 belonging to the enumerated classes of differential forms.

The consistency of the differential systems for $s = 0$ may be investigated by the same method.

BIBLIOGRAPHY

[1] P. I. Petrov, *Invariants and classification of differential quadratic forms in four variables*, Izv. Akad. Nauk SSSR Ser. Mat. 23 (1959), 387–420. (Russian) MR 22 #2947.

Translated by:
A. West

Amer. Math. Soc. Transl.
(2) Vol. 92, 1970

A CHARACTERIZATION OF RIEMANNIAN MANIFOLDS*

P. I. PETROV

1. A Riemannian space V_n is a differentiable manifold on which is defined a nondegenerate quadratic differential form

$$ds^2 = g_{ij} du^i du^j, \qquad g \neq 0. \tag{1}$$

Two forms ds^2 and $d\bar{s}^2$ in the same coordinate system are said to be equivalent if there is a point transformation $x \to \bar{x}$, with nonzero Jacobian, that sends one form into the other.

The collection of facts common to equivalence classes of forms of the type (1) constitutes the subject matter of Riemannian geometry. On this basis we can consider the problem of classifying the Riemannian manifolds V_n as the problem of separating the differential quadratic forms in n variables into nonequivalent classes. Gauss [1] and Riemann [2] used the structure of the curvature tensor to distinguish the particular types of geometries based upon a metric defined by means of ds^2.

This idea of the founders of Riemannian geometry, in a somewhat modernized form, has found an application and concretization in [3], [4] and [5]. In these articles the manifolds V_3 and V_4 have been classified by using the arithmetic invariants of the fundamental forms of these spaces.

The aim of the present article is to characterize analytically the metric tensors $g_{ij}(u)$ of the various classes of the Riemannian manifolds V_3 and C_3, and of the 4-dimensional Einstein spaces.**

2. In the theory of invariants of affine transformations one considers covariant and contravariant vectors.

A linear vector-function A for which the argument and the function itself are vectors of the same form is said to be a vector-function of the first kind.

To each vector-function

$$\bar{y} = A(\bar{x})$$

corresponds the matrix

* Translation of Ukrain. Mat. Ž. 17 (1965), no. 4, 55–62.

** *Editor's note.* In a later paper, *Characterization of types of conformally flat four-dimensional Riemannian space*, Ukrain. Mat. Ž. 18 (1966), no. 2, 42–49; English transl., Amer. Math. Soc. Transl. (2) 73 (1968), 251–260; MR 34 #752 the author considers a similar problem.

$$A = \begin{pmatrix} a_1^1 & a_2^1 \dots a_n^1 \\ a_1^2 & a_2^2 \dots a_n^2 \\ \cdot & \cdot \quad \cdot \quad \cdot \\ a_1^n & a_2^n \dots a_n^n \end{pmatrix}, \tag{2}$$

which establishes the linear transformation

$$y^i = a^i_\alpha x^\alpha. \tag{3}$$

The algebra of vector-functions is the algebra of the matrices which correspond to these vector-functions. In 3-dimensional space these are six types of linear vector-functions with the Weierstrass characteristics [6]

$$[111],\ [21],\ [(11)1],\ [3],\ [(21)],\ [(111)]. \tag{4}$$

3. By the characteristic polynomial of a vector-function A (or matrix A), as is usual, we mean the determinant $|\lambda E - A|$:

$$\varphi(\lambda) = \lambda^3 - \sigma_1\lambda^2 + \sigma_2\lambda - \sigma_3. \tag{5}$$

The discriminant of the cubic equation (5),

$$D = \sigma_1^2\sigma_2^2 - 4\sigma_1^3\sigma_3 + 18\sigma_1\sigma_2\sigma_3 - 4\sigma_2^3 - 27\sigma_3^2, \tag{6}$$

after using the formulas

$$\begin{cases} \sigma_1 E = 3A - \dfrac{1}{2}\varphi''(A), \\ \sigma_2 E = 3A^2 + \varphi'(A) - A\varphi''(A), \\ \sigma_3 E = A^3 + A\varphi'(A) - \dfrac{A^2}{2}\varphi''(A), \end{cases} \tag{7}$$

can be written as

$$H\varphi'(A) = -\frac{D}{4}E, \tag{8}$$

where we have put

$$H = [\varphi'(A)]^2 - \frac{1}{16}[\varphi''(A)]^2\varphi'(A). \tag{9}$$

We denote the rank of the matrix $\phi'(A)$ by ρ_2 and let ρ_1 denote the greatest of the values of the ranks of the matrices $\phi'(A)$ and $\phi''(A)$. G. B. Gurevič has suggested [7] the problem of proving the

Lemma. *The linear vector-functions of the first kind of a 3-dimensional*

space have the following arithmetic $(\rho_1\,\rho_2)$ *and algebraic characteristics*:

$$\begin{array}{ll} (33) & H\varphi'(A)\neq 0, \\ (32) & H\varphi'(A)=0, \quad H\neq 0. \\ (31) & H=0, \quad \varphi'^2(A)\neq 0. \\ (21) & \varphi'^2(A)=0, \quad \varphi'(A)\neq 0. \\ (10) & \varphi''^2(A)=0, \quad \varphi''(A)\neq 0. \\ (00) & \varphi''(A)=0. \end{array} \tag{10}$$

Proof. 1°. Suppose that a linear vector-function of the first kind has the Weierstrass characteristic [111]. This means that, by choosing the contravariant coordinate vectors in the principal directions of the vector-function, its matrix can be reduced to the diagonal form

$$A=\begin{pmatrix}\lambda_1 & 0 & 0\\ 0 & \lambda_2 & 0\\ 0 & 0 & \lambda_3\end{pmatrix}, \ \lambda_1,\lambda_2,\lambda_3 \text{ not equal.} \tag{11}$$

Hence

$$\varphi'(A)=\begin{pmatrix}\varphi'(\lambda_1) & 0 & 0\\ 0 & \varphi'(\lambda_2) & 0\\ 0 & 0 & \varphi'(\lambda_3)\end{pmatrix} \tag{12}$$

Because the λ_i are simple roots of $\phi(\lambda)$ (that is, $\phi(\lambda_i)=0$ and $\phi'(\lambda_i)\neq 0$), we have $\rho_2=3$. We conclude from here that $\rho_1=3$. Since λ_1, λ_2, λ_3 are distinct, the expression $D=\Pi_{\alpha>\beta}(\lambda_\alpha-\lambda_\beta)^2$ is nonzero.

Thus, if the Weierstrass characteristic of A is [111], then

$$(\varrho_1\varrho_2)=(33), \qquad H\varphi'(A)\neq 0. \tag{13}$$

Conversely, in view of (8) it follows from (13) that $D\neq 0$; hence the characteristic numbers λ_1, λ_2, λ_3 of A are distinct, and therefore its Jordan form coincides with the matrix (11) which has characteristic [111].

2°. We can always reduce a matrix with the Weierstrass characteristic [21] to the normal form

$$A=\begin{pmatrix}\lambda_1 & 1 & 0\\ 0 & \lambda_1 & 0\\ 0 & 0 & \lambda_2\end{pmatrix}, \lambda_1,\lambda_2 \text{ not equal.} \tag{14}$$

For the matrix (14) we have

$$\varphi'(A) = \begin{pmatrix} 0 & \varphi''(\lambda_1) & 0 \\ 0 & 0 & 0 \\ 0 & 0 & \varphi'(\lambda_2) \end{pmatrix}, \tag{15}$$

$$\varphi''(A) = \begin{pmatrix} \varphi''(\lambda_1) & 0 & 0 \\ 0 & \varphi''(\lambda_1) & 0 \\ 0 & 0 & \varphi''(\lambda_2) \end{pmatrix} \tag{16}$$

Since $\phi(\lambda_1) = 0$, $\phi'(\lambda_1) = 0$, $\phi''(\lambda_1) \neq 0$, $\phi'(\lambda_2) \neq 0$ and $\phi''(\lambda_2) \neq 0$, we have $\rho_2 = 2$ and $\rho_1 = 3$. By substituting (15) and (16) into (9) we find that

$$H = -\frac{1}{16}\begin{pmatrix} 0 & \varphi''^3(\lambda_1) & 0 \\ 0 & 0 & 0 \\ 0 & 0 & 0 \end{pmatrix}. \tag{17}$$

Hence, if a linear vector-function A of the first kind has the Weierstrass characteristic [21], then

$$(\varrho_1\varrho_2) = (32), \quad H\varphi'(A) = 0, \; H \neq 0. \tag{18}$$

Conversely, if we are given a matrix A satisfying the conditions (18), then the discriminant of its characteristic polynomial is zero, and so $\phi(\lambda)$ has a multiple root λ_1. The order of multiplicity of this root is two, for otherwise, from the condition for the polynomial (5) to have a triple root:

$$\begin{cases} \varphi'(A) - \dfrac{1}{12}\varphi''^2(A) = 0, \\ \varphi'(A)\,\varphi''(A) = 0 \end{cases} \tag{19}$$

it would follow that $H = 0$, thus contradicting our assumption that $H \neq 0$. Because $\rho_2 = 2$, to this double root corresponds only one principal direction. Hence A has the canonical form (14) and its Weierstrass characteristic is expressed by the symbol [21].

3°. We take the matrix A with characteristic [(11)1]. We can always consider it as

$$\begin{pmatrix} \lambda_1 & 0 & 0 \\ 0 & \lambda_1 & 0 \\ 0 & 0 & \lambda_2 \end{pmatrix}, \; \lambda_1 \neq \lambda_2. \tag{20}$$

It is not difficult to see that for the matrix (20)

$$(\varrho_1\varrho_2) = (31), \; H = 0, \; \varphi'^2(A) \neq 0. \tag{21}$$

The converse is also true, because the conditions (21) express, first and foremost, the fact that $|\lambda E - A| = 0$ has a multiple root λ_1. If λ_1 were a triple root, then on the basis of the equation

$$\varphi'^2(A) = (\sigma_1^2 - 3\sigma_2) A^2 + (9\sigma_3 - \sigma_1\sigma_2) A + (\sigma_2^2 - 3\sigma_1\sigma_3) E \qquad (22)$$

it would follow that $\phi^{12}(A) = 0$ notwithstanding one of the conditions (21). Since $\rho_2 = 1$, to the root λ_1 correspond two principal directions. Hence the matrix A is of simple type and its characteristic is represented by the symbol [(11)1].

4°. We next assume that A has the characteristic [3]. We transform A to the Jordan form

$$\begin{pmatrix} \lambda_1 & 1 & 0 \\ 0 & \lambda_1 & 1 \\ 0 & 0 & \lambda_1 \end{pmatrix}. \qquad (23)$$

For the matrix (23) it is not difficult to establish that

$$(\varrho_1\varrho_2) = (21), \ \varphi'^2(A) = 0, \ \varphi'(A) \neq 0. \qquad (24)$$

The characteristic polynomial $\phi(\lambda)$ of a matrix A satisfying (24) has a multiple root λ_1 because the equation $\phi'^2(A) = 0$, together with the formula

$$\frac{D}{4} E = \left(\frac{\varphi''^2(A)}{16} - \varphi'(A)\right)\varphi'^2(A),$$

shows that $D = 0$. If λ_1, λ_1 and λ_2 were the roots of the polynomial $\phi(\lambda)$, where $\lambda_1 \neq \lambda_2$, then we would have $\phi'(\lambda_1) = 0$ and $\phi'(\lambda_2) = 0$ as the characteristic numbers of the nilpotent vector-function $\phi'(A)$. But this contradicts our assumption that $\phi(\lambda_2) = 0$ and $\phi'(\lambda_2) \neq 0$. We conclude from here that $\lambda_1 = \lambda_2$. Since $\rho_1 = 2$ and $\rho_2 = 1$, to the triple root λ_1 corresponds only one principal direction of the matrix A. This means that it has the normal form (23) and characteristic [3].

5°. If A has the characteristic [(21)]; that is, if it can be represented by the matrix

$$\begin{pmatrix} \lambda_1 & 1 & 0 \\ 0 & \lambda_1 & 0 \\ 0 & 0 & \lambda_1 \end{pmatrix}, \qquad (25)$$

then the conditions

$$(\varrho_1\varrho_2) = (10), \ \varphi''^2(A) = 0, \ \varphi''(A) \neq 0 \qquad (26)$$

hold. The converse is also true: together with the equation $\phi'(A) = 0$ we have $[\phi''(A)]^2 = 0$; hence $|\lambda E - A| = (\lambda - \lambda_1)^3$; of the three possibilities for the structure of A only (25) corresponds to the values $\rho_1 = 1$, $\rho_2 = 0$ of the arithmetic invariants.

6°. The matrix

$$A = \begin{pmatrix} \lambda_1 & 0 & 0 \\ 0 & \lambda_1 & 0 \\ 0 & 0 & \lambda_1 \end{pmatrix} \tag{27}$$

has the characteristic [(111)] and the properties

$$(\varrho_1\varrho_2) = (00), \ \varphi''(A) = 0. \tag{28}$$

Equations (28) completely characterize the matrix (27): $\phi'(A) = 0$ and $\phi''(A) = 0$, thus, at the same time as $\phi(\lambda_1) = 0$, we have $\phi'(\lambda_1) = 0$ and $\phi''(\lambda_1) = 0$. Among the matrices with a triad of characteristic numbers that have the canonical forms (23), (25), (27), only the latter corresponds to the conditions $\rho_1 = \rho_2 = 0$.

4. The tensor-differential invariant of the second order of the form (1) (more often called the curvature tensor of this form):

$$R_{hijk} = \frac{\partial \Gamma_{ik,h}}{\partial u^j} - \frac{\partial \Gamma_{ij,h}}{\partial u^k} + g^{lm}\left(\Gamma_{ij,l}\,\Gamma_{hk,m} - \Gamma_{ik,l}\Gamma_{jh,m}\right), \tag{29}$$

where $\Gamma_{ij,k}$ is the Christoffel symbol of the first kind

$$\Gamma_{ij,k} = \frac{1}{2}\left(\frac{\partial g_{ik}}{\partial u^j} + \frac{\partial g_{jk}}{\partial u^i} - \frac{\partial g_{ij}}{\partial u^k}\right),$$

can be expressed in terms of $g_{ij}(u)$, $\partial g_{ij}/\partial u^k$ and $\partial^2 g_{ij}/\partial u^k \partial u^l$.

Equation (29) shows how to calculate the components R_{hijk} for a given $g_{ij}(u)$. When $n = 3$ it allows us to determine, by integrating a system of second order partial differential equations, the six unknown functions $g_{ij}(u^1 u^2 u^3)$ for a given $R_{hijk}(u)$ (see [3]. The result to which we have just referred reduces the problem of classifying the 3-dimensional Riemannian manifolds V_3 to the algebraic problem of investiagting the inequivalent types of curvature tensor R_{hijk} in a 3-dimensional space. When $n = 3$ the latter problem is simplified, because in this special case we can always represent R_{hijk} linearly in terms of g_{ij} and L_{ij}:

$$R_{hijk} = 4g_{[h[j}L_{i]k]}. \tag{30}$$

The tensor L_{ij} of valency two is defined by

$$L_{ij} = -R_{ij} + \frac{R}{4} g_{ij}, \tag{31}$$

where R_{ij} is the Ricci tensor and R is the scalar curvature.

By setting

$$L_j^i = g^{i\alpha} L_{\alpha j}, \tag{32}$$

we can reduce the problem of classifying the V_3 to that of finding the types of the linear vector-functions

$$\mathfrak{L} = (L_j^i) \qquad (i,j = 1,2,3). \tag{33}$$

As is well known, there are six of these, with the Weierstrass characteristics $[111]$, $[21]$, $[(11)1]$, $[3]$, $[(21)]$, $[(111)]$. Following the method described above, we see that with each nonsingular form ds^2 in three variables there is uniquely associated an arithmetic invariant $[e_1 e_2 e_3]$; this is the characteristic of the vector-function $\mathfrak{L}$ that corresponds to this form. For brevity we call it the *characteristic of the form*. Two equivalent forms have the same characteristic. Let us agree to include in the same class not only forms that are equivalent to one another, but also those forms for which the matrices (33) can be reduced to matrices similar in structure to the Jordan canonical forms. As a result we can classify the V_3 according to their differential invariants.

Let $\phi(\lambda) = |\lambda E - \mathfrak{L}|$, let ρ_2 be the rank of the matrix $\phi'(\mathfrak{L})$ and let ρ_1 be the greatest value of the ranks of $\phi'(\mathfrak{L})$, $\phi''(\mathfrak{L})$. If we bear in mind the lemma of § 3 concerning a new characterization of vector-functions, we obtain the following

Theorem I. *The set of nonsingular ternary differential quadratic forms*

$$ds^2 = g_{ij}(u)\, du^i du^j \qquad (ij = 1,2,3) \tag{34}$$

is divided into six nonempty, distinct classes, which are characterized respectively by the conditions

$$\begin{array}{lll} \text{I.}\ (\varrho_1\varrho_2) = (33) & H\varphi'(\mathfrak{L}) \neq 0, & \\ \text{II.}\ (\varrho_1\varrho_2) = (32) & H\varphi'(\mathfrak{L}) = 0, & H \neq 0. \\ \text{III.}\ (\varrho_1\varrho_2) = (31) & H = 0, & \varphi'^2(\mathfrak{L}) \neq 0. \\ \text{IV.}\ (\varrho_1\varrho_2) = (21) & \varphi'^2(\mathfrak{L}) = 0, & \varphi'(\mathfrak{L}) \neq 0. \\ \text{V.}\ (\varrho_1\varrho_2) = (10) & \varphi''^2(\mathfrak{L}) = 0, & \varphi''(\mathfrak{L}) \neq 0. \\ \text{VI.}\ (\varrho_1\varrho_2) = (00) & \varphi''(\mathfrak{L}) = 0. & \end{array} \tag{35}$$

This theorem gives necessary and sufficient tests that allow us to decide when a form of the type (34) belongs to one or another of the classes I – VI. Any criterion from (35) is an explicit, compatible differential system for investigating the fundamental tensor $g_{ij}(u)$ of a space of a definite class. For this reason Theorem I provides a new means of investigating the differential properties of the 3-dimensional Riemannian manifolds V_3.

5. As is well known, the conformally flat Riemannian spaces C_3 are obtained from the set of all V_3 by the additional condition

$$L_{i[j,k]} = 0. \tag{36}$$

Because of this, the method of classifying the manifolds V_3 in terms of the characteristics of their metric forms ds^2, in the case of the spaces C_3, is applied to the elements of the matrix $\mathfrak{L}$ with the conditions (36) taken into account; it leads to the following

Theorem II. *There are six types of spaces C_3; they are characterized respectively by the conditions*

$$\boxed{L_{i[j,k]} = 0} \tag{37}$$

I II III IV V VI

According to Theorem II each class of spaces C_3 is defined analytically by (36) and by one of the conditions I – VI of the series (35). In its form this deduction is analogous to the preceding one, although we do not have the right, in this case, to use the theorem on the mutual determinacy, when $n = 3$, of g_{ij} and R_{hijk}; the arguments in § 4 were essentially based on this theorem. But from where does it follow that any of the six differential systems of the series (37) has a solution? In other words, from where do we know that each of the six series of conditions (37) is not only necessary but is also sufficient to construct the tensor g_{ij} of the corresponding type of conformally flat 3-dimensional Riemannian space? This question has a positive solution in view of the examples, constructed in [4], that illustrate the existence of all of the six types of spaces C_3 enumerated in Theorem II. Obviously, this again implies the existence of the class of spaces V_3 mentioned in Theorem I.

6. The Riemannian space V_n for which

$$R_{ij} - \frac{R}{n} g_{ij} = 0, \tag{38}$$

is said to be an Einstein space [8]. We have reduced [5] the problem of classifying Einstein spaces when $n = 4$ to that of finding, in euclidean space, the inequivalent types of two symmetric matrices with zero trace: $T = (T_{ij})$, $\overline{T} = (T_{\bar{i}\bar{j}})$ $(i, j = 1, 2, 3)$. When the signature s of the fundamental form ds^2 of an Einstein space satisfies $s = -2$, the elements T_{ij}, $T_{\bar{i}\bar{j}}$ turn out to be complex conjugate linear homogeneous functions that form the curvature tensor relative to a local orthonormal coordinate system. On this basis, in Theorem 28 of [5] we asserted the existence of three types of 4-dimensional Einstein spaces of singnature $s = -2$; these correspond, respectively, to the following characteristics of the λ-matrices $\lambda E - T$ and $\lambda\overline{E} - \overline{T}$:

$$\text{I. } [111], [\overline{111}]. \quad \text{II. } [12], [\overline{12}]. \quad \text{III. } [3], [\overline{3}]. \tag{39}$$

This statement, in conjuction with the lemma in § 3, leads to a qualitatively new theorem that finally resolves, in a well-known sense, the problem of classifying 4-dimensional Einstein spaces. By using the notation obtained from that used in § 3 above by simply replacing A by T, $\overline{T}$ respectively, we can state this theorem as

Theorem III. *When $n = 4$ and $s = -2$ there are three types of Einstain spaces. The arithmetic and analytic characteristics of these types are given respectively by the scheme:*

$$\boxed{R_{ij} - \frac{R}{4} g_{ij} = 0}$$

— I
- (33) $H\varphi'(T) \neq 0$, *if* $[111]$, $[\overline{1}\overline{1}\overline{1}]$;
- (31) $H = 0$, $\varphi'^2(T) \neq 0$, *if* $[(11)1]$, $[(\overline{1}\overline{1})\overline{1}]$;
- (00) $\varphi''(T) = 0$, *if* $[(111)]$, $[(\overline{1}\overline{1}\overline{1})]$. (40.1)

— II
- (32) $H\varphi'(T) = 0$, *if* $[12]$, $[\overline{1}\overline{2}]$;
- (10) $\varphi''^2(T) = 0$, $\varphi''(T) \neq 0$, *if* $[(21)]$, $[(\overline{2}\overline{1})]$. (40.2)

— III (21) $\varphi'^2(T) = 0$, $\varphi'(T) \neq 0$, *if* $[3]$, $[\overline{3}]$. (40.3)

The following examples of forms show the compatibility of the differential conditions (40.1), (40.2), (40.3) imposed on the fundamental tensors:

$$ds^2 = -(du')^2 - \left(\frac{\operatorname{sh} \frac{3u'}{2}}{\sqrt[3]{\operatorname{ch} \frac{3u'}{2}}} \right)^2 (du^2)^2 + 2\left(\operatorname{ch} \frac{3u^1}{2}\right)^{\frac{4}{3}} du^3 du^4,$$

$$ds^2 = -\cos^2 u^4 du'^2 - e^{2u^4} du^{2^2} + 2du^3 du^4, \tag{41}$$

$$ds^2 = -e^{\alpha u^4 + \beta}(2du' du^3 + Adu^{2^2}) + 2(B_1 e^{\alpha u^4} + B_2 e^{-\frac{\alpha}{2} u^4}) du^2 du^3 +$$

$$+\left(C_1e^{\alpha u^4}+C_2e^{-\frac{\alpha}{2}u^4}-\frac{B_2^2}{2A}e^{-2\alpha u^4-\beta}\right)du^{3^2}-du^{4^2},$$

where α, β, A, B_i, C_i are constants, and

$$\alpha>0, \qquad A>0, \qquad C_1e^{\alpha u^4}+C_2e^{-\frac{\alpha}{2}u^4}-\frac{B_2^2}{2A}e^{-2\alpha u^4-\beta}<0,$$

which have, respectively, characteristics from the series (39). Each of the three compatible differential systems given in Theorem III is a necessary and sufficient condition for the metric form ds^2 of an Einstein space to belong to one of these three classes; it serves to investigate the fundamental tensor $g_{ij}(u)$ of a particular class of Einstein spaces. This situation gives rise to the possibility of studying individually any of the three classes of spaces lying at the foundations of the construction of the principle of relativity.

None of the memoirs devoted to the problem of classifying Einstein spaces contains a characterization of the metric tensors of the spaces of a selected class; thus an exhaustive solution of the problem has not been given.

BIBLIOGRAPHY

[1] C. F. Gauss, *Disquisitiones generales circa superficies curvas*, Comment Soc. Sci. Gottingen Recent. Comment. Cl. Math. 6 (1823–1827), 99–146.

[2] G. F. B. Riemann, *Über die Hypothesen, welche der Geometrie zu Grunde liegen*, Abh. Ges. Wiss. Göttingen 13 (1868), 1–20.

[3] P. I. Petrov, *A classification of 3-dimensional Riemannian spaces according to their differential invariants*, Učen. Zap. Kazan. Gos. Univ. 117 (1957), no. 9, 119–121. (Russian)

[4] ———, *Classification des espaces riemanniens à trois dimensions conformément plats*, Czechoslovak Math. J. 11 (86) (1961), 161–164. (Russian) MR 23 #A599.

[5] ———, *Invariants and classification of differential quadratic forms in four variables*, Izv. Akad. Nauk SSSR Ser. Mat. 23 (1959), 387–420. (Russian) MR 22 #2947.

[6] P. A. Širokov, *Tensor calculus*. Chapter III, Izdat. Kazan. Univ., Kazan, 1934; 2nd ed., 1961. (Russian) MR 29 #5187.

[7] G. B. Gurevič, *Foundations of the theory of algebraic invariants*, OGIZ, Moscow, 1948; p. 338, problem no. 11; English transl., Noordhoff, Groningen, 1964. MR **11**,413; MR 32 #1211.

[8] L. P. Eisenhart, *Riemannian geometry*, Princeton Univ. Press, Princeton, N. J., 1926; 2nd ed., 1949; Russian transl., IL, Moscow, 1948, p. 117. MR 11,687.

Translated by:
L. W. Longdon

Amer. Math. Soc. Transl.
(2) Vol. 92, 1970

AN INVARIANT TENSOR CRITERION FOR CONFORMALLY REDUCIBLE RIEMANNIAN SPACES*

UDC 513.813

A. M. ANČIKOV

In the present paper we find an invariant criterion for the conformally reducible Riemannian spaces V_n, and as a consequence of this we obtain criteria for reducible and semireducible spaces. We state the invariant criterion for reducible Riemannian spaces given in [1]:

Theorem. *For the Riemannian space* V_n *to be reducible it is necessary and sufficient that there exist a covariantly constant, symmetric, idempotent tensor* $c_{\alpha\beta}$ *that is not equal to the metric tensor:*[1)]

$$1)\ c_{\alpha\beta,\gamma}=0,\quad 2)\ c_{\alpha\beta}\neq g_{\alpha\beta},\quad 3)\ c_{[\alpha\beta]}=0,\quad 4)\ c_{\alpha\gamma}c^{\gamma}_{\beta}=c_{\alpha\beta}. \qquad \text{(I)}$$

In the system of coordinates in which the metric is written in the form

$$\begin{gathered} ds^2=ds_1^2+ds_2^2,\\ ds_1^2=g_{i_1j_1}(x^{k_1})\,dx^{i_1}\,dx^{j_1}\quad (i_1,\ j_1=1,\ 2,\dots,\ p),\\ ds_2^2=g_{i_2j_2}(x^{k_2})\,dx^{i_2}\,dx^{j_2}\quad i_2,\ j_2=p+1,\ p+2,\dots,\ n), \end{gathered} \qquad \text{(II)}$$

$c_{\alpha\beta}$ *is the fundamental tensor of one of the families of parallel geodesic surfaces.*

This criterion is generalized in [2] to the case of semireducible Riemannian spaces:

$$ds^2=ds_1^2+\sigma ds_2^2, \qquad \text{(III)}$$

where

$$\begin{gathered} ds_1^2=g_{i_1j_1}(x^{k_1})\,dx^{i_1}\,dx^{j_1}\quad (i_1,\ j_1=1,\ 2,\dots,\ p),\\ ds_2^2=g_{i_2j_2}(x^{k_2})\,dx^{i_2}\,dx^{j_2}\quad (i_2,\ j_2=p+1,\ p+2,\dots,\ n).\\ \sigma=\sigma(x^1,\ x^2,\dots,\ x^p). \end{gathered}$$

In this case the following theorem holds

Theorem. *For the Riemannian space* V_n *to be semireducible it is necessary and sufficient that there exist a symmetric tensor* $c_{\alpha\beta}$, *not proportional*

---

* Translation of Izv. Vysš. Učebn. Zaved. Matematika 1965, no. 5 (48), 13–18.

1) We can take it to be not proportional to the metric tensor.

to the metric tensor, that satisfies, jointly with some gradient $u_{,\alpha}$, *the equations*

$$c_{\alpha\beta,\gamma} = -\frac{1}{2}(u_{,\alpha}c_{\beta\gamma} + u_{,\beta}c_{\alpha\gamma}) \tag{IV}$$

and the condition

$$c_{\alpha\gamma}c^{\gamma}_{\beta} = c_{\alpha\beta}. \tag{V}$$

In [2] the sufficiency of this theorem was proved only for spaces with a definite metric, and the author considered criteria for certain special semi-reducible spaces. In the present article we are going to consider an invariant criterion for conformally reducible Riemannian spaces. First we give a definition of these spaces.

Definition. The Riemannian space V_n is said to be *conformally reducible* if, in some system of coordinates, its metric has the form

$$ds^2 = u^2(ds_1^2 + ds_2^2), \tag{VI}$$

where $u = u(x^1, x^2, \cdots, x^n)$, and ds_1^2 and ds_2^2 are defined as in (II).

§1. Conditions for conformal reducibility

We are going to consider the conformally reducible space (VI) and look for an invariant criterion that defines this space. Because the space with the metric (VI) is conformal to the reducible space (II), in finding a criterion for conformal reducibility we can use a conformal mapping of the conditions for reducibility (I); that is, we will write these conditions in the space (VI).

Theorem 1.1. *For the Riemannian space* V_n *to be conformally reducible it is necessary and sufficient that these exist a symmetric tensor* $c_{\alpha\beta}$, *not proportional to the metric tensor, that satisfies, jointly with some scalar* u, *the equations*

$$c_{\alpha\beta,\gamma} = -2c_{\alpha\beta}\sigma_\gamma - (c_{\alpha\gamma}\sigma_\beta + c_{\beta\gamma}\sigma_\alpha) + \sigma_\tau(c^\tau_\alpha g_{\beta\gamma} + c^\tau_\beta g_{\alpha\gamma}), \tag{1.1}$$

$$u^2 c_{\alpha\gamma}c^\gamma_\beta = c_{\alpha\beta}, \tag{1.2}$$

where $\sigma_\alpha = \partial_\alpha \ln u$.

Proof. Necessity. Let us take a symmetric tensor $c_{\alpha\beta}$ (not proportional to the metric tensor) whose covariant derivative in (VI) has the form

$$\begin{aligned} c_{\alpha\beta,\gamma} = c_{\alpha\beta;\gamma} &- 2c_{\alpha\beta}\sigma_\gamma - (c_{\alpha\gamma}\sigma_\beta + c_{\beta\gamma}\sigma_\alpha) \\ &+ \sigma_\tau(c^\tau_\alpha g_{\beta\gamma} + c^\tau_\beta g_{\alpha\gamma}), \end{aligned} \tag{1.3}$$

where $\sigma_\alpha = \partial_\alpha \ln u$ and the semicolon denotes the covariant derivative with respect to the metric tensor of the reducible space. We assume that in the reducible space (II) $c_{\alpha\beta}$ satisfies the conditions (I). Hence $c_{\alpha\beta;\gamma} = 0$; that is, we obtain (1.1). In our case the idempotency conditions (4) in (I) can be written as

$$u^2 c_{\alpha\gamma} c^{\gamma}_{\beta} = c_{\alpha\beta}. \tag{1.4}$$

In fact $g_{\alpha\beta} = u^2 \overset{*}{g}_{\alpha\beta}$ ($\overset{*}{g}_{\alpha\beta}$ is the metric tensor of the reducible space) and $\overset{*}{g}{}^{\alpha\beta} = u^2 g^{\alpha\beta}$. By using the last relation we obtain (1.4).

Sufficiency. It follows from (1.2) that $g_{\alpha\beta} = u^2 \overset{*}{g}_{\alpha\beta}$. Next, by writing down the covariant derivative of the tensor $c_{\alpha\beta}$ with respect to the metric tensor $g_{\alpha\beta}$, we obtain (1.3), where $c_{\alpha\beta;\gamma}$ is the covariant derivative in the space whose metric tensor is $\overset{*}{g}_{\alpha\beta}$. In view of (1.1) it follows from here that

$$c_{\alpha\beta;\gamma} = 0; \tag{1.5}$$

in addition, $c_{\alpha\beta}$ satisfies the idempotency condition in the space with the metric tensor $\overset{*}{g}_{\alpha\beta}$. By a theorem of P. A. Širokov, it follows, from (1.5) and the idempotency condition $c_{\alpha\gamma} \overset{*}{c}{}^{\gamma}_{\beta} = c_{\alpha\beta}$ for a symmetric tensor $c_{\alpha\beta}$ not proportional to the metric tensor $\overset{*}{g}_{\alpha\beta}$, that $\overset{*}{g}_{\alpha\beta}$ is the metric tensor of a reducible space. Hence V_n with the fundamental tensor $g_{\alpha\beta}$ is a conformally reducible space. The sufficiency is proved.

It is easy to show that, if we use the tensor $\overset{*}{c}_{\alpha\beta} = u^2 c_{\alpha\beta}$ instead of $c_{\alpha\beta}$, we can state Theorem 1.1 in the following way:

Theorem 1.2.[1] *For the Riemannian space V_n to be conformally reducible it is necessary and sufficient that these exist a symmetric tensor $c_{\alpha\beta}$, not proportional to the metric tensor, that satisfies, jointly with some scalar u, the conditions*

$$c_{\alpha\beta,\gamma} = -(c_{\alpha\gamma}\sigma_{\beta} + c_{\beta\gamma}\sigma_{\alpha}) + \sigma_{\tau}(c^{\tau}_{\alpha} g_{\beta\gamma} + c^{\tau}_{\beta} g_{\alpha\gamma}), \tag{1.6}$$

$$c_{\alpha\gamma} c^{\gamma}_{\beta} = c_{\alpha\beta}, \tag{1.7}$$

where $\sigma_{\alpha} = \partial_{\alpha} \ln u$.

The equations (1.6) under the condition (1.7) will be said to be the equations for the conformal reducibility of the Riemannian space V_n. The integrability conditions for (1.5) are written as

$$c_{\alpha\tau} R^{\tau}_{\beta\gamma\delta} + c_{\tau\beta} R^{\tau}_{\alpha\gamma\delta} = \sigma_{\beta\gamma} c_{\alpha\delta} - \sigma_{\beta\delta} c_{\alpha\gamma}$$
$$+ \sigma_{\alpha\gamma} c_{\beta\delta} - \sigma_{\alpha\delta} c_{\beta\gamma} + \sigma_{\tau\delta}(g_{\alpha\gamma} c^{\tau}_{\beta} + g_{\beta\gamma} c^{\tau}_{\alpha}) -$$

---

[1] Here we have omitted the sign * over the tensor $\overset{*}{c}_{\alpha\beta}$.

$$- \sigma_{\tau\gamma} (c^{\tau}_{\alpha} g_{\beta\delta} + c^{\tau}_{\beta} g_{\alpha\delta}) +$$

$$+ \Delta_1 \sigma [g_{\beta\delta} c_{\alpha\gamma} - g_{\beta\gamma} c_{\alpha\delta} + g_{\alpha\delta} c_{\beta\gamma} - g_{\alpha\gamma} c_{\beta\delta}], \tag{1.8}$$

where $\Delta_1 \sigma = \sigma_\tau \sigma^\tau = g^{\alpha\beta} \sigma_\alpha \sigma_\beta$ is the Beltrami first differential parameter, $\sigma_{\alpha\beta} = \sigma_{,\alpha\beta} + \sigma_\alpha \sigma_\beta$, and $R^{\alpha}_{\beta\gamma\delta}$ is the curvature tensor of V_n.

As is well known ([3], Russian p. 45), if $g_{\alpha\beta}$ and $c_{\alpha\beta}$ are the components of two symmetric tensors and if

$$g_{\beta\delta} c_{\alpha\gamma} - g_{\beta\gamma} c_{\alpha\delta} + g_{\alpha\delta} c_{\beta\gamma} - g_{\alpha\gamma} c_{\beta\delta} = 0$$

$$(\alpha, \beta, \gamma, \delta = 1, 2, \ldots, n), \tag{1.9}$$

then these tensors are proportional; that is,

$$c_{\alpha\beta} = \rho g_{\alpha\beta}. \tag{1.10}$$

The expression in the square brackets at (1.8) cannot be equated to zero, for otherwise, by (1.9) and (1.10), we would have a tensor $c_{\alpha\beta}$ proportional to the metric tensor, and this is excluded from the considerations in Theorems 1.1 and 1.2.[1)]

§2. Criteria for certain special conformally reducible Riemannian spaces

We are going to consider the special type of conformally reducible spaces when one of the ds_1^2 and ds_2^2 is the metric of a one-dimensional space; for example, when $ds_1^2 = e_1 dx^{1^2}$, where $e_1 = \pm 1$; that is, we are going to investigate spaces with the metric

$$ds^2 = u^2 (e_1 dx^{1^2} + g_{ij}\, dx^i dx^j),$$

$$e_1 = \pm 1, \quad u = u(x^1, x^2, \ldots, x^n),$$

$$g_{ij} = g_{ij}(x^2, x^3, \ldots, x^n) \quad (i, j = 2, 3, \ldots, n). \tag{2.1}$$

In this case we can prove a theorem that is analogous to Theorem 1.2 but different from it.

Theorem 2. *For V_n to be a conformally reducible space of the form (2.1) it is necessary and sufficient that there exist a nonisotropic vector field c_α that, jointly with some scalar u, satisfies the equations*

$$c_{\alpha,\beta} = -(c_\alpha \sigma_\beta + c_\beta \sigma_\alpha) + g_{\alpha\beta} c^\tau \sigma_\tau. \tag{2.2}$$

---

1)Let us note that the integrability conditions for equations (1.1) also coincide with (1.8).

In proving this theorem we will use the following well-known fact: for the Riemannian space V_n to be a reducible space of the form

$$d\bar{s}^2 = e_1 dx^{1^2} + g_{ij} dx^i dx^j,$$
$$g_{ij} = g_{ij}(x^2, x^3, \dots, x^n) \quad (i, j = 2, 3, \dots, n), \tag{2.3}$$

it is necessary and sufficient that there exist a nonisotropic, covariantly constant vector field c_α: $c_{\alpha;\beta} = 0$.

Proof. Necessity. We take a nonisotropic vector field c_α and write down the covariant derivative with respect to the metric (2.1):

$$c_{\alpha,\beta} = \partial_\beta c_\alpha - c_\tau \bar{\Gamma}^\tau_{\alpha\beta} - c_\tau \delta^\tau_\alpha \sigma_\beta - c_\tau \delta^\tau_\beta \sigma_\alpha + c^\tau \sigma_\tau g_{\alpha\beta}, \tag{2.4}$$

where $\bar{\Gamma}^\tau_{\alpha\beta}$ is the Christoffel symbol for the metric (2.3) and $\sigma_\alpha = \partial_\alpha \ln u$. The first two terms on the right-hand side of (2.4) represent the covariant derivative of c_α with respect to the metric (2.3); this derivative is zero. Hence we obtain (2.2).

Sufficiency. It follows from (2.2) that $c_{\alpha,\beta} = c_{\beta,\alpha}$; that is, that c_α is a gradient vector: $c_\alpha = \partial_\alpha \lambda$. Consequently the vector field c_α is orthogonal to the surfaces $\lambda = \text{const}$, which cannot be isotropic since the vector field c_α is nonisotropic. Next, by introducing a system of coordinates in which $\lambda = \text{const}$ represent the surfaces $x^1 = \text{const}$ and the orthogonal trajectories are the coordinate lines x^1, we can write the metric in the form

$$ds^2 = \overset{*}{g}_{11}(x^1, x^2, \dots, x^n)\, dx^{1^2} + \overset{*}{g}_{ij}(x^1, x^2, \dots, x^n)\, dx^i dx^j, \quad (i, j = 2, 3, \dots, n), \tag{2.5}$$

or

$$ds^2 = u^2 (e_1 dx^{1^2} + g_{ij}\, dx^i dx^j), \tag{2.6}$$

where

$$e_1 u^2 = \overset{*}{g}_{11}, \; g_{ij} = \frac{\overset{*}{g}_{ij} e_1}{\overset{*}{g}_{11}} \quad (i, j = 2, 3, \dots, n),$$

g_{ij} and u^2 depend on all the variables $x^1, x^2, \cdots, x^n$.

We write the covariant derivative of c_α with respect to the metric (2.6) in the form (2.4). From here, by (2.2), we have

$$c_{\alpha;\beta} = \partial_\beta c_\alpha - c_\tau \bar{\Gamma}^\tau_{\alpha\beta} = 0.$$

This is equivalent to saying that the expression in brackets on the right-hand side of (2.6) represents the metric of a reducible space; that is, $\partial_1 g_{ij} = 0$. Hence (2.6) is the metric of a conformally reducible space.

The integrability conditions for equations (2.2) have the form

$$c_\tau R^\tau_{\alpha\beta\gamma} = c_\gamma \sigma_{\alpha\beta} - c_\beta \sigma_{\alpha\gamma} + c^\tau (g_{\alpha\beta} \sigma_{\tau\gamma} - g_{\alpha\gamma} \sigma_{\tau\beta})$$
$$+ \Delta_1 \sigma (g_{\alpha\gamma} c_\beta - g_{\alpha\beta} c_\gamma), \quad \Delta_1 \sigma = \sigma_\tau \sigma^\tau .$$

§3. On semireducible spaces V_n

In general, the coefficients of conformality u^2 in the metric (VI) depends on all the coordinates $x^1, x^2, \cdots, x^n$. If we assume that u is a function of the variables of only one of the metrics ds_1^2 and ds_2^2, then we obtain a particular case of conformally reducible spaces; these spaces are said to be semireducible [2]. Let us take, for example, the case when $u = u(x^1, x^2, \cdots, x^p)$ and the metric is written in the form

$$ds^2 = u^2 (ds_1^2 + ds_2^2), \tag{3.1}$$

where ds_1^2 and ds_2^2 depend on $x^1, x^2, \cdots, x^p$ and $x^{p+1}, x^{p+2}, \cdots, x^n$ respectively. This metric coincides with (III) if $\sigma = u^2$, and the ds_1^2 from (III) is equal to the $u^2 ds_1^2$ in (3.1).

In this case the conditions (1.6) and (1.7) for conformal reducibility reduce to

$$c_{\alpha\beta,\gamma} = -(c_{\alpha\gamma}\sigma_\beta + c_{\beta\gamma}\sigma_\alpha), \tag{3.2}$$

$$c_{\alpha\gamma} c^\gamma_\beta = c_{\alpha\beta} \tag{3.3}$$

and, in addition, in the system of coordinates in which the metric is written in the form (3.1), $c_{\alpha\beta}$ coincides with the "part" of the metric tensor corresponding to the metric $u^2 ds_2^2$. Hence we have

Theorem 3.1. *For V_n to be a conformally reducible space of the form (3.1) (or semireducible), where*

$$ds_1^2 = g_{i_1 j_1} (x^1, x^2, \ldots, x^p)\, dx^{i_1} dx^{j_1} \quad (i_1, j_1 = 1, 2, \ldots, p),$$
$$ds_2^2 = g_{i_2 j_2} (x^{p+1}, x^{p+2}, \ldots, x^n)\, dx^{i_2} dx^{j_2} \quad (i_2, j_2 = p+1, p+2, \ldots, n),$$

and u is a function of the variables x^{i_1}, it is necessary and sufficient that there exist a symmetric tensor $c_{\alpha\beta}$, not proportional to the metric tensor, that satisfies, jointly with some scalar u, the conditions (3.2) *and* (3.3), *where $\sigma_\alpha = \partial_\alpha \ln u$.*

We next consider corollaries of Theorem 2. We make the following assumptions regarding the function u in (2.1):

$$\text{A)} \quad u = u(x^1),$$
$$\text{B)} \quad u = u(x^2, x^3, \ldots, x^n). \tag{3.4}$$

For case A), if we introduce the new vector $\overset{*}{c}_\alpha = u^2 c_\alpha$, the following is a corollary of Theorem 2.

Theorem 3.2.[1)] *For V_n to be a conformally reducible space of the form* (2.1) *with* $u = u(x^1)$ (*or semireducible*), *it is necessary and sufficient that there exist scalars ϕ and u satisfying the equations*

$$\varphi_{,\alpha\beta} = g_{\alpha\beta}\varphi^{\tau}\sigma_{\tau}, \tag{3.5}$$

where $\sigma_\alpha = \partial_\alpha \ln u$. [2].

In case B) the equations (2.2) are rewritten in the form

$$c_{\alpha,\beta} = -(c_\alpha \sigma_\beta + c_\beta \sigma_\alpha), \tag{3.6}$$

or, setting $\overset{*}{c}_\alpha = u^2 c_\alpha$, reduce to

$$\overset{*}{c}_{\alpha,\beta} = \overset{*}{c}_\alpha \sigma_\beta - \overset{*}{c}_\beta \sigma_\alpha. \tag{3.7}$$

From here we obtain the following as a second corollary of Theorem 2.

Theorem 3.3. *For V_n to be a conformally reducible space of the form* (2.1) *with* $u = u(x^2, x^3, \cdots, x^n)$ (*or semireducible*), *it is necessary and sufficient that there exist a vector field c_α (or $\overset{*}{c}_\alpha$) which, jointly with some scalar u, satisfies* (3.6) (*or* (3.7)), *where* $\sigma_\alpha = \partial_\alpha \ln u$.

In (3.6) c_α is a gradient. Therefore in Theorem 3.3 we can talk about a gradient instead of the vector c_α.

In conclusion let us note that the last three theorems were considered in [2] (in another form); however, as we mentioned earlier, the sufficiency of the theorem on semireducibility is proved in [2] only for a definite metric and, moreover, the proof is clumsy. In the present article we have obtained the theorems proved in [2] as corollaries of a criterion for the conformal reducibility of V_n.

BIBLIOGRAPHY

[1] P. Širokov, *Symmetric conformally euclidean spaces*, Izv. Fiz.-Mat. Obšč. Naučn.-Issled. Inst. Mat. Meh. Kazan. Univ. (3) 11 (1938), 9–27.

[2] G. I. Kručkovič, *On a class of Riemannian spaces*, Trudy Sem. Vektor. Tenzor. Anal. 11 (1961), 103-128; erratum, ibid. 12 (1963), 95. MR 28 #5401.

[3] L. P. Eisenhart, *Riemannian geometry*, Princeton Univ. Press, Princeton, N. J., 1926; 2nd ed., 1949; Russian transl., IL, Moscow, 1948. MR 11, 687.

Translated by L. W. Longdon

---

1)a) We have taken $\phi_{,\alpha}$ instead of $\overset{*}{c}_\alpha$ in the theorem, because $\overset{*}{c}_\alpha$ is a gradient. b) If, in the case of an indefinite metric, the gradients ϕ_α and σ_β are orthogonal ($\phi^\alpha\sigma_\alpha = 0$), we require that ϕ_α be nonisotropic.

Amer. Math. Soc. Transl.
(2) Vol. 92, 1970

CONFORMALLY EUCLIDEAN GENERALIZED RIEMANNIAN SPACES*

UDC 513.813

S. M. BAHRAH

1°. By a generalized Riemannian space W_n we mean an n-dimensional elementary manifold in which there is defined the field of a covariant tensor g_{ij} of valency two that, in general, is nonsymmetric $(g_{ij} \neq g_{ji})$ and is such that $g_{ij}x^ix^j > 0$ for any nonzero vector x^i. The interest in these spaces arose in connection with a variant of the unified field theory proposed by Einstein in 1950.

A one-to-one correspondence between the points of two generalized Riemannian spaces W_n and $\tilde{W}_n$ is said to be conformal [1] if the metric tensors g_{ij} and $\tilde{g}_{ij}$ at corresponding points are connected by the equation

$$\tilde{g}_{ij} = e^{2\sigma} g_{ij}, \tag{1}$$

where the scalar σ depends only on the point. Equation (1) is equivalent to the two equations

$$\tilde{g}_{\underline{ij}} = e^{2\sigma} g_{\underline{ij}}, \quad \tilde{g}_{\underset{\smile}{ij}} = e^{2\sigma} g_{\underset{\smile}{ij}}, \tag{2}$$

where

$$g_{\underline{ij}} = \frac{g_{ij} + g_{ji}}{2}, \quad g_{\underset{\smile}{ij}} = \frac{g_{ij} - g_{ji}}{2}. \tag{3}$$

A generalized Riemannian space that admits a conformal mapping onto a generalized euclidean space will be said to be conformally euclidean [2]. The following theorem, established in [2], plays an essential role in what is to follow.

Theorem 1. *For $\tilde{W}_n$ with the metric tensor g_{ij} to be a generalized euclidean space it is necessary and sufficient that*

$$\tilde{\rho}_{ijkl} = 0, \quad \tilde{g}_{ij\|k} = 0, \tag{4}$$

where $\underset{\sim}{\|}$ denotes absolute differentiation with the help of the Christoffel symbol $\Gamma^l_{\underset{\sim}{ij}}$, and $\tilde{\rho}_{ijkl}$ is the curvature tensor corresponding to the symmetric tensor $\tilde{g}_{\underline{ij}}$.

2°. We are going to establish necessary and sufficient conditions for the space W_n to be conformally euclidean. Suppose that W_n is conformally

* Translation of Izv. Vysš. Učebn. Zaved. Matematika 1965, no. 6(49), 23–26.

euclidean; that is, there exists a tensor $\tilde{g}_{ij}$, obtained from g_{ij} according to (1) and satisfying (4). From (1) and (4_2) we obtain

$$g_{\underset{\smile}{ij} \| k} + 2\sigma_k g_{ij} = 0, \quad \sigma_k = \frac{\partial \sigma}{\partial u^k}. \tag{5}$$

As is well known [3],

$$\tilde{\Gamma}^l_{ij} = \Gamma^l_{ij} + \delta^l_j \sigma_i + \delta^l_i \sigma_j - \overset{ls}{g^{-}} g_{\underset{-}{ij}} \sigma_s; \tag{6}$$

here Γ^l_{ij} is the Christoffel symbol corresponding to the symmetric tensor g_{ij}. By using (6) we can present (5) in the following form:

$$g_{\underset{\smile}{ij} | k} = g_{\underset{\smile}{ik}} \sigma_j - g_{\underset{\smile}{jk}} \sigma_i + g_{\underset{\smile}{jl}} \overset{ls}{g^{-}} g_{\underset{-}{ik}} \sigma_s - g_{\underset{\smile}{il}} \overset{ls}{g^{-}} g_{\underset{-}{jk}} \sigma_s, \tag{7}$$

where $|$ denotes absolute differentiation with the help of Γ^l_{ij}. Condition (4) is equivalent [3] to the following:

$$\rho^{\cdot}_{ijkl} = S_{jl} g_{\underset{-}{ik}} + S_{ik} g_{\underset{-}{jl}} - S_{il} g_{\underset{-}{jk}} - S_{jk} g_{\underset{-}{il}},$$

$$S_{jk} = \sigma_{j|k} - \sigma_j \sigma_k + \frac{1}{2} g_{\underset{-}{jk}} \overset{sl}{g^{-}} \sigma_s \sigma_l. \tag{8}$$

Thus, if W_n is conformally euclidean, then there exists a scalar field σ satisfying (7) and (9). It is easy to establish the sufficiency of our condition.

Theorem 2. *For a generalized Riemannian space to be conformally euclidean it is necessary and sufficient that there exists a scalar field σ satisfying* (7) *and* (8).

3°. Theorem 2 does not give an effective means of solving the problem of the conformal euclidean character of W_n. It is desirable to have conditions for a space to be conformally euclidean in the form of explicit restrictions on the tensor g_{ij} and on tensors constructed from it. The problem of finding these conditions is substantially easier in the case of a generalized Riemannian space of maximal rank. Let W^0_n denote an even-dimensional generalized Riemannian space if the rank of the skew-symmetric part of the metric tensor is equal to the dimension of the space; for W^0_n we thus have that $\mathrm{Det}\, \|g_{\underset{\smile}{ij}}\| \neq 0$.

We introduce the contravariant tensor $\overset{kp}{g^{\smile}}$ defined by

$$g_{\underset{\smile}{il}} \overset{lj}{g^{\smile}} = \delta^j_i . \tag{9}$$

To find conditions for W^0_n to be conformally euclidean we contract both sides of (7) with respect to i and k by using the tensor $\overset{ij}{g^{\smile}}$. Leaving the case when $n = 2$ for special consideration, we obtain

$$\sigma_j = \frac{1}{n-2} g_{sj\,|\,l} \overset{sl}{\underset{\smile}{g}}. \tag{10}$$

We form the integrability conditions for the system (10):

$$g_{si\,|\,lj} \overset{sl}{\underset{\smile}{g}} + g_{si\,|\,l} \overset{sl}{\underset{\smile}{g}}{}_{|\,j} = 0 \quad [i, j], \tag{11}$$

where $[i, j]$ denotes alternation with respect to the indices i and j.

Equations (11) together with the equations obtained by substituting the covector σ_j, constructed according to (10), in (7) and (8) (we will call these equations the conditions A and B) are a necessary condition for W_n^0 to be conformally euclidean. It is easy to establish that these conditions are sufficient.

Theorem 3. *For the generalized Riemannian space* W_n^0 $(n > 2)$ *of maximal rank to be conformally euclidean it is necessary and sufficient that its metric tensor* g_{ij} *satisfies conditions* (11), (A) *and* (B).

The space W_2^0 can be conformally euclidean only when it is a generalized euclidean space.

4°. Returning to the general case let us note that it follows from (7) that

$$\varphi_{ijk} = \underset{\smile}{g}_{[ik}\sigma_{j]}, \tag{12}$$

where

$$\varphi_{ijk} = \underset{\smile}{g}_{ij\,|\,k} + \underset{\smile}{g}_{jk\,|\,i} + \underset{\smile}{g}_{ki\,|\,j}. \tag{13}$$

We multiply both sides of (12) by g_{pq} and alternate with respect to k, p and q; we multiply both sides of the resulting equation by g_{fl} and alternate with respect to i, f and l. By using (12) we obtain

$$\sigma_j \omega_\alpha = \pi_{j\beta}, \tag{14}$$

where

$$\omega_\alpha \equiv \omega_{flikpq} = \underset{\smile}{g}_{[fl}\underset{\smile}{g}_{i][k}\underset{\smile}{g}_{pq]}, \tag{15}$$

$$\pi_{j\beta} \equiv \pi_{jflikpq} = \underset{\smile}{g}_{[fl}\varphi_{i]j[k}\underset{\smile}{g}_{pq]} + \underset{\smile}{g}_{j[k}\underset{\smile}{g}_{pq]}\varphi_{fil} + \underset{\smile}{g}_{[fl}\underset{\smile}{g}_{i]j}\varphi_{pkq}.^{1)} \tag{16}$$

We consider a space W_n such that

$$\pi_{j\beta} \neq 0 \tag{17}$$

(a nonsingular generalized Riemannian space). Then $\omega_\alpha \neq 0$, since otherwise the space would not be conformally euclidean. In this case, when the follow-

1) Here and in what follows the Greek indices are collective; they replace the group of six Latin indices.

ing condition holds:

$$\pi_{i\alpha}\pi_{j\beta} = \pi_{i\beta}\pi_{j\alpha} \tag{18}$$

the covector σ_j is determined uniquely from (14) [4]. We can write the integrability conditions for the system (14) as

$$\omega_{\alpha\,|\,[j}\pi_{i]\beta} = (\pi_{i\alpha\,|\,j} - \pi_{j\alpha\,|\,i})\,\omega_\beta\,. \tag{19}$$

By multiplying both sides of (7) by the tensor ω_α and by using (14), we obtain

$$\omega_\alpha g_{\underset{\smile}{ij}\,|\,k} = \pi_{j\alpha} g_{\underset{\smile}{ik}} - \pi_{i\alpha} g_{\underset{\smile}{jk}} + g_{\underset{\smile}{jl}} \overset{ls}{\underline{g}} g_{\underline{ik}}\pi_{s\alpha} - g_{\underset{\smile}{il}} \overset{ls}{\underline{g}} g_{\underline{jk}}\omega_{s\alpha}\,. \tag{20}$$

It follows from (14) that

$$\omega_\alpha\omega_\beta\sigma_{i\,|\,j} = \omega_\beta\pi_{i\alpha\,|\,j} - \omega_{\alpha\,|\,j}\pi_{i\beta}\,. \tag{21}$$

We use (14) and (21). Then

$$S_{jk}\omega_\alpha\omega_\beta = \omega_\beta\pi_{j\alpha\,|\,k} - \omega_{\alpha\,|\,k}\pi_{j\beta} - \pi_{j\alpha}\pi_{k\beta} + \frac{1}{2} g_{\underline{jk}} \overset{sp}{\underline{g}} \pi_{s\alpha}\pi_{p\beta}\,. \tag{22}$$

It follows from (8) and (22) that

$$\begin{aligned} \rho_{ijkl}\omega_\alpha\omega_\beta = {} & \omega_\beta g_{\underline{l[i}}\pi_{j]\alpha\,|\,k} - g_{\underline{l[i}}\pi_{j]\beta}\,\omega_{\alpha\,|\,k} - g_{\underline{l[i}}\pi_{j]\alpha}\pi_{k\beta} + g_{\underline{l[i}} g_{\underline{j]k}} \overset{sp}{\underline{g}} \pi_{s\alpha}\pi_{p\beta} \\ & + \omega_\alpha g_{\underline{k[i}}\pi_{j]\alpha\,|\,l} - g_{\underline{k[i}}\pi_{j]\beta}\,\omega_{\alpha\,|\,l} - g_{\underline{k[i}}\pi_{j]\alpha}\pi_{l\beta}\,. \end{aligned} \tag{23}$$

Thus (18), (19), (20) and (23) are necessary conditions for the nonsingular generalized Riemannian space to be conformally euclidean. We can show that the stated conditions are also sufficient.

5°. We make a remark concerning singular Riemannian spaces. A generalized Riemannian space appears, in particular, on surfaces in the generalized euclidean space H_n. It was established in [5] that the metric tensor g_{ij} of a surface in H_n is subject to the condition $g_{\check{ij}|k} + g_{\check{jk}|i} + g_{\check{ki}|j} = 0$.

Thus, the generalized Riemannian spaces that appear on surfaces in H_n are singular. If here the tensor $\omega_\alpha \neq 0$, then the spaces W_n can be conformally euclidean only when they are flat.

BIBLIOGRAPHY

[1] V. Murgescu, *Propriétés conformes des espaces de Riemann généralisés*, I, Bull. Inst. Politehn. Iași 4 (8) (1958), no. 3/4, 77–80. MR 22 #1858.

[2] S. M. Bahrah, *The theory of hypersurfaces of a generalized euclidean space and generalized Riemannian spaces of first class*, Izv. Vysš. Učebn. Zaved. Matematika 1960, no. 3 (16), 54–61. (Russian) MR 24 #A1674.

[3] P. K. Raševskiĭ, *Riemannian geometry and tensor analysis*, 2nd. ed., "Nauka," Moscow, 1964. (Russian) MR 28 #4437.

[4] A. M. Lopšic, *Some problems of tensor algebra in linear dimensionless spaces*, Trudy Sem. Vektor. Tenzor. Anal. 6 (1948), 365–419. (Russian) MR 15,283.

[5] S. M. Bahrah, *The geometry of generalized Euclidean spaces*, Candidate's Dissertation, Moscow, 1961. (Russian).

Translated by:
L. W. Longdon

Amer. Math. Soc. Transl.
(2) Vol. 92, 1970

COMPLETELY REDUCIBLE CONNECTIONS *

V. I. VEDERNIKOV

The main content of the present article arose as a result of explaining the algebraic basis of the concept of the conjugate connections introduced by A. P. Norden [1]. It turns out that this concept is closely related to the concept of the connections that arise naturally in defining connections in principal fiber spaces of a certain type, which we call completely reducible. Only linear connections are considered in detail in the present article, although the results can be carried over to the general case of an infinitesimal connection. The reason for this is that we intend to apply our results to problems of classical differential geometry: affine, projective, conformal, etc.; therefore we restrict ourselves to considerations in the small; that is, we assume that the fiber spaces being considered are represented by the direct product of a neighborhood of a point and a fiber. However, this restriction is not necessary.

1. We are going to consider a group G that is a subgroup of the direct product $G_1 \times G_2$ of groups G_1 and G_2. The elements of G are pairs (g_1, g_2), where $g_1 \in G_1$ and $g_2 \in G_2$. We consider only those groups G for which the condition of "essentiality" holds for G_1 and G_2: for each $g_1 \in G_1$ $(g_2 \in G_2)$ there is an element of G of the form (g_1, g_2') $((g_1', g_2))$. The set H_1 of all elements of the form (g_1, e_2), where e_2 is the unit element of G_2, forms a subgroup of G because $(g_1^{-1}, e_2)(g_1', e_2) = (g_1^{-1}g_1', e_2)$. Analogously the set H_2 of all elements of the form (e_1, g_2) forms a subgroup of G; here e_1 is the unit element of G_1. Accordingly we isolate subgroups H_i' of G_i but we do not distinguish them from H_i. It is obvious that the H_i are normal divisors of the groups G_i. In fact, for any $g_1 \in G_1$, since G_1 and G_2 are essential, we can find a $(g_1, g_2) \in G$. Then $(g_1, g_2)(H_1, e_2)(g_1^{-1}, g_2^{-1}) = (g_1 H_1 g_1^{-1}, e_2) \subset H_1$ and so H_1 is a normal divisor of G. We define the factor-groups G_1/H_1 and G_2/H_2. If $\alpha = (g_1, g_2)$ and $\beta = (g_1', g_2')$ are such that g_1 and g_1' belong to the same coset; that is, if $g_1^{-1}g_1' = h_1 \in H_1$, then $\alpha^{-1}\beta(h_1^{-1}, e_2) = (e_1, g_2^{-1}g_2') \in H_2$ and hence g_2 and g_2' belong to the same coset of G_2/H_2. Since the

* Translation of Izv. Vysš. Učebn. Zaved. Matematika 1965, no. 1 (44), 33–34.

converse is obviously true, g_2 and g_2' belong to the same coset of G_2/H_2 if and only if g_1 and g_1' belong to the same coset of G_1/H_1, where $(g_1, g_2), (g_1', g_2') \in G$.

Thus there is a one-to-one correspondence ϕ between the cosets of the spaces G_1/H_1 and G_2/H_2; ϕ is determined by the following correspondence between G_1 and G_2: g_2 corresponds to g_1 if $(g_1, g_2) \in G$. This correspondence is an isomorphism. In fact, if we are given $(g_1, g_2), (g_1', g_2') \in G$, then their product also belongs to G; this means that they are simultaneously corresponding elements of the cosets g_1H_1 and g_2H_2, $g_1'H_1$ and $g_2'H_2$, $g_1g_1'H_1$ and $g_2g_2'H_2$. Then $\phi(g_1g_1'H_1) = \phi(g_1H_1g_1'H_1) = g_2g_2'H_2 = g_2H_2 \times g_2'H_2 = \phi(g_1H_1)\phi(g_1'H_1)$ and ϕ is a one-to-one homomorphism; that is, an isomorphism. Thus we have explained the structure of a subgroup G of the direct product $G_1 \times G_2$ (under the condition that G_1 and G_2 are essential): in the groups G_1 and G_2 there are normal divisors H_1 and H_2 to which correspond factor-groups G_1/H_1 and G_2/H_2 that are isomorphic under the isomorphism ϕ. Then to G belong those and only those pairs (g_1, g_2) that contain elements g_1 and g_2 that lie in cosets of G_1/H_1 and G_2/H_2 which correspond under ϕ. Conversely, the set of elements (g_1, g_2) formed in this way is a subgroup of the direct product (with the requirement that G_1 and G_2 be essential, as is obvious). In fact, if we have (g_1, g_2) and (g_1', g_2') such that $g_2H_2 = \phi(g_1H_1)$ and $g_2'H_2 = \phi(g_1'H_1)$, then $g_2^{-1}g_2'H_2 = g_2^{-1}H_2g_2'H_2 = \phi(g_1^{-1}H_1)\phi(g_1'H_1) = \phi(g_1^{-1}g_1'H_1)$, and consequently $(g_1, g_2)^{-1}(g_1', g_2')$ belongs to the same set; that is, the indicated set is a subgroup of $G_1 \times G_2$.

Let us note some particular cases. a) $H_2 = e_2$. Here the factor-group $G_2/H_2 = G_2$ is isomorphic to G_1/H_1. Hence G_1 is homomorphic to G_2. The kernel of the corresponding homomorphism ϕ_1 is the group H_1. G consists of pairs of the form $(g_1, \phi(g_1))$ and is isomorphic to G_1. b) $H_1 = e_1$, $H_2 = e_2$. The homomorphism is an isomorphism. c) $H_1 = G_1$. In this case G_1 is the only coset of G_1/H_1 and, since G_2/H_2 is isomorphic to G_1/H_1, we have $G_2 = H_2$. This brings us to the case of the direct product $G_1 \times G_2$. The case when $G_1 = G_2$, or, more generally, when $G_1/H_1 = G_2/H_2$, is of particular importance for us; then ϕ is an automorphism of the group G_1/H_1. The involutory automorphism, when $\phi^{-1} - \phi$, is even more interesting. Let us also note the case when the groups G_i are direct products: $G_i = H_i \times K_i$. Here the isomorphism of the factor-spaces G_1/H_1 and G_2/H_2 reduces to an isomorphism between K_1 and K_2. In this case we will say that G is determined by the partial isomorphism ϕ.

We next consider a group G that can be represented as a group of transformations of the space $F = F_1 \times F_2$ (the direct product of the manifolds F_1 and F_2). We denote this representation by ϕ. Let us make the following assumptions: 1) $\phi(G)$ acts effectively in $F_1 \times F_2$; that is, the identity transformation of $F_1 \times F_2$ corresponds only to the unit element of G [2]. 2) The spaces F_i are domains of intransitivity for $\phi(G)$. In view of this assumption, the representation ϕ induces a representation $\phi_i(G)$ in F_i; we denote the groups of transformations $\phi_i(G)$ by G_i and form their direct product $G_1 \times G_2$. We show that G is isomorphic to a subgroup of $G_1 \times G_2$ for which the G_i are essential. In fact, the mapping ϕ: $g \to (\phi_1(g), \phi_2(g))$ is a homomorphism because ϕ_i is a representation of G. In view of assumption 1) the homomorphism ϕ is one-to-one and so is an isomorphism (if $g \neq e$, then $\phi_1(g)$ or $\phi_2(g)$ is not the identity transformation). The pairs of the form $(\phi_1(g), E_2)$, where E_2 is the identity transformation in F_2, form a group H_1. Thus, in this case, H_1 is the stationary subgroup of R_2 (R_2 is the frame in F_2 that is the kernel of noneffectiveness of $\phi_2(G)$). Analogously H_2 is the kernel of noneffectiveness of $\phi_1(G)$ and the stationary group of the frame R_1 of F_1. The pair (R_1, R_2) is a frame in $F = F_1 \times F_2$ and the stationary group of this geometrical object reduces to the unit element of G. As before, H_i are normal divisors of G_i and the factor-groups G_1/H_1 and G_2/H_2 are isomorphic. By recalling that G_1 and G_2 act transitively in F_1 and in F_2, we can regard the space F_i as the homogeneous space G_i/G_i', where G_i' is the stationary group of the point F_i.

When G is a linear group, $g \in G$, can be written in a special coordinate system, as

$$g = \begin{pmatrix} A & O' \\ O & B \end{pmatrix},$$

where the dimensions of the matrices A and B may or may not be the same. The set of matrices of G for which B is the unit matrix forms a subgroup H_1; similarly we obtain a subgroup H_2 when $A = E_1$. The elements of H_1 and H_2 commute, and their product is the subgroup $G' = H_1 \times H_2$. Thus, when the groups G under consideration are linear, they are completely reducible. Hence in the general case we will call them completely reducible.

It is obvious that the considerations can be easily generalized to the case when the group acts effectively in the direct product F of several spaces: $F = F_1 \times F_2 \times \cdots \times F_m$, where each F_i is a set of intransitivity of

G. Let G_i denote the group induced by G in the space F_i, and let H_i denote the kernel of noneffectiveness of G acting in F_i. If G acts effectively, for example, in $F_1 \times F_2 \times \cdots \times F_k$, then we can restrict our discussion to the pair $F_1 \times F_2 \times \cdots \times F_k$ and $F_{k+1} \times \cdots \times F_m$. When this is not so we must first consider the action of the group on the product $F_1 \times F_1'$, where $F_1' = F_2 \times \cdots \times F_m$. Then, from what has been said, the factor-spaces G_1/H_1' and G_1'/H_1 are isomorphic. Here G_1' is the group induced by G on F_1' and H_1' is the corresponding kernel of noneffectiveness. By further consideration of the action of G_1'/H_1 on F_1' we can study the structure of the group G in this case.

2. When considering completely reducible Lie groups we must describe the Lie algebra of such a Lie group. As is well known [2], the ideals h_i of the Lie algebra l_i of the group G_i correspond to the normal divisors H_i, and the factor-algebras l_i/h_i correspond to the factor-groups G_i/H_i. Hence, the form of the structure of a completely reducible group G, studied in the previous section, leads us to the conclusion that the Lie algebra l of this group consists of all pairs (ω_1, ω_2), where the ω_i are elements of l_i and belong to those cosets of the factor-algebra l_i/h_i that correspond under the isomorphism f of these algebras; f is determined by the isomorphism of the groups G_1/H_1 and G_2/H_2. We can also define the Lie algebra l as a subalgebra of the direct product $l_1 \times l_2$ for which these l_i are essential. The particular cases of completely reducible groups we considered earlier reduce to the following particular cases of their subalgebras: a) the algebra l consists of pairs $(\omega_1, \psi'(\omega_1))$, where ψ' denotes a homomorphism of l_1 onto l_2; b) the previous case, but where ψ' is an isomorphism or, more particularly, an automorphism that can be involutory; c) l is the direct product $l_1 \times l_2$ of l_1 and l_2.

In what follows we consider principal fiber spaces $P(M, G)$ with completely reducible Lie groups G as the fiber; that is, the fiber is a subgroup of the direct product of the Lie groups G_1 and G_2 with the condition that G_1 and G_2 are essential. Such a space $P(M, G)$ will be said to be completely reducible. We next assume that a connection is defined in $P(M, G)$ by the form ω. Then, as is well known [3], this form takes values from the Lie algebra of the structural group G. When the group is completely reducible its algebra has the structure indicated above and consists of pairs (ω_1, ω_1), where ω_i takes values from the Lie algebra of the group G_i. In the above particular cases we have: a) ω_1 and ω_2 are homomorphic connections; b) ω_1 and ω_2 are isomorphic connections. In a particular case these can be automorphic connections and, moreover, the automorphism can be involutory. If (case c)) the group G is the direct product of G_1 and G_2, then the connection in $P(M, G)$ is the set of two independent connections ω_1 and ω_2.

We are especially interested in the case when the base manifold M is the topological product of two manifolds: $M = M_1 \times M_2$. We denote the coordinates of points of cubic neighborhoods of M_1 and M_2 by $u = \{u^1, u^2, \cdots, u^k\}$ and $v = \{v^1, v^2, \cdots, v^m\}$ respectively, and the coordinates of points of $M = M_1 \times M_2$ by $x = \{u^1, \cdots, u^k, v^1, \cdots, v^m\} = \{x^1, \cdots, x^n\}$. In this case, for brevity, we write $\omega = \{\omega_1(x, dx, dG_1), \omega_2(x, dx, dG_2)\} = \{\omega_1(u, v, du, dv, dG_1), \omega_2(u, v, du, dv, dG_2)\}$. The forms ω_1 and ω_2 define a connection in the principal fiber spaces $P_1'(M, G_1)$ and $P_2'(M, G_2)$ respectively. The following special cases are important in applications: 1) $\omega = \{\omega_1(u, v, dx, dG_1), \omega_2(u, v, dv, dG_2)\}$; that is, the form $\omega_1(u, v, du, dG_1)$ can be considered as defining a set of connections on the principal fiber space $P_1(M_1, G_1)$, the set depending on the parameter v. Similarly $\omega_2(u, v, dv, dG_2)$ defines a set of connections on $P_2(M_2, G_2)$ that depends on the parameter u. 2) $\omega = \{\omega_1(u, du, dG_1), \omega_2(v, dv, dG_2)\}$. Here the connection ω is a set of two connections defined on the manifolds $P_1(M_1, G_1)$ and $P_2(M_2, G_2)$ respectively. These connections can be dependent (as we noted above); for example, they can be isomorphic or automorphic connections. Examples of such connections have been given by A. P. Norden [1], [4], [5]. Suitable examples will be given in the present article.

3. In this section we briefly present the theory of linear equi-affine connections in a form that is convenient for future use. By definition an equi-affine connection is an infinitesimal connection in the principal fiber space $P(M, U(n))$, where M is the n-dimensional base manifold and $U(n)$ is the structural group of unimodular matrices whose algebra, as is well known [2], consists of matrices whose trace is zero. We define this connection by the form

$$\tilde{\omega} = dA_0 A_0^{-1} + A_0 \omega A_0^{-1}, \tag{1}$$

where $A \in U(n)$ and $\operatorname{sp} \omega = 0$. By a linear equi-affine connection associated with an equi-affine connection we mean a linear connection that is obtained from an equi-affine connection by "extending" the group to the full linear group. Thus, if an equi-affine connection is defined by (1), the associated linear equi-affine connection is given by the form

$$\tilde{\omega}_1 = dAA^{-1} + A\omega A^{-1}, \tag{2}$$

defined on the principal fiber manifold $P(M, GL(n, R))$. Here $A \in GL(n, R)$ (the full linear group), and so $A = \sigma A_0$, where $\sigma^n = |A|$. It is easy to calculate that $\tilde{\omega}_1 = d \ln \sigma E + \tilde{\omega}$ (E is the unit matrix). For a linear equi-affine connection we have that $\operatorname{sp} \tilde{\omega}_1 = nd \ln \sigma$ is an exact differential; that is, that

$$d(\operatorname{sp} \tilde{\omega}_1) = 0, \tag{3}$$

where d denotes exterior differentiation. Condition (3) is a necessary condition for a linear connection to be equi-affine; we show that it is also sufficient. In fact, if (2) holds, then we can set $\operatorname{sp} \tilde{\omega}_1 = -nd \ln p$. We replace ω by the form $\omega_1 = d(pE)(pE)^{-1} + (pE)\omega(pE)^{-1}$; it is obvious that when we replace ω by ω_1 in (2) we obtain the same linear connection. But then it is easy to calculate that $\operatorname{sp} \omega_1 = 0$; that is, the linear connection is, in fact, equi-affine. Thus, (3) is both necessary and sufficient for a linear connection to be equi-affine. By using the expressions $\tilde{\Omega} = d\tilde{\omega}_1 + \tilde{\omega}_1 \wedge \tilde{\omega}_1$ for the curvature forms of the linear connection [6], it is easy to see that (3) is equivalent to

$$\operatorname{sp} \tilde{\Omega} = 0. \tag{4}$$

Because the form $\tilde{\Omega}$ takes values from the space of affinors ($\tilde{\Omega} = A\Omega A^{-1}$), it is sufficient to check that (4) holds for some (local) cross section $P(M, GL)$: $A = Ax$, $x \in M$ (the trace is invariant under the transformation $\Omega \dashrightarrow A\Omega A^{-1}$). Let us note that, in tensor form, (4) is equivalent to $R^{\alpha}_{\alpha,jk} = 0$; when the connection is without torsion it is equivalent to the symmetry of the Ricci tensor.

If the linear connection $\tilde{\omega}_1$ is not equi-affine, then, since the full linear group is the direct product of the unimodular group and the group of scalar matrices ($GL(n, R) = U(n) \times D_1$), the Lie algebra of this group is the pair $(\tilde{\omega}_e, \tilde{\omega}_0^0 E)$, where $\tilde{\omega}_0^0 E$ is an element of the Lie algebra of the group of scalar matrices and $\operatorname{sp} \tilde{\omega}_e = 0$. For a given cross-section $A(x)$ we obtain a form on M:

$$\omega_1 = \omega_e + \omega_0^0 E; \quad \operatorname{sp} \omega_e = 0, \tag{5}$$

from which it follows that

$$\omega_0^0 = \frac{1}{n} \operatorname{sp} \omega_1. \tag{6}$$

The form $\tilde{\omega}_1$ of the linear connection is determined from (2), in which we must replace ω by ω_1. If we set $A = |A|^{1/n} A_0$, then

$$\tilde{\omega} = \left(\omega_0^0 + \frac{1}{n} d \ln |A|\right) E + dA_0 A_0^{-1} + A_0 \omega_e A_0^{-1} = \tilde{\omega}_0^0 + \tilde{\omega}_e. \tag{7}$$

Here $\tilde{\omega}_e$ is the form of an equi-affine connection and $\tilde{\omega}_0^0 = \omega_0^0 + d \ln \sigma = \omega_0^0 + (1/n) d \ln |A|$ is the object of the connection in the principal fiber space $P(M, D)$; this object allows us to construct the covariant derivative for scalar and tensor densities [7]. In this case, if $\tilde{\omega}_0^0$ is an exact differential, at the expense of "renormalizing" we can derive an equi-affine connection. In what follows we call $\tilde{\omega}_e$ the equi-affine part of the linear connection. We must bear in mind that the preceding arguments imply that it is not single-valued.

4. We are going to give some examples of connections in the principal fiber space $P(M, G)$, where G is a completely reducible group. It follows from the previous discussion that to do this it is sufficient to give examples of pairs of connections, independent in the case when $G = G_1 \times G_2$, or related in a specific way (isomorphic, homomorphic, automorphic, etc. connections). The most simple example of isomorphic connections is a pair of coincident connections. A more interesting example of isomorphic connections is a pair of conjugate connections; this pair is constructed in terms of an automorphism ϕ of the full linear group defined by: for $A \in GL(n, R)$, $\phi(A)$ is the matrix $(A^{-1})'$.

As is known from the theory of Lie algebras [8], to ϕ there corresponds an automorphism f of the Lie algebra of the form

$$\omega_2 = f(\omega_1) = -\omega_1', \tag{8}$$

where the matrix $\tilde{\omega}_1$ is considered as an element of the Lie algebra of $GL(n, R)$. We call (8) the conjugacy condition for the connections. In what follows we will consider automorphic connections of a general form; that is, the case when f is an arbitrary automorphism of the Lie algebra. In the cases we are considering this is an automorphism of the Lie algebra of the full linear group and $\tilde{\omega}_1$ and $\tilde{\omega}_2 = f(\tilde{\omega}_1)$ are elements of the same Lie algebra. The case when f is an involutory automorphism is of particular interest; then $\tilde{\omega}_2 = f(\tilde{\omega}_1)$ and $\tilde{\omega}_1 = f(\tilde{\omega}_2)$ simultaneously and we say that the connections are mutually automorphic. In particular, conjugate connections are mutually automorphic.

It is easy to give an example of homomorphic connections. To do this it is sufficient to consider the pair $(\tilde{\omega}, \tilde{\omega}_e)$, where $\tilde{\omega}$ is a given linear connection and $\tilde{\omega}_e$ is its equi-affine part. This pair is determined by a homomorphism ϕ: $GL(n, R) \longrightarrow U(n)$ under which the unimodular matrices are mapped into themselves. Under the corresponding homomorphism f of the Lie algebra the matrix $\tilde{\omega}$ is mapped onto the matrix $\tilde{\omega}_e = f(\tilde{\omega}) = \tilde{\omega} - (1/n)\,\mathrm{sp}\,\tilde{\omega} E$. Here $\omega_0 = (1/n)\mathrm{sp}\,\tilde{\omega} E$ is an arbitrary connection in $P(M, D_1)$ and the pair $(\tilde{\omega}, \tilde{\omega}_e) = (\tilde{\omega}_e + \omega_0, \tilde{\omega}_e)$ is the form of a connection in $P(M, G)$, where G is the subgroup of the direct product $GL(n, R) \times U(n)$ distinguished by the homomorphism ϕ. We must also note the case of the pair $(\tilde{\omega}_1 + \tilde{\omega}_2, \tilde{\omega}_1)$, where $\tilde{\omega}_1 + \tilde{\omega}_2$ is the form of a connection in $P(M, G)$ $(G = G_1 \times G_2)$. The transition to the homomorphic connection consists of "rejecting" one of the two independent connections. The pair $(\tilde{\omega} + \tilde{\omega}_0^0 E, \tilde{\omega})$, where $\tilde{\omega}$ is a euclidean connection and $\tilde{\omega} + \tilde{\omega}_0^0 E$ is a metrical connection [9], is a particular example. A somewhat more complicated and, for us, more interesting example is one that is related to the general case. To construct it we consider the group G as the subgroup of the direct product of two full linear groups that is defined by the isomorphism ϕ of $GL(n, R)/D_1$ onto $GL(n, R)/D_1$. Because $GL(n, R)/D_1$ is isomorphic to the unimodular group, ϕ is an automorphism of the unimodular group. We are especially interested in the case when ϕ denotes a transition to the conjugate group: ϕ: $A \longrightarrow (A^{-1})'$. It is ovbious that the Lie algebra of this group consists of matrices of the form $(\tilde{\omega}_1, \tilde{\omega}_2) = (\tilde{\omega}_e + \tilde{\omega}_0^0 E, f(\tilde{\omega}_e) + \tilde{\omega}_n^n E)$, where $\tilde{\omega}_0^0 E$ and $\tilde{\omega}_n^n E$ are arbitrary connections in the principal fiber space $P(M, D_1)$ and f is an automorphism of the algebra of the unimodular group.

In our particular case we have that $f(\tilde{\omega}_e) = -\tilde{\omega}_e'$; the corresponding linear connections $\tilde{\omega}_1$ and $\tilde{\omega}_2$ will be said to be semiconjugate. It is obvious that the condition for semiconjugacy is

$$\tilde{\omega}_2 + \tilde{\omega}_1' = (\tilde{\omega}_n^n - \tilde{\omega}_0^0)\, E. \tag{9}$$

In the general case, this equation corresponds to

$$\tilde{\omega}_2 - \hat{f}\,(\tilde{\omega}_1) = (\tilde{\omega}_n^n - \tilde{\omega}_0^0)\, E. \tag{10}$$

Here we suppose that f is the restriction of an automorphism $\hat{f}$ of the full linear group and bear in mind that, under a homomorphism of the Lie algebra, the scalar matrices are mapped into themselves.

Conversely, if for some form $\tilde{p}$ we have

$$\tilde{\omega}_2 - \hat{f}(\tilde{\omega}_1) = \tilde{p}E, \tag{11}$$

then we see directly that the equi-affine parts of the linear connections $\tilde{\omega}_1$ and $\tilde{\omega}_2$ are isomorphic.

Thus, in the fiber space $P(M, G)$, where G is the group of pairs of matrices of the form $(\sigma A, \tau\phi(A))$, $|A| = 1$, $|\phi(A)| = 1$, the connection is a pair of "semi-isomorphic" connections. More precisely, there is a "semi-isomorphism" between a linear connection and a nonlinear one (it would be better to say, linear in the space that has the group of matrices $\phi(A)$ as a fiber). A complete expression for the forms of the connections $\tilde{\omega}_1$ and $\tilde{\omega}_2$ is given by

$$\tilde{\omega}_1 = (\omega_0^0 + d\ln\sigma)E + (dAA^{-1} + A\omega_1 A^{-1}),$$

$$\tilde{\omega}_2 = (\omega_n^n + d\ln\tau)E + \{[d\varphi(A)][\varphi(A)]^{-1} + \varphi(A)f(\omega_1)[\varphi(A)]^{-1}\}. \tag{12}$$

Let us note the more general case when $\tilde{\omega} = (\tilde{\omega}_1 + \tilde{\omega}_2,\ f(\tilde{\omega}_1) + \tilde{\omega}_2^*)$, where $\tilde{\omega}_1 + \tilde{\omega}_2$ is the form of a connection in $P(M, G_1 \times G_2)$ and f is an automorphism of the Lie algebra of the group G_1.

We must stress that the conjugate connections we have introduced are not the same as those introduced by A. P. Norden [1], because linear connections defined on the same fiber space are considered in the latter case. We will indicate a method of finding a linear connection from an existing automorphic connection; then to find a conjugate (automorphic) linear connection it is necessary first to determine a conjugate (automorphic) connection and then to use it to construct a linear connection. We call the latter a conjugate (automorphic) linear connection and, in the general case, a semi-automorphic connection.

5. Let us suppose that a linear connection is defined in $P(M, GL(n, R))$ by the form $\tilde{\omega} = dAA^{-1} + A\omega A^{-1}$. The connection is completely defined by specifying ω as a form on the base manifold M with values in the space of matrices (the Lie algebra of the full linear group). The same connection can be defined by the form

$$\omega(B) = dBB^{-1} + B\omega B^{-1}, \tag{13}$$

where $B(x)$ is a local cross section defined in a coordinate neighborhood. Let us note that we can formally go from $\tilde{\omega} = dAA^{-1} + A\omega A^{-1}$ to $\tilde{\omega}(B) = dAA^{-1} + A\omega(B)A^{-1} = dAA^{-1} + A(dBB^{-1} + B\omega B^{-1})A^{-1}$ by replacing the matrix A by $AB(x)$ in $\tilde{\omega}$; that is, by taking the right translation by $B(x)$ in a fiber of $P(M, GL)$. Suppose then that we are given an automorphism ϕ of the group and the corresponding automorphic connection, defined by the form of the connection $f(\tilde{\omega})$. Since $f(\tilde{\omega})$, defined for the cross section $C = C(x)$ where we have put $B = \phi(C)$, belongs to the Lie algebra of $GL(n, R)$, we can use it to construct the linear connection $\hat{\tilde{\omega}} = dAA^{-1} + Af(\tilde{\omega})A^{-1}$; here we can replace $f(\omega)$ by $\tilde{\omega} = dBB^{-1} + Bf(\omega)B^{-1}$ and must assume that $B = \phi(C) = B(x)$.

Thus, we can define a linear connection by the form

$$\hat{\tilde{\omega}} = dAA^{-1} + A[dBB^{-1} + Bf(\omega)B^{-1}]A^{-1}, \tag{14}$$

where $B = B(x)$ is a local cross section in $P(M, GL)$. Our definition is linked with the specification of a cross section in $P(M, GL)$; however, we can give a definition of this connection that does not depend on the choice of the cross section in $P(M, GL)$ and that does not require this cross section to be prescribed.

We replace the requirement for a cross section by demanding that we have an operator $\tilde{B}$ of a special form; the sign $\sim$ indicates that it is defined in a principal fiber space of a definite kind, in the given case, in the space $P(M, G)$, where G is the subgroup of the direct product $GL \times GL$ defined by the automorphism ϕ. Let $\tilde{B}$ be the operator that maps elements of the group $\phi(GL)$ into elements of $GL(n, R)$ invariant with respect to left translations; that is, the operator that maps $\phi(A) \in \phi(GL)$ into $A_1 \in GL$ so that the correspondence is preserved under a left translation in $GL(n, R)$ and the corresponding translation in $\phi(GL)$. Then we have that $A_1 = B\phi(A)$ and, in addition, that

$$TA_1 = \tilde{B}\varphi(TA) = \tilde{B}\varphi(T)\varphi(A). \tag{15}$$

This means that we must assume that $\tilde{B}$ is a matrix operator (that is, that B is an element of the group GL) and that it depends on the element generating the translation in the following way:

$$\tilde{B} = B(T) = TB[\varphi(T)]^{-1}. \tag{16}$$

Here $\tilde{B}$ is defined in each fiber of $P(M, G)$, and so we must assume that $B = B(x)$.

The operator $\tilde{B}$ acts as an operator of left translation (that is, as a matrix, $\tilde{B}$ multiplies on the left). If we consider that, under the left translation $A \longrightarrow TA$, the form $\tilde{\omega} = dAA^{-1} + A\omega A^{-1} \longrightarrow dTT^{-1} + T\tilde{\omega}T^{-1}$, and that an analogous change holds for $f(\tilde{\omega})$, then we must expect that

$$\tilde{\omega} = [dB(T)][B(T)]^{-1} + B(T)\, f(\tilde{\omega})\, [B(T)]^{-1};\quad B(T) = \tilde{B} \tag{17}$$

is the form of a linear connection and that this form corresponds to the operator $\tilde{B}$ which maps elements of $\phi(GL)$ into elements of GL. This can be verified by calculations; in fact, by using (16), we have

$$\begin{aligned} \tilde{\omega} &= d\tilde{B}\tilde{B}^{-1} + \tilde{B}f(\tilde{\omega})\,\tilde{B}^{-1} = d\{TB[\varphi(T)]^{-1}\}\{TB[\varphi(T)]^{-1}\}^{-1} \\ &+ TB[\varphi(T)]^{-1}\{d\varphi(T)[\varphi(T)]^{-1} + \varphi(T)f(\omega)[\varphi(T)]^{-1}\}\{TB[\varphi(T)]^{-1}\}^{-1} \\ &= dTT^{-1} + T(aBB^{-1} + B\{d[\varphi(T)]^{-1}\varphi(T) + [\varphi(T)]^{-1}d\varphi(T)\}B^{-1})\,T^{-1} \\ &+ TBf(\omega)\,B^{-1}T^{-1} = dTT^{-1} + T(dBB^{-1} + Bf(\omega)\,B^{-1})\,T^{-1}, \end{aligned}$$

that is, apart from a change of notation we have obtained (14). The connection defined by the form $\hat{\omega}$ will be said to be the linear connection automorphic to the connection $\tilde{\omega}$. In particular we can define the matrix $B(x)$ to be constant on the base manifold, and for B we can take the unit matrix E. The condition for the connections $\hat{\omega}$ and $\tilde{\omega}$ to be automorphic can be written as at (17), or, equivalently, as

$$d\tilde{B} + \tilde{B}f(\tilde{\omega}) - \tilde{\omega}\tilde{B} = 0. \tag{18}$$

If ϕ is an involutory automorphism we can assume that the connections $\tilde{\omega}$ and $f(\tilde{\omega})$ are mutually automorphic, for, if otherwise, by using the same operator $\tilde{B}$ we could obtain a third connection automorphic to $\tilde{\omega}$. If $\phi(A) = (A^{-1})'$, then $f(\omega) = -\omega'$ and (16) yields

$$\tilde{B} = TBT'. \tag{19}$$

This equation shows that $\tilde{B}$ is a twice covariant tensor. Condition (18) reduces to the condition for the connections to be conjugate:

$$d\tilde{B} - \tilde{B}\,\tilde{\omega}' - \hat{\tilde{\omega}}\tilde{B} = 0, \tag{20}$$

which can be written as

$$db_{ij} - b_{ik}\omega_j^k - b_{kj}\,\omega_i^k = 0.$$

and (when the tensor b_{ij} is symmetric) yields the well-known condition for linear connections to be conjugate in the sense of A. P. Norden [1]. Let us note that, if $B = E$, then $\tilde{B} = TT'$, and consequently the tensor $\tilde{B}$ is positive definite. We obtain the general case if we put

$$B = \begin{bmatrix} 1 & & & & \\ & \ddots & & & \\ & & 1 & & \\ & & & -1 & \\ & & & & \ddots \\ & & & & & -1 \end{bmatrix}$$

Although all the time we have considered a linear connection $\tilde{\omega}$, it was intentionally not indicated in the definitions that GL is the full linear group, because everything can be generalized, without change, to nonlinear connections: euclidean, equi-affine, pseudolinear, etc. Formula (1) is formed on the basis of these connections but we take a subgroup of the linear group. Let us also note the case $\phi(A) = A$. Here $\tilde{B}$ is an affinor and the connection $\hat{\tilde{\omega}}$ is equivalent to the connection $\tilde{\omega}$; more precisely, we can obtain $\hat{\tilde{\omega}}$ from $\tilde{\omega}$ by means of a right translation.

We next consider two connections for which only the equi-affine parts are automorphic; that is, connections for which (10) holds, or, what is the same, (11) holds. We construct a transformation of the equi-affine part of the connection $f(\tilde{\omega}_1)$ into an automorphic connection of the same type by using the operator $\tilde{B} = AB[\phi(A)]^{-1}$, where B is an equi-affine matrix; that is, we introduce the form

$$\hat{\tilde{\omega}}_1^{\mathrm{e}} = d\tilde{B}\tilde{B}^{-1} + \tilde{B}\,f(\tilde{\omega}_1^{\mathrm{e}})\,\tilde{B}^{-1}. \tag{21}$$

By extending the group of transformations to the full linear group, that is, by replacing $\tilde{B}$ by $\tilde{C} = \lambda\tilde{B}$, and by extending (21) to the linear connections $\tilde{\omega}_1 = \tilde{\omega}_1^{\mathrm{e}} + \tilde{\omega}_0^0 E$ and $\tilde{\omega}_2 = f(\tilde{\omega}_1^{\mathrm{e}}) + \tilde{\omega}_n^n E$, in view of (10) we obtain

$$\hat{\tilde{\omega}}_1 = d\tilde{C}\tilde{C}^{-1} + \tilde{C}\,\tilde{\omega}_2\tilde{C}^{-1} = d\tilde{C}\tilde{C}^{-1} + \tilde{C}f(\tilde{\omega}_1)\,\tilde{C}^{-1} + (\tilde{\omega}_n^n - \tilde{\omega}_0^0)\,E. \tag{22}$$

We put $\hat{\tilde{\omega}}_1 = \hat{\tilde{\omega}}_1^e + \hat{\tilde{\omega}}_0^0 E$. Then

$$\hat{\tilde{\omega}}_0^0 = \tilde{\omega}_n^n + d\ln\lambda. \tag{23}$$

We assume that the connections $\hat{\tilde{\omega}}_1$ and $\tilde{\omega}_1$ are of the same kind; that is, that $\tilde{\omega}_0^0 = \omega_0^0 + d\ln\sigma$ and $\hat{\tilde{\omega}}_0^0 = \tilde{\omega}_0^0 + d\ln\sigma$ simultaneously, and that $\tilde{\omega}_n^n = \omega_n^n + d\ln\tau$. Then from (23) we obtain that $\lambda = C_0\sigma/\tau$, where we can obviously have $C_0 = 1$. Thus,

$$\tilde{C} = \frac{\sigma}{\tau} AB\,[\varphi(A)]^{-1}.$$

It is obvious that this operator maps the elements of the group of matrices $\{\tau\phi(A)\}$ into the elements of the group of matrices $\{\sigma A\}$. We can derive the same results if we start from a pseudolinear connection. Here, for simplicity, we assume that $\phi(A) = (A^{-1})'$ and $f(\omega) = -\omega'$. Starting from the definition of a pseudolinear connection, we can form the group of matrices of the form

$$A_1 = \begin{pmatrix} \sigma & \\ & \sigma A \end{pmatrix} = \sigma \begin{pmatrix} 1 & \\ & A \end{pmatrix},$$

where A is an n-dimensional matrix. The form for a pseudolinear connection is

$$\begin{pmatrix} \tilde{\omega}_0^0 & \\ & \tilde{\omega}_1 \end{pmatrix} = dA_1A_1^{-1} + A_1 \begin{pmatrix} \omega_0^0 & \\ & \omega_1 \end{pmatrix} A_1^{-1}$$
$$= \begin{pmatrix} \omega_0^0 + d\ln\sigma & \\ & dAA^{-1} + A\omega_1A^{-1} + d\ln\sigma E \end{pmatrix}.$$

The automorphism $A \to \phi(A)$ of the linear group generates the following automorphism of the group of matrices:

$$\begin{pmatrix} 1 & \\ & A \end{pmatrix} \to \begin{pmatrix} 1 & \\ & \varphi(A) \end{pmatrix},$$

and the latter automorphism generates a group of pairs of matrices of the form

$$\left\{\sigma \begin{pmatrix} 1 & \\ & A \end{pmatrix},\ \tau \begin{pmatrix} 1 & \\ & \varphi(A) \end{pmatrix}\right\}.$$

As in the previous case we introduce the operator

$$\tilde{C} = \frac{\sigma}{\tau}\begin{pmatrix} 1 & \\ & A \end{pmatrix}\begin{pmatrix} 1 & \\ & B \end{pmatrix}\begin{pmatrix} 1 & \\ & (A^{-1})' \end{pmatrix} = \begin{pmatrix} \frac{\sigma}{\tau} & \\ & \frac{\sigma}{\tau} ABA' \end{pmatrix},$$

defined by the "mixed" pseudotensor $\tilde{B} = (\sigma/\tau)ABA'$.

As before, the conjugate linear connection is defined by (22), where the form of the connection $\tilde{\omega}_2$ is defined by

$$\begin{pmatrix} \overset{*}{\tilde{\omega}}{}^n_n & \\ & \overset{*}{\tilde{\omega}} \end{pmatrix} = \begin{pmatrix} \omega^n_n + d\ln\tau & \\ & [d(A^{-1})']A' + (A^{-1})'(-\omega_1') A' + d\ln\tau E \end{pmatrix}.$$

By going over to the associated linear connections, which we define by the forms $\tilde{\theta} = \tilde{\omega}_1 - \tilde{\omega}^0_0 E$ and $\overset{*}{\tilde{\theta}} = \overset{*}{\tilde{\omega}}{}^* - \tilde{\omega}^n_n E$, we obtain from (20) that

$$d\tilde{B} - \tilde{B}\tilde{\theta}' - \tilde{\theta}\tilde{B} = (\tilde{\omega}^0_0 + \tilde{\omega}^n_n)\tilde{B} = 2\tilde{l}\tilde{B},$$

which, obviously, coincides with (22)—it is only necessary to replace C by B. In tensor form the last equation becomes

$$db_{ij} - b_{ik}\theta^k_j - b_{kj}\theta^j_i = 2l_k b_{ij},$$

where we have put $l = l_k du^k$. The condition we have derived is the same as the general condition of A. P. Norden [1] for linear connections to be conjugate.

Remark 1. General conjugate connections arise in the study of problems in projective geometry.

It is well known that the full linear group functions in projective geometry, but the kernel of noneffectiveness of this group consists of scalar matrices (a renormalization). Thus, in fact, the group of unimodular matrices functions in projective geometry, and the equi-affine part of the corresponding linear connections that arise in projective geometry is essential. Hence under these conditions we have to consider an automorphism of only the equi-affine parts of the linear connections. The geometrical meaning of the conjugacy of linear connections associated with pseudolinear connections has been explained by A. P. Norden [1].

Remark 2. Pseudolinear connections allow us to define the concept of the parallel transfer of pseudovectors; that is, of directions. By the same analogy, the conjugate connection allows us to define the parallel transfer of pseudovectors; that is, of $(n-1)$-dimensional directions defined in the tangent plane M by the equation $v_k x^k = 0$, where v_k is a given pseudovector and x^k is a variable pseudovector belonging to a given $(n-1)$-direction.

With the help of $\tilde{B} = (b_{ij})$ we can go from the pseudovector y^i to the pseudovector $v_k = b_{ki} y^i$ and define an $(n-1)$-direction. We can find the pseudovectors of this direction from the equation $b_{ik} x^i y^k = 0$ which defines a polarity in the space of pseudovectors. It is clear (and this is easily verified by calculations) that the equation $b_{ik} x^i y^k = 0$ remains unchanged under a parallel transfer of y^i in a given linear connection and of x^i in the conjugate linear connection.

BIBLIOGRAPHY

[1] A. P. Norden, *Affinely connected spaces*, GITTL, Moscow, 1950. (Russian) MR 12, 441.

[2] L. S. Pontrjagin, *Continuous groups*, 2nd ed., GITTL, Moscow, 1954; English transl., Gordon and Breach, New York, 1966. MR 17, 171; 34 #1439.

[3] A. Lichnerowicz, *Théorie globale des connexions et des groupes d'holonomie*, Edizioni Cremonese, Rome, 1955; Russian transl., IL, Moscow, 1960. MR 19, 453; 23 #A1320.

[4] A. P. Norden, *Cartesian composition spaces*, Izv. Vysš. Učebn. Zaved. Matematika 1963, no. 4 (35), 117–128. (Russian) MR 29 #6417.

[5] ———, *On some possible directions for the development of line geometry*, Kazan. Gos. Univ. Učen. Zap. 123 (1963), no. 1, 141–151. (Russian) MR 33 #653.

[6] J. Favard, *Cours de géométrie différentielle locale*, Gauthier-Villars, Paris, 1957; Russian transl., IL, Moscow, 1960. MR 18, 668; 22 #11317.

[7] J. A. Schouten and D. J. Struik, *Einführung in die neueren Methoden der Differentialgeometrie*. Vol. 2, Noordhoff, Groningen, 1938; Russian transl., IL, Moscow, 1948. MR 12, 128.

[8] C. Chevalley, *Theory of Lie groups*. Vol. I, Princeton Math. Series, vol. 8, Princeton Univ. Press, Princeton, N. J., 1946; Russian transl., IL, Moscow, 1948. MR 7, 412.

[9] V. I. Vedernikov, *A generalization of A. P. Norden's method of normalization for the case of a fibered space*, Kazan. Gos. Univ. Učen. Zap. 123 (1963), no. 1, 3–23. (Russian) MR 33 #669.

Translated by L. W. Longdon

Amer. Math. Soc. Transl.
(2) Vol. 92, 1970

CONNECTIONS IN HOMOGENEOUS BUNDLES*

UDC 513.73

Ju. G. LUMISTE

Introduction

Suppose we are given a G/H-bundle E^*; that is, a fiber bundle with differentiable bundle space E^* for which fibers $p^{*-1}(x)$ over x, arising from the projection $p^*: E^* \to B$, are diffeomorphic to a homogeneous space G/H with an effective, connected structural group G. Suppose that on E^* there is defined a differentiable distribution Δ^*, with dimension $n = \dim B$, that is complementary to the vertical distribution Θ^* of planes $\Theta^*_{z^*}$ tangential to the fibers; that is, at an arbitrary point $z^* \in E^*$, $\Theta^*_{z^*} \cap \Delta^*_{z^*} = \{0\}$. If a path $\lambda(x_0, x_1)$ is defined in B, then a field of 1-directions belonging to Δ^* is defined on $p^{*-1}(\lambda)$. The integral paths of this field, which exist in view of Vinogradov's theorem [1], generate a diffeomorphism of $p^{*-1}(x_1)$ onto $p^{*-1}(x_0)$. The set of diffeomorphisms of $p^{*-1}(x)$ onto itself that correspond to loops at the point x is endowed in a natural way with the structure of a topological group $\Psi_x(\Delta^*)$ that acts in $p^{*-1}(x)$.

If, for some point $x \in B$, $\Psi_x(\Delta^*)$ is isomorphic to a subgroup of the structural group G, which, for simplicity, we also denote by $\Psi_x(\Delta^*)$, then, for any $x' \in B$, $\Psi_{x'}(\Delta^*)$ is the subgroup of G conjugate to $\Psi_x(\Delta^*)$. In this case the distribution Δ^* on E^* is said to be a *horizontal distribution* or a *connection* in E^*, and the subgroup $\Psi_x(\Delta^*) \subset G$, defined with accuracy up to inner automorphisms, is said to be the *holonomy group* of this connection.

At the present time there is a well developed theory of connections in principal G-bundles E. (A G/H-bundle for which $H = \{e\}$, and, consequently, the standard fiber is the manifold G, is said to be principal.) It is well known that the distribution Δ on E that is complementary to the vertical distribution Θ is horizontal if and only if it is invariant relative to the invariant action of G on E (see, for example, [2], [3]). An effective necessary and sufficient condition for this, which is suitable for concrete examples, is given by a theorem of G. F. Laptev [4], [5].

The theory of connections Δ^* in general G/H-bundles with $H \neq \{e\}$ has been developed to a lesser extent, although these connections occur more

*Translation of Mat. Sb. 69 (111) (1966), 434–469.

frequently in practice. It is well known that there is a one-to-one correspondence between the connections Δ in principal G-bundles E and the connections Δ^* in those G/H-bundles E^* obtained from E under the projection $\pi\colon E \longrightarrow E^* = E/H$ of the manifold E into the manifold E^* of orbits of a subgroup H under the action of G on E [2]. Corresponding horizontal distributions are related by $\Delta^* = \pi'(\Delta)$, where π' is the differential of the projection π and the holonomy groups $\Psi_x(\Delta)$ and $\Psi_x(\Delta^*)$ are isomorphic.

Next, it is well known that we can derive each G/H-bundle E^* from some principal G-bundle E by the above factorization π [6], [5]. This suggests the following program to investigate the connections Δ^* in G/H-bundles E^*: it is necessary to determine the corresponding connection Δ in the corresponding principal G-bundle E, to find the objects and groups (the curvature object, holonomy group, etc.) characterizing it and to study their relation with the connection Δ^* in E^*.

To start to realize this program it is necessary to solve problems of principal significance. Firstly we require the establishment of effective necessary and sufficient conditions for a distribution Δ^* defined on E^* to be horizontal. Then we must study the action of the holonomy group $\Psi_x(\Delta^*)$ on $p^{*-1}(x)$, and introduce the objects at points $z^* \in E^*$ that define it. It is desirable to give the constructions in such a form that the new objects are enveloped naturally by the objects characterizing a given horizontal distribution Δ^* on E^*; that is, so that we obviate reverting to objects connected with Δ that, in general, belong to the complementary distribution.

In the present article we use the invariant method of Cartan-Laptev-Vasil'ev [7], [4], [8] when solving these problems; this method is characterized by the application of free frames and, chiefly, of coframes only by necessity compatible with the structure of the manifold under consideration. All geometrical objects are given by systems of differential equations that define the change in the components of an object under the infinitesimal transformations of this free coframe. The investigations are developed by means of differential extensions of these systems, and of the objects defined by them, by using the inclusion of certain objects in others. The large degree of freedom of the system of reference allows us, when we wish, to use the methods of tensor or of global differential geometry which gives us a number of advantages.

The first three sections of the present article are mainly preparatory. In §1 we derive the structure equations of G. F. Laptev for a principal G-bundle E. In §2 we obtain from these the structure equations of a G/H-bundle E^*.

In this connection we take into account the structure of a solvable Lie group H_1 in the stationary subgroup H of the standard fiber G/H that is determined by a sequence $H_1 \supset H_2 \supset \cdots \supset H_p = \{e\}$ of kernels of natural homomorphisms of H into the nonholonomy differential groups that act in tangent spaces to G/H of higher order. The system of structure equations acquires a step structure; each step corresponds to a certain order of a differential group. In §3 we present the relevant information from the theory of connections in principal G-bundles. We develop a proof of a theorem of G. F. Laptev; the global character of this theorem becomes clear from the proof. We indicate a method of calculating the infinitesimal holonomy group of the connection Δ in E.

The connections Δ^* in general G/H-bundles E^*, which form the main subject material of this article, are considered in the last three sections. In §4 we given a condition for a distribution Δ^* on E^*, complementary to the vertical distribution Θ, to be horizontal. We obtain this in the form of a compatibility condition for a certain system of linear equations that contain the structure constants of G in the coefficients, and whose free terms contain, besides these, the partial derivatives, up to some order p, of the functions determining Δ^*; here the order p depends (in the above manner) only on the properties of the standard fiber G/H. This condition is considerably simplified in the case of reductive (in particular, symmetric and locally grouped) G/H related to the class $p = 1$. In §5 we study the curvature and torsion objects of various orders into which the single curvature object of the corresponding connection Δ in E splits, and which characterize the action of the infinitesimal holonomy group on the fibers of E^*.

Finally, in §6, we give examples of applications of our results. Let us note that the most important of these applications is connected with the problem of rigging (or normalizing) the embedded G/H-bundles of certain concrete classes, and of studying the induced connections; however, this problem lies outside the framework of the present article (see [9]). In our first example in §6 we give possible ways of interpreting a linear connection as a horizontal distribution on a vector bundle or on an affine bundle with a chosen intersecting surface (see [10], [11], [12], [13], [14]). In subsequent examples we consider connections in projective and conformal bundles (see [10], [12], [15], [16]).

The contents of the present article can be regarded as a generalization of the theory of connections in these two bundles [12] to the general case when the standard fiber can be any homogeneous space G/H and we do not require that $\dim B = \dim G/H$. The possibilities that this opens up will not be exhausted for a long time.

§1. The structure of a principal G-bundle

1. Suppose that we are given a principal G-bundle; that is, a principal fiber bundle consisting of a bundle space E with a structural Lie group G acting on E and a projection $p\colon E \to E/G = B$ onto the base (see, for example, [2]). There is always an open covering $\{U\}$ of B such that for each domain U of the covering there is a cross-section in E; that is, a diffeomorphism $\mu_U\colon U \to p^{-1}(U)$ with the property that $p\mu_U(x) = x$ for each $x \in U$. We have a diffeomorphism $\phi_U\colon U \times G \to p^{-1}(U)$ defined by $\phi_U(x, g) = \mu_U(x) \circ g$. Each vector field X on U determines, in a natural way, a vector field on $U \times G$ which, under ϕ_U, is mapped into some vector field Z on $p^{-1}(U)$. Suppose that on U there are n $(= \dim B)$ vector fields $X_1, \cdots, X_n$ that form a basis in the module of vector fields on U. Then, corresponding to these, we have vector fields $Z_1, \cdots, Z_n$ defined in $p^{-1}(U)$.

The action of G on E allows us, at each point $z \in E$, to define a diffeomorphism γ_z of G into the fiber through z by $\gamma_z(g) = z \circ g$. Its differential $(d\gamma_z)_e$ at the identity element $e \in G$ maps the Lie algebra G' of G onto the tangent space to the fiber at the point z. A vector field Z on E is said to be fundamental [2], and is denoted by $Z = (d\gamma)_e\alpha$, if it consists of the images $Z_z = (d\gamma_z)_e\alpha$ of the same element $\alpha \in G'$.

2. We choose a basis in G' consisting of $r (= \dim G)$ linearly independent elements $\epsilon_1, \cdots, \epsilon_r$. Then the corresponding fundamental vector fields $Z_{n-1} = (d\gamma)_e\epsilon_1, \cdots, Z_{n+r} = (d\gamma)_e\epsilon_r$, together with the vector fields $Z_1, \cdots$ $\cdots, Z_n$, form a complete basis in the module of vector fields on $p^{-1}(U)$. The following structure equations hold:

$$[Z_iZ_j] = Q_{ij}^kZ_k + R_{ij}^\rho Z_\rho, \tag{1.1}$$

$$[Z_iZ_\sigma] = S_{i\sigma}^kZ_k + T_{i\sigma}^\rho Z_\rho, \tag{1.2}$$

$$[Z_\sigma Z_\tau] = C_{\sigma\tau}^\rho Z_\rho \tag{1.3}$$

$$(i, j, \ldots = 1, \ldots, n; \quad \rho, \sigma, \ldots = n+1, \ldots, n+r),$$

where the coefficients in the first two equations are differentiable functions on $p^{-1}(U)$, and $C_{\sigma\tau}^\rho = \text{const}$ are the structure constants of the Lie algebra G'.

The structure of equations (1.1)–(1.3) is preserved under the transformations

$$Z_{i'} = \Gamma^i_{i'} Z_i + \Gamma^\rho_{i'} Z_\rho, \tag{1.4}$$

$$Z_{\rho'} = A^\rho_{\rho'} Z_\rho, \tag{1.5}$$

where $\Gamma^\rho_{i'}$ and $\Gamma^i_{i'}$ are differentiable functions on $p^{-1}(U)$ with $\det|\Gamma^i_{i'}| \neq 0$, and $A^\rho_{\rho'}$ are constants with $\det|A^\rho_{\rho'}| \neq 0$. Hence, in what follows, by $Z_1, \cdots, Z_n$ we mean an arbitrary system of n vector fields on $p^{-1}(U)$ that supplement the r linearly independent fundamental fields $Z_{n+1}, \cdots, Z_{n+r}$ to form a complete basis in the module of vector fields on $p^{-1}(U)$.

3. On going over to the dual statement of equations (1.1)–(1.3) we obtain the structure equations of G. F. Laptev for a principal G-bundle E. Let the 1-forms $\omega', \cdots, \omega^{r+n}$ form a basis in the module of 1-forms on $p^{-1}(U)$, dual to the basis $Z_1, \cdots, Z_{n+r}$. Then

$$\omega^i(Z_j) = \delta^i_j, \quad \omega^i(Z_\rho) = 0, \quad \omega^\sigma(Z_i) = 0, \quad \omega^\sigma(Z_\rho) = \delta^\sigma_\rho. \tag{1.6}$$

By using there relations and the well-known formula (see, for example, [2])

$$d\omega(Z, Z') = Z\omega(Z') - Z'\omega(Z) - \frac{1}{2}\omega([ZZ']), \tag{1.7}$$

that links the operation of exterior differentiation of a 1-form with the bracket operation over vector fields, it is not difficult, with the help of (1.1), (1.2) and (1.3), to derive the equations

$$d\omega^i = -\frac{1}{2}(Q^i_{jk}\omega^j \wedge \omega^k + S^i_{j\rho}\omega^j \wedge \omega^\rho),$$

$$d\omega^\rho = -\frac{1}{2}(R^\rho_{ij}\omega^i \wedge \omega^j + T^\rho_{i\sigma}\omega^i \wedge \omega^\sigma + C^\rho_{\sigma\tau}\omega^\sigma \wedge \omega^\tau).$$

By writing

$$\omega^i_j = -\frac{1}{2}(Q^i_{jk}\omega^k + S^i_{j\rho}\omega^\rho), \tag{1.8}$$

$$\omega^\rho_i = -\frac{1}{2}(R^\rho_{ij}\omega^j + T^\rho_{i\sigma}\omega^\sigma), \tag{1.9}$$

we can present the above equations in the following more compact form:

$$d\omega^i = \omega^j \wedge \omega^i_j, \tag{1.10}$$

$$d\omega^\rho = -\frac{1}{2}C^\rho_{\sigma\tau}\omega^\sigma \wedge \omega^\tau + \omega^i \wedge \omega^\rho_i, \tag{1.11}$$

in which they were given by G. F. Laptev [17].

4. It is obvious that the structure of the equations (1.10) and (1.11) is preserved under the transformations

$$\omega^{i'} = \Gamma_i^{i'} \omega^i, \tag{1.12}$$

$$\omega^{\rho'} = A_\rho^{\rho'} \omega^\rho - \Gamma_i^{\rho'} \omega^i, \tag{1.13}$$

that follow directly from (1.4) and (1.5). In addition, we can regard the coefficients $\Gamma_i^{\rho'}$ and $\Gamma_i^{i'}$ not only as functions on $p^{-1}(U) \subset E$ (as follows from (1.4) and (1.5)), but also, when desired, as new parameters that determine the position of a free linear frame in the tangent space $T_z(E)$ to E at z; the last r vectors of this frame are tangential to a fiber. In other words, the coefficients $\Gamma_i^{\rho'}$ and $\Gamma_i^{i'}$ in (1.12) and (1.13) can be considered as functions on a domain of the principal bundle space P of the above linear frames. This possibility gives equations (1.10) and (1.11) certain advantages over equations (1.1)–(1.3). In what follows we will regard the forms occurring in them as 1-forms on P. When we do this it is important to note the following.

In view of (1.11) the Pfaffian system $\omega^i = 0$ is completely integrable; hence the forms ω^i can be expressed in terms of the differentials of n parameters that play the role of the local coordinates of a point x in a domain $U \subset B$. When x is fixed the forms ω^i are zero. The system $\omega^i = 0$, $\omega^\rho = 0$ is also completely integrable; hence we can express ω^ρ in terms of the differentials of $n + r$ local coordinates of a point $z \in E$; when z is fixed, both ω^i and ω^ρ vanish.

We can fix the local coordinates of z only in the coefficients of the forms ω^i and ω^ρ. We then obtain a coframe $\{\omega_z^i, \omega_z^\rho\}$ at the point z; this coframe is an element of the space P^* conjugate to P. It is not difficult to derive an expression for the infinitesimal transformation of this coframe at z.

To do this we must choose two commuting vector fields W and δ on P, of which δ is vertical in the fibering of the manifold P; that is, it corresponds to taking the point $z \in E$ to be fixed. Then $\omega^i(\delta) = \omega^\rho(\delta) = 0$ and, if we write out (1.10) and (1.11) for the values of the 2-forms on W and δ, on using (1.7) these equations give

$$\delta\omega^i (W) = - \omega^j (W) \omega_j^i (\delta), \quad \delta\omega^\rho (W) = - \omega^i (W) \omega_i^\rho (\delta).$$

By considering these equations at a fixed point z and assuming that W_z is arbitrary, we obtain that

$$\delta\omega_z^i = -\omega_z^j \omega_j^i(\delta), \tag{1.14}$$

$$\delta\omega_z^\rho = -\omega_z^i \omega_i^\rho(\delta). \tag{1.15}$$

5. By using the calculus of G'-valued 1-forms (see [3]) we can write (1.11) in the following more compact form:

$$d\omega = -\frac{1}{2}[\omega\omega] + \omega^i \wedge \omega_i, \tag{1.11'}$$

where

$$\omega = \omega^\rho \otimes \varepsilon_\rho, \quad \omega_i = \omega_i^\rho \otimes \varepsilon_\rho. \tag{1.16}$$

Here (1.2) becomes

$$\omega' = \omega - \Gamma_i \omega^i \tag{1.17}$$

with $\omega = \omega^{\rho'} \otimes \epsilon_{\rho'}$ and $\Gamma_i = \Gamma_i^{\rho'} \otimes \epsilon_{\rho'}$.

6. The particular principal fiber bundle whose structure equations arise from successive extensions of (1.10) is of important significance in what follows. After u steps we obtain (cf. [17])

$$d\omega^i = \omega^j \wedge \omega_j^i,$$

$$d\omega_j^i = \omega_j^k \wedge \omega_k^i + \omega^k \wedge \omega_{jk}^i,$$

$$\cdots\cdots\cdots\cdots\cdots\cdots$$

$$d\omega^i_{j_1\cdots j_u} = \sum_{\lambda(s,t,u)}^{0 \leqslant s < t \leqslant u} \omega^k_{j_{s+1} j_{\lambda_{s+2}} \cdots j_{\lambda_t}} \wedge \omega^i_{j_1 \cdots j_s k j_{\lambda_{t+1}} \cdots j_{\lambda_u}} + \omega^k \wedge \omega^i_{j_1 \cdots j_u k}, \tag{1.18}$$

where $\lambda_{(s,t,u)}$ denotes a permutation $(\lambda_{s+2}, \cdots, \lambda_t, \cdots, \lambda_u)$ of the numbers $s+2, \cdots, t, \cdots, u$ and that $\lambda_{s+2} < \cdots < \lambda_t$, $\lambda_{t+1} < \cdots < \lambda_u$, and the summation is taken over all such permutations for $s = 0, \cdots, u-1$ and for $t = s + 1, \cdots, u$.

This system has the structure of the system (1.10) (1.11), where now the forms $\omega_j^i, \cdots, \omega^i_{j_1, \cdots, j_u}$ take the part of the forms ω^ρ. If we put $\omega^i = 0$ we obtain the structure equations of a Lie group, called the nonholonomy differential group $\widetilde{D}_u(n, R)$ of order u. We can represent this group in the linear group $GL(N, R)$ of transformations of an N-dimensional vector space $(N = n + n^2 + \cdots + n^u)$ as the group of transformations of the linear frame formed from the vectors

$$e_j, \dots, e_{j_1 \dots j_u}, \tag{1.19}$$

which are linearly independent under the various arrangements (with repetitions) of the indices. The structure equations of the group $\widetilde{D}_u(n, R)$ (that is, equations (1.18) with $\omega^i = 0$) are the integrability conditions of the expressions for an infinitesimal transformation of this frame:

$$de_j = \omega_j^i e_i,$$
$$\cdots\cdots\cdots\cdots\cdots\cdots\cdots\cdots$$
$$de_{j_1 \dots j_u} = \sum_{\lambda_{(s,t,u)}}^{0 \leqslant s < t \leqslant u} \omega^k_{j_{s+1} j_{\lambda_{s+2}} \dots j_{\lambda_t}} e_{j_1 \dots j_s k j_{\lambda_{t+1}} \dots j_{\lambda_u}}.$$

This completely integrable system

$$\omega^i_{j_{\lambda_1} \dots j_{\lambda_s}} = \omega^i_{j_1 \dots j_s} \quad (s = 1, \dots, u)$$

distinguishes a Lie subgroup $D_u(n, R) \subset \widetilde{D}_u(n, R)$, which is known as the (holonomy) differential group of order u [18]. We can represent it in the group $GL(2^n - 2, R)$ as the subgroup of transformations of a linear frame consisting of vectors (1.19) that are symmetric with respect to the lower indices [17].

§2. The structure of a homogeneous fiber bundle

1. Suppose we are given a homogeneous G/H-bundle; that is, a fiber bundle with bundle space E^* whose standard fiber is a homogeneous space G/H with an effective structural Lie group G and stationary Lie subgroup H. It is well known that there is a one-to-one correspondence between the G/H-bundles E^* and the pairs (E, H), where E is a principal G-bundle and $H \subset G$ is a Lie subgroup that does not contain normal divisors of G. Under this correspondence E^* can be identified (with accuracy up to an isomorphism) with the manifold E/H of orbits of H under the action of G on E; that is, the correspondence is given by the projection $\pi: E \to E/H$ $(= E^*)$. It is obvious that the base spaces of corresponding bundles E and E^* can be identified (with accuracy up to a diffeomorphism), and that the projections p: $E^* \to B$ are connected by the equation $p = p^* \pi$.

2. The structure equations for the G/H-bundle $E^* = E/H$ follow from the structure equations (1.10), (1.11) for the corresponding principal G-bundle E. For, suppose that the basis in G' is chosen so that $\epsilon_{n+m+1}, \dots, \epsilon_{n+r}$ belong to the Lie algebra H' of the stationary subgroup H. Then

$$[\varepsilon_\alpha \varepsilon_\beta] = C^\gamma_{\alpha\beta}\varepsilon_\gamma; \quad \alpha, \beta, \ldots = n+m+1, \ldots, n+r,$$

that is,

$$C^a_{\alpha\beta} = 0; \quad a, b, \ldots = n+1, \ldots, n+m. \tag{2.1}$$

Equations (1.10) and (1.11) give

$$d\omega^i = \omega^j \wedge \omega^i_j, \tag{2.2}$$

$$d\omega^a = -\frac{1}{2} C^a_{bc}\omega^b \wedge \omega^c - C^a_{b\alpha}\omega^b \wedge \omega^\alpha + \omega^i \wedge \omega^a_i, \tag{2.3}$$

$$d\omega^\alpha = -\frac{1}{2} C^\alpha_{\sigma\tau}\omega^\sigma \wedge \omega^\tau + \omega^i \wedge \omega^\alpha_i. \tag{2.4}$$

It follows from here that the Pfaffian system $\omega^i = 0$, $\omega^a = 0$ is completely integrable; hence the forms ω^a can be expressed in terms of the differentials of $n+m$ parameters, n of which (the complete integrals of the system $\omega^i = 0$) are the local coordinates in the domain U of the base B. These $n+m$ parameters play the role of the local coordinates on the G/H-bundle E^*. When the fiber $p^{*-1}(x)$ in E^* is fixed the forms ω^i vanish, and when we fix the point z^* in the fiber, the forms ω^a also vanish.

3. If we consider the 1-forms ω^i and ω^a at a fixed point $z^* \in E^*$, then at this point they form a linear coframe $\{\omega^1_{z^*}, \cdots, \omega^{n+m}_{z^*}\}$ tangential to E^*; that is, a basis in the linear space $T^*_{z^*}(E^*)$ conjugate to the linear tangent space $T_{z^*}(E^*)$. (This corresponds to fixing the local coordinates of z^* only in the coefficients of the forms ω^i and ω^a.) It is not difficult to find expressions for an infinitesimal transformation of this coframe at z^*. To do this we take an arbitrary element α of the Lie algebra H' of the stationary subgroup H and consider the fundamental vector field on E that corresponds to it as a symbol δ^* of differentiation. For each vector $Z_z \in T_z(E)$ we can find a vector field Z on the inverse image $p^{-1}(U)$ of some domain $U \subset B$ such that Z contains Z_z and commutes with δ^*, so that $[\delta^* Z] = 0$. By calculating the values of both sides of the equations (2.2) and (2.3) on the vector fields δ^* and Z, and using (17) and the obvious equations $\omega^a(\delta^*) = 0$ and $\omega^i(\delta^*) = 0$, we obtain

$$\delta^*\omega^i(Z) = -\omega^j(Z)\,\omega^i_j(\delta^*), \quad \delta^*\omega^a(Z) = \omega^b(Z)\,C^a_{b\alpha}\omega^\alpha(\delta^*) - \omega^i(Z)\,\omega^a_i(\delta^*).$$

If we consider these equations at the point $z \in E$ (let us recall that the vector $Z_z \in T_z(E)$ is arbitrary) then, after passing to E^* by using the differential

π' of the projection π, we obtain the following expressions for an infinitesimal displacement of our coframe at z^*:

$$\delta^*\omega^i_{z^*} = -\omega^j_{z^*}\omega^i_j(\delta^*), \tag{2.5}$$

$$\delta^*\omega^a_{z^*} = -\omega^b_{z^*}\omega^a_b(\delta^*) - \omega^i_{z^*}\omega^a_i(\delta^*). \tag{2.6}$$

From these we obtain the following results for the dual frame $\{e_i, e_a\}$ at $z^* \in E^*$:

$$\delta^* e_i = e_j\omega^j_i(\delta^*) + e_a\omega^a_i(\delta^*), \tag{2.5'}$$

$$\delta^* e_a = e_b\omega^b_a(\delta^*). \tag{2.6'}$$

4. Here the 1-forms

$$\omega^a_b = -C^a_{b\alpha}\omega^\alpha \tag{2.7}$$

form the matrix $\|\omega^a_b\|$ which is the image of an arbitrary element of the Lie algebra of the so-called isotropy group $\widetilde{H}$ of a standard fiber G/H. Note that these forms allow us to write (2.3) as

$$d\omega^a = \omega^b \wedge \omega^a_b - \frac{1}{2}C^a_{bc}\omega^b \wedge \omega^c + \omega^i \wedge \omega^a_i, \tag{2.8}$$

and that they satisfy the equations

$$d\omega^a_b = \omega^c_b \wedge \omega^a_c + \omega^c \wedge \omega^a_{bc} + \frac{1}{2}C^a_{b\alpha}C^\alpha_{cd}\omega^c \wedge \omega^d + \omega^i \wedge \widetilde{\omega}^a_{bi}, \tag{2.9}$$

which stem from (2.4) in view of the Jacobi identity; here

$$\omega^a_{bc} = C^a_{b\beta}C^\beta_{c\alpha}\omega^\alpha, \tag{2.10}$$

$$\widetilde{\omega}^a_{bi} = -C^a_{b\alpha}\omega^\alpha_i. \tag{2.11}$$

When the natural homomorphism of the stationary subgroup of the point $y_0 \in G/H$ into the group $GL(m, R)$ of linear transformations of the tangent space $T_{y_0}(G/H)$ is an isomorphism (that is, when the isotropy group $\widetilde{H}$ of the standard fiber G/H is isomorphic to the stationary subgroup H), then the rank of the system of forms ω^a_b is the same as the rank of the system of forms ω^α, which is $r - m = \dim H$, and, in view of (2.7), the ω^α can be expressed in terms of the ω^a_b. In this case (2.9) completely replaces the equations (2.4), and the local structure of the G/H-bundle E^* is defined by (2.2), (2.8) and (2.9).

5. In the general case when the above representation of H in the group $GL(m, R)$ of transformations of the space $T_{y_0}(G/H)$ is not faithful, we must go further and consider representations of H in the differential groups that transform the tangent spaces of higher order to G/H at the point y_0. We will investigate some natural representations of H in the nonholonomy differential groups $\widetilde{D}_u(m, R)$ below.

Let us set

$$\omega^a_{b_1\dots b_u} = -C^a_{b_1\dots b_u\alpha}\omega^\alpha, \tag{2.12}$$

where

$$C^a_{b_1\dots b_u\alpha} = (-1)^{u-1}C^a_{b_1\beta_1}C^{\beta_1}_{b_2\beta_2}\ \dots\ C^{\beta_{u-1}}_{b_u\alpha}. \tag{2.13}$$

We note that when $u = 1$ this result yields the forms (2.7). Our aim is to derive structure equations similar to (2.9) for the system of forms (2.12).

Lemma 1. *The following holds:*

$$C^a_{b_1\dots b_u\alpha}C^\alpha_{\beta\gamma} = 2\sum_{\lambda(s,t,u)}^{0\leqslant s<t\leqslant u} C^c_{b_{s+1}b_{\lambda_{s+2}}\dots b_{\lambda_t}[\beta}C^a_{|b_1\dots b_s\, c\, b_{\lambda_{t+1}}\dots b_{\lambda_u}|\gamma]}. \tag{2.14}$$

Proof. It follows from the Jacobi identity and from (2.1) that

$$C^a_{b\alpha}C^\alpha_{\beta\gamma} = 2C^c_{b[\beta}C^a_{|c|\gamma]};$$

hence (2.14) holds when $u = 1$.

Let us assume that (2.14) holds when $u = q$. We obviously have

$$C^a_{b_1\dots b_u\beta}C^\beta_{b_{u+1}\alpha} = -C^a_{b_1\dots b_{u+1}\alpha},$$

and, in view of the Jacobi identity and (2.14),

$$\begin{aligned} C^a_{b_1\dots b_{q+1}\alpha}C^\alpha_{\beta\gamma} &= \{C^a_{b_1\dots b_q\delta}C^\delta_{\beta\alpha}C^\alpha_{\gamma b_{q+1}} + C^c_{b_{q+1}\beta}C^a_{b_1\dots b_q c\gamma}\}\,[\beta,\gamma] \\ &= \left\{\left(\left\{\sum_{\lambda(s,t,q)}^{0\leqslant s<t\leqslant q} C^c_{b_{s+1}b_{\lambda_{s+2}}\dots b_{\lambda_t}\beta}C^a_{b_1\dots b_s\, c\, b_{\lambda_{t+1}}\dots b_{\lambda_q}\alpha}\right\}[\beta,\alpha]\right)C^\alpha_{\gamma b_{q+1}}\right. \\ &\qquad \left. + C^c_{b_{q+1}\beta}C^a_{b_1\dots b_q c\gamma}\right\}[\beta,\gamma], \end{aligned}$$

where $[\beta, \gamma]$ denotes alternation (without division by 2) of the preceding expression $\{\cdots\}$ with respect to the indices β and γ. The last equations show that (2.14) also holds when $u = q + 1$. The lemma is proved.

It is now not difficult to derive the structure equations for the forms (2.12). It follows from (2.4) that

$$d\omega^a_{b_1\ldots b_u} = -C^a_{b_1\ldots b_u\alpha}\Big(-\frac{1}{2}C^\alpha_{\beta\gamma}\omega^\beta\wedge\omega^\gamma - C^\alpha_{c\beta}\omega^c\wedge\omega^\beta - \frac{1}{2}C^\alpha_{cd}\omega^c\wedge\omega^d + \omega^i\wedge\omega^\alpha_i\Big);$$

hence, in view of (2.14).

$$d\omega^a_{b_1\ldots b_u} = \sum_{\lambda_{(s,t,u)}}^{0\leqslant s<t\leqslant u} \omega^c_{b_{s+1}b_{\lambda_{s+2}}\ldots b_{\lambda_t}}\wedge\omega^a_{b_1\ldots b_s c\, b_{\lambda_{t+1}}\ldots b_{\lambda_u}} + \omega^c\wedge\omega^a_{b_1\ldots b_u c} + \frac{1}{2}C^a_{b_1\ldots b_u\alpha}C^\alpha_{cd}\omega^c\wedge\omega^d + \omega^i\wedge\tilde\omega^a_{b_1\ldots b_u i}, \quad (2.15)$$

where

$$\tilde\omega^a_{b_1\ldots b_u i} = -C^a_{b_1\ldots b_u\alpha}\omega^\alpha_i.$$

Note that equations (2.9) are a particular case of the structure equations (2.15) and correspond to the value $u = 1$.

6. By taking $\omega^i = 0$ and $u = 1, 2, \cdots$ in the equations (2.8) and (2.15) we obtain a number of consequences concerning the local structure of the homogeneous standard fiber G/H.

It is not difficult to observe that, when $\omega^i = 0$ and $\omega^a = 0$, equations (2.15) are the same as the structure equations of the nonholonomy differential group $\tilde D_u(m, R)$, which are obtained from (1.18) with $\omega^i = 0$ (and by changing the notation of the indices). Hence equations (2.12) determine a natural representation of the stationary subgroup H in $\tilde D_u(m, R)$.

Lemma 2. *We can always find a faithful representation in the sequence of natural representations of the stationary subgroup H of the homogeneous space G/H in the nonholonomy differential groups $\tilde D_u(m, R)$* ($m = \dim G/H$, $u = 1, 2, \cdots$) *defined by* (2.12); *the order of the group $\tilde D_p(m, R)$ in which this faithful representation is first achieved can be arbitrarily large.*

Proof. The kernel H_u of the homomorphism of H onto $\tilde D_u(m, R)$ defined by (2.12), is distinguished by the completely integrable system

$$\omega^a = 0, \quad \omega^a_b = 0, \quad \ldots, \quad \omega^a_{b_1\ldots b_u} = 0. \quad (2.16)$$

We have the sequence

$$H_1 \supset H_2 \supset \ldots \supset H_u \supset \ldots \tag{2.17}$$

of embeddings of the normal divisors of the subgroup H. The Lie algebra H'_u of the subgroup H_u is characterized by the condition

$$[H'_u G'] \subset H'_{u-1}, \quad u = 1, 2, \ldots, \tag{2.18}$$

because, in view of (2.16), (2.12), (2.13) and (2.14), H'_u consists of those elements $\xi = \xi^\alpha \epsilon_\alpha \subset H'_{u-1}$ for which

$$C^a_{b_1 \ldots b_{u-1}\beta} C^\beta_{b_u \alpha} \xi^\alpha = 0. \tag{2.19}$$

It follows from here that, when $\dim H'_{u-1} \neq 0$, we have $\dim H'_{u-1} > \dim H'_u$, because $H'_{u-1} \supset H'_u$ and, if $\dim H'_{u-1} = \dim H'_u$, the algebra H'_{u-1} would be a nontrivial ideal in G', which is impossible (G acts effectively on G/H). Thus

$$\dim H > \dim H_1 > \ldots > \dim H_u > \ldots .$$

Hence there is a first value $u = p$ such that $\dim H_p = 0$; the representation of H in $\widetilde{D}_p(m, R)$ is faithful.

The above reasoning shows that p can be an arbitrarily large natural number. Hence the lemma is proved.

The number p is an important characteristic of the homogeneous space G/H; we call it the *order of the isotropy* of G/H. The number $s_u = \dim H_u$ is said to be the *character of the isotropy of order* u of G/H.

Remark 1. The sequence (2.17) defines the structure of a solvable group in H_1; that is, the factor-groups H_{u-1}/H_u ($u = 2, 3, \cdots$) are abelian.

In fact, in the Jacobi identity

$$[[H'_{u-1} H'_{u-1}] G'] + [[G' H'_{u-1}] H'_{u-1}] + [[H'_{u-1} G'] H'_{u-1}] = 0$$

the last two terms belong to the subalgebra H'_{u-1} because $[G' H'_{u-1}] \subset H'_{u-2} \subset H'$ ($u \geq 2$) and H'_{u-1} is an ideal in H'. Hence

$$[H'_{u-1} H'_{u-1}] \subset H'_u \tag{2.20}$$

and so H_{u-1}/H_u is abelian.

Let us note that, in an analogous way, it is easy to verify by induction that H'_u is in fact an ideal in H'; this follows from the Jacobi identity

$$[[H'_u H']\,G'] + [[G'H'_u]\,H'] + [[H'G']\,H'_u] = 0,$$

where the last two terms belong to H'_{u-1} because $[G'H'_u] \subset H'_{u-1}$ and, by the inductive hypothesis, $[H'_{u-1}H'] \subset H'_{u-1}$ (this is obviously true when $u = 1$ if we assume that $H'_0 = H'$). Hence $[H'_u H'] \subset G$, $u = 1, 2, \cdots$.

Remark 2. It is important to note for what is to follow that, in view of the condition $s_1 > s_2 > \cdots > s_{p-1} > 0$, $s_p = 0$, when $u < p$ the system

$$C^a_{b\alpha}\xi^\alpha = 0, \quad C^a_{b_1 b_2 \alpha}\xi^\alpha = 0, \quad \ldots, \quad C^a_{b_1 \ldots b_u \alpha}\xi^\alpha = 0 \tag{2.21}$$

has a nontrivial solution, and when $u \geq p$ it has only the trivial solution $\xi^\alpha = 0$. In other words, p is the first value of $u = 1, 2, \cdots$ for which the rank of the matrix of the system (2.24) is equal to the number $r - m$ of unknowns.

It follows, in particular, that the local structure of the G/H-bundle E^* is completely determined by (2.2), (2.8) and (2.15) with the values $u = 1, 2, \cdots, p$, because, in view of (2.12) and Remark 2, the latter equations completely replace those at (2.4).

Example. Because the Maurer-Cartan forms $\omega^i_{j_1 \cdots j_u}$ of the holonomy differential group $D_p(n, R)$ (see §1.6) are symmetric with respect to the lower indices, we can write the structure equations of $D_p(n, R)$ in the form [17]:

$$d\omega^i_j = \omega^k_j \wedge \omega^i_k,$$
$$d\omega^i_{j_1 j_2} = \omega^k_{j_1 j_2} \wedge \omega^i_k + \omega^k_{j_1} \wedge \omega^i_{j_2 k} + \omega^k_{j_2} \wedge \omega^i_{j_1 k},$$
$$\cdots\cdots\cdots\cdots\cdots\cdots$$
$$d\omega^i_{j_1 \ldots j_p} = \sum_{u=1}^{p} \frac{p!}{u!\,(p-u)!}\, \omega^k_{(j_1 \ldots j_u} \wedge \omega^i_{j_{u+1} \ldots j_p) k}.$$

The completely integrable Pfaffian system

$$\omega^i_1 = 0, \quad \omega^i_{11} = 0, \quad \ldots, \quad \omega^i_{\underbrace{11\ldots1}_{p}} = 0$$

distinguishes a Lie subgroup $H \subset D_p(n, R)$. It is not difficult to check that the order of the isotropy of the homogeneous space $D_p(n, R)/H$ (the space of hypervectors [18] of order p) is equal to p.

In fact, of all the Maurer-Cartan forms for $D_p(n, R)$, apart from $\omega^i_1, \cdots, \omega^i_{\underbrace{11\cdots 1}_{p}}$, only $\omega^i_{j'_1}, \omega^i_{1j'_1}, \cdots, \omega^i_{\underbrace{1\cdots 1}_{p} j'_1}$ $(j'_1, j'_2, \cdots = 2, \cdots, n)$ occur in the expression for $d\omega^i_{\underbrace{11\cdots 1}_{p}}$; next, of the remaining forms, only $\omega^i_{j'_1 j'_2}, \omega^i_{1j'_1 j'_2}, \cdots, \omega^i_{\underbrace{1\cdots 1}_{p-2} j'_1 j'_2}$ occur the expression for $d\omega^i_{\underbrace{1\cdots 1}_{p-1} j'_1}$, etc. The last Maurer-Cartan forms $\omega^i_{j'_1\cdots j'_p}$ of $D_p(n, R)$ only enter at the pth stage. In other words, by using the terminology and notation of subsections 4 and 5, we can assert that only when $u = p$ is the rank of the system of forms $\omega^a_b, \omega^a_{b_1 b_2}, \cdots, \omega^a_{b_1\cdots b_u}$ equal to the rank of the system of forms ω^a; that is, that the index of the isotropy of the space $D_p(n, R)/H$ is p.

§3. Connections in principal bundles

1. The concept of a connection in a G/H-bundle E^* was introduced in the Introduction: a differentiable distribution Δ^* on E^*, complementary to the vertical distribution Θ^*, is said to be a *horizontal distribution*, or a *connection*, in E^* if the group $\Psi_x(\Delta^*)$ generated by it is a subgroup of the structural Lie group G. As we remarked in the Introduction, the connections Δ^* in E^* are in a one-to-one correspondence with the connections Δ in the principal G-bundle E, the factorization $\pi\colon E \longrightarrow E/H$ of which by the action of the subgroup H on it leads to a G/H-bundle E/H, isomorphic to E^*. Here Δ^* is the image of Δ under the differential π' of the projection π (see [2]): $\Delta^* = \pi'(\Delta)$. Hence, connections in principal fiber bundles, which up to the present have been the main object of investigation, play a decisive role in the theory of connections.

2. Suppose that a distribution Δ, complementary to the vertical distribution Θ, is defined on the principal G-bundle E. We can take a basis in the module of vector fields on $p^{-1}(U)$ consisting of $n\,(=\dim B)$ fields Z_i and $r\,(=\dim G)$ fundamental fields Z_ρ (see §1.1), so that the Z_i belong to Δ.

Lemma 3. *A distribution Δ on the principal G-bundle E, complementary to the vertical distribution Θ, is a horizontal distribution, or connection, in E if and only if the Lie derivative of any vector field $\xi^i Z_i$ from Δ in the direction of any fundamental vector field $\lambda^\rho Z_\rho$* ($\lambda^\rho =$ const) *belongs to Δ (in other words, if we have $T^\rho_{i\sigma} = 0$ in* (1.2)).

Proof. Necessity. This follows from the well-known result [2] that the bracket of two vector fields on E, one of which is horizontal and the other is fundamental, is a horizontal vector field.

Sufficiency. We take an intersecting surface $\mu(U)$ in the inverse image $p^{-1}(U) \subset E$ of a sufficiently small domain $U \subset B$, and consider the plane $\Delta_{\mu(x)}$ of the distribution Δ at an arbitrary point $\mu(x)$ of this surface. Let Δ'_z denote the image of $\Delta_{\mu(x)}$ under the action on E of an element $g \in G$ such that $z = \mu(x) \circ g$. We aim to show that Δ is invariant; that is, that $\Delta'_z = \Delta$.

Suppose that Δ'_z is spanned by the vectors

$$(Z'_i)_z = (Z_i)_z + \Gamma_i^\sigma \, (Z_\sigma)_z, \tag{3.1}$$

where the Γ_i^σ are differentiable functions on the fiber through $\mu(x)$. The vectors $(Z'_i)_z$ on $p^{-1}(U)$ form a vector field Z'_i. In view of (1.1), (1.2) and (3.1) we have

$$[Z'_i Z_\rho] = S^j_{i\rho} Z_j + \Gamma_i^\sigma C^\tau_{\rho\sigma} Z_\tau + (Z_\rho \Gamma_i^\sigma)\, Z_\sigma.$$

On the other hand, because the field Z'_i is invariant we have

$$[Z'_i Z_\rho] = {}'S^j_{i\rho} Z'_j = {}'S^j_{i\rho} (Z_j + \Gamma_j^\tau Z_\tau).$$

Hence

$${}'S^j_{i\rho} = S^j_{i\rho}, \quad Z_\rho \Gamma_i^\tau = \Gamma_j^\tau S^j_{i\rho} - \Gamma_i^\sigma C^\tau_{\rho\sigma}.$$

Since we can regard each vector field Z_ρ as a global differential operator on $p^{-1}(U)$, the last equations form a homogeneous system of first order differential equations on $p^{-1}(U)$. This system has the trivial solution $\Gamma_i^\sigma \equiv 0$ which, in view of the initial conditions $(\Gamma_i^\sigma)_{\mu(x)} = 0$, coincides with the system of functions being considered. Thus $Z'_i = Z_i$, and so $\Delta' = \Delta$ on $p^{-1}(U)$. We can repeat this argument for each domain of a sufficiently fine covering $\{U\}$ of B. Hence $\Delta' = \Delta$ on the whole of E, and the lemma is proved.

It is not difficult to observe that this proof is only applicable when the structural group G is connected.

3. Theorem 1 (G. F. Laptev [4]). *If a 1-form ω is defined on the principal G-bundle E whose structural group is a connected Lie group G, and if ω has values in the Lie algebra G' of G, where*

1) $$\omega_z(Z_z) = (d\gamma_z^{-1}) Z_z \tag{3.2}$$

for each vertical vector $Z_z \subset \Theta_z$, *and*

2) *the* G'*-valued* 2*-form*

$$\Omega = d\omega + \frac{1}{2}[\omega\omega] \tag{3.3}$$

is semibasic on E (*that is,* $\Omega_z(Z'_z, Z''_z) = 0$ *if at least one of the vectors* Z'_z, Z''_z *is vertical*), *then a distribution* Δ *on* E, *for which the plane* Δ_z *at any point* $z \in E$ *annihilates* ω_z, *is a connection. Conversely, any connection* Δ *in* E *annihilates some* G'*-valued* 1*-form* ω *on* E *that satisfies conditions* 1) *and* 2).

Proof. It follows from (3.2) that Δ is complementary to a vertical distribution. Because, in view of (3.3), (1.11′), (1.16) and (1.9),

$$\Omega = -\frac{1}{2}\omega^i \wedge (R^\rho_{ij}\omega^j + T^\rho_{i\sigma}\omega^\sigma) \otimes \varepsilon_\sigma,$$

it follows that Ω is semibasic if and only if $T^\rho_{i\sigma} = 0$; that is, if and only if Δ is horizontal. The theorem is proved.[1)]

A G'-valued 1-form ω on E satisfying the conditions of Theorem 1 is called a *connection form*. The 2-form Ω, formed from ω and defined by (3.3), is called the *curvature form* of the corresponding connection. On each $p^{-1}(U)$ it is given by

$$\Omega = -\frac{1}{2} R^\rho_{ij}\omega^i \wedge \omega^j \otimes \varepsilon_\rho. \tag{3.4}$$

By taking the exterior derivative of (3.3) we obtain the Bianchi identity [19]

$$d\Omega = [\Omega\omega], \tag{3.5}$$

from which it follows that

$$dR^\rho_{ij} - R^\rho_{kj}\omega^k_i - R^\rho_{ik}\omega^k_j + R^\sigma_{ij}\omega^\rho_\sigma = \widetilde{R}^\rho_{ijk}\omega^k, \tag{3.6}$$

---

1) A global formulation of the theorem is given in [5], but it is somewhat imprecise: 1) the requirement that G be connected is omitted; 2) in a number of assumptions it is required that Ω be a form of the type ad G, which, in fact, is a consequence of the theorem.

where $\widetilde{R}^{\rho}_{[ijk]} = 0$ and

$$\omega^{\rho}_{\sigma} = -C^{\rho}_{\sigma\tau}\omega^{\tau}. \tag{3.7}$$

Hence the quantities R^{ρ}_{ij} form a differential-geometric object on E which is called the *curvature object* of the given connection.

If each domain $p^{-1}(U) \subset E$ is related to an arbitrarily chosen cobasis $\{\omega^1, \cdots, \omega^{r+n}\}$ satisfying the structure equations (1.10) and (1.11′), then, in the general case, the form of the connection defined in E is given by (1.17):

$$\omega' = \omega - \Gamma^{\rho}_{i}\omega^{i} \otimes \varepsilon_{\rho}.$$

Since

$$d\omega' = -\frac{1}{2}[\omega'\omega'] + \omega^{i} \wedge (d\Gamma^{\rho}_{i} - \Gamma^{\rho}_{j}\omega^{j}_{i} + \Gamma^{\sigma}_{i}\omega^{\rho}_{\sigma} + \omega^{\rho}_{i}) \otimes \varepsilon_{\rho} \tag{3.8}$$
$$+ \frac{1}{2}\omega^{i} \wedge \omega^{j} \otimes [\Gamma_{i}\Gamma_{j}],$$

ω' is a connection form if and only if

$$d\Gamma^{\rho}_{i} - \Gamma^{\rho}_{j}\omega^{j}_{i} + \Gamma^{\sigma}_{i}\omega^{\rho}_{\sigma} + \omega^{\rho}_{i} = \Gamma^{\rho}_{ij}\omega^{j}. \tag{3.9}$$

Hence the system of quatities Γ^{ρ}_{i} forms a differential-geometric object, which we call the *object of the connection* [17]. It follows from (3.8), (3.9) and (3.4) that the curvature object is given by

$$R^{\rho}_{ij} = -(2\Gamma^{\rho}_{[ij]} + C^{\rho}_{\sigma\tau}\Gamma^{\sigma}_{i}\Gamma^{\tau}_{j}). \tag{3.10}$$

4. R^{ρ}_{ij} and its extended objects allow us to calculate the infinitesimal holonomy group of the connection Δ at any point $z \in E$ [20].

Let R^{0}_{z} denote the linear subspace of G' spanned by the elements $\epsilon_{ij} = R^{\rho}_{ij}\epsilon_{\rho}$ taken at the point $z \in E$. It follows from (3.6) that $\delta\epsilon_{ij} = \epsilon_{kj}\omega^{k}_{i}(\delta) + \epsilon_{ik}\omega^{k}_{j}(\delta)$; that is, R^{0}_{z} is invariantly associated to the point z.

Next, an extension of the system (3.6) with the help of the structure equations (1.11) and (1.18) leads to the equation

$$d\widetilde{R}^{\rho}_{ijk} = \widetilde{R}^{\rho}_{ljk}\omega^{l}_{i} + \widetilde{R}^{\rho}_{ilk}\omega^{l}_{j} + \widetilde{R}^{\rho}_{ijl}\omega^{l}_{k} + R^{\rho}_{li}\omega^{l}_{ik} + R^{\rho}_{il}\omega^{l}_{jk} + R^{\sigma}_{ij}C^{\rho}_{\sigma\tau}\omega^{\tau}_{k}$$
$$+ \widetilde{R}^{\sigma}_{ijk}C^{\rho}_{\sigma\tau}\omega^{\tau} + \overline{R}_{ijkl}\omega^{l}. \tag{3.11}$$

If we form the quantities

$$R^{\rho}_{ijk} = \widetilde{R}^{\rho}_{ijk} + R^{\sigma}_{ij} C^{\rho}_{\sigma\tau} \Gamma^{\tau}_{k}, \tag{3.12}$$

then, in view of (3.6), (3.11) and (3.9), we obtain the following equations for them:

$$\begin{aligned} dR^{\rho}_{ijk} = R^{\rho}_{ljk}\omega^{l}_{i} + R^{\rho}_{ilk}\omega^{l}_{j} + R^{\rho}_{ijl}\omega^{l}_{k} + R^{\rho}_{lj}\omega^{l}_{ik} + R^{\rho}_{il}\omega^{l}_{jk} \\ + R^{\sigma}_{ijk} C^{\rho}_{\sigma\tau}\omega^{\tau} + \widetilde{R}^{\rho}_{ijkl}\omega^{l}. \end{aligned} \tag{3.13}$$

Hence the linear subspace $R^{1}_{z} \subset G'$ spanned by the elements $\epsilon_{ij} = R^{\rho}_{ij}\epsilon_{\rho}$ and $\epsilon_{ijk} = R^{\rho}_{ijk}\epsilon_{\rho}$, taken at the point $z \in E$, is also invariantly associated to z. This process of constructing subspaces of G' can be continued. Suppose we have a collection of the quantities $R^{\rho}_{ijk_1\dots k_u}$ satisfying

$$\begin{aligned} dR^{\rho}_{ijk_1\dots k_u} = \sum^{0\leqslant t\leqslant u}_{\lambda_{(-1,\,t,\,u)}} \left(R^{\rho}_{ilk_{\lambda_1}\dots k_{\lambda_t}}\omega^{l}_{jk_{\lambda_{t+1}}\dots k_{\lambda_u}} + R^{\rho}_{ljk_{\lambda_1}\dots k_{\lambda_t}}\omega^{l}_{ik_{\lambda_{t+1}}\dots k_{\lambda_u}}\right) \\ + \sum^{0\leqslant s<t\leqslant u}_{\lambda_{(s,t,u)}} R^{\rho}_{ijk_1\dots k_s lk_{\lambda_{s+2}}\dots k_{\lambda_t}}\omega^{l}_{k_{s+1}k_{\lambda_{t+1}}\dots k_{\lambda_u}} + R^{\sigma}_{ijk_1\dots k_u} C^{\rho}_{\sigma\tau}\omega^{\tau} + \widetilde{R}^{\rho}_{ijk_1\dots k_u k_{u+1}}\omega^{k_{u+1}}. \end{aligned} \tag{3.14}$$

By extending (3.11) with the help of (1.11) and (1.18) we can immediately check that the quantities

$$R^{\rho}_{ijk_1\dots k_{u+1}} = \widetilde{R}^{\rho}_{ijk_1\dots k_{u+1}} + R^{\sigma}_{ijk_1\dots k_u} C^{\rho}_{\sigma\tau}\Gamma^{\tau}_{k_{u+1}} \tag{3.15}$$

satisfy equations analogous to (3.14), only formed for the value $u+1$. This means that the linear subspace $R^{u+1}_{z} \subset G'$ spanned by the elements

$$\varepsilon_{ij} = R^{\rho}_{ij}\varepsilon_{\rho}, \quad \varepsilon_{ijk} = R^{\rho}_{ijk}\varepsilon_{\rho}, \quad \dots, \quad \varepsilon_{ijk_1\dots k_{u+1}} = R^{\rho}_{ijk_1\dots k_{u+1}}\varepsilon_{\rho},$$

taken at $z \in E$ is invariantly associated to z.

Theorem 2. *The linear space in the Lie algebra G' that is spanned by the elements*

$$\varepsilon_{ij}, \quad \varepsilon_{ijk}, \quad \dots, \quad \varepsilon_{ijk_1\dots k_u}, \quad \dots,$$

taken at some point $z \in E$, is a Lie subalgebra of G'.

Proof. By applying the vector fields $[Z_l Z_m]$ to the functions $R^{\rho}_{ijk_1\dots k_u}$ defined on E we obtain, in view of (1.1), that

$$[Z_l Z_m] R^{\rho}_{ijk_1\dots k_u} = Q^k_{lm} Z_k R^{\rho}_{ijk_1\dots k_u} + R^{\tau}_{lm} Z_\tau R^{\rho}_{ijk_1\dots k_u}. \tag{3.16}$$

Here $Z_k R^{\rho}_{ijk_1\dots k_u}$ is the value of the differential $dR^{\rho}_{ijk_1\dots k_u}$ on the horizontal vector field Z_k. According to (3.14) and (3.15) $Z_k R^{\rho}_{ijk_1\dots k_u}\epsilon_\rho \in G'$ belongs to R^{u+1}_z, because for the horizontal field Z_k we have $\omega^\tau(Z_k) = \Gamma^\tau_i \omega^i(Z_k)$. Similarly we see that the element

$$[Z_l Z_m] R^{\rho}_{ijk_1\dots k_u} \varepsilon_\rho = \frac{1}{2}(Z_l Z_m R^{\rho}_{ijk_1\dots k_u} \varepsilon_\rho - Z_m Z_l R^{\rho}_{ijk_1\dots k_u} \varepsilon_\rho)$$

of G' belongs to R^{u+2}_z.

When considering the value of the differential $dR^{\rho}_{ijk_1\dots k_u}$ for the fundamental vector field Z_τ, we must take into account that $\omega^\phi(Z_\tau) = \delta^\phi_\tau$ and $\omega^i(Z_\tau) = 0$. Hence $Z_\tau R^{\rho}_{ijk_1\dots k_u}\epsilon_\rho$ is given by (3.14) as the sum of an element of the subspace R^u_z and of the element $R^{\sigma}_{ijk_1\dots k_u} C^{\rho}_{\sigma\tau}\epsilon_\rho$. Taking all this into account, on the basis of (3.16) we can say that

$$C^{\rho}_{\sigma\tau} R^{\sigma}_{ijk_1\dots k_u} R^{\rho}_{lm} \varepsilon_\rho \tag{3.17}$$

belongs to R^{u+2}_z; briefly, $[R^u_z R^0_z] \subset R^{u+2}_z$. If we apply the horizontal vector field Z_q to the expression for the element (3.17) and take the above remarks into consideration, we find that $C^{\rho}_{\sigma\tau} R^{\sigma}_{ijk_1\dots k_u} R^{\tau}_{lmq}\epsilon_\rho$ belongs to R^{u+3}_z; briefly, $[R^u_z R^1_z] \subset R^{u+3}_z$. By repeating this process we find that $[R^u_z R^v_z] \subset R^{u+v+2}_z$. The theorem is proved.

The Lie subalgebra of G' referred to in Theorem 2 coincides with the Lie algebra of the infinitesimal holonomy group $\psi'_z(\Delta)$ of the connection Δ at the point $z \in E$; this group was introduced in [21] for a linear connection and in [20] for the general case of a connection in a principal G-bundle E (see also [3]). In the main our proof of Theorem 2 repeats the arguments used in [20] and [21].

5. The dimensions $\dim R^u_z$ $(u = 1, 2, \cdots)$ form a nondecreasing sequence of natural numbers bounded from above by $\dim \psi'_z(\Delta) \leq \dim G = r$. Hence there is a first value $u = q$ such that $\dim R^q_z = \dim \psi'_z(\Delta)$. The Lie algebra of $\psi'_z(\Delta)$ obviously coincides with R^q_z; consequently, R^q_z determines a unique subgroup $\psi'_z(\Delta)$ in G.

Let us recall some well-known results concerning holonomy groups (see [20], [21] and also [3]); the extent to which our algorithm for calculating these groups can be applied will be clear from these results.

We can define the holonomy group $\Psi_z(\Delta)$ of a connection Δ in E at $z \in E$ as the set of $g \in G$ such that the points z and $z \circ g^{-1}$ can be joined by a horizontal path (that is, by the integral path of a horizontal distribution Δ). If the base B satisfies the second axiom of countability, then the connected component of the identity of $\Psi_z(\Delta)$ coincides with the restricted holonomy group $\psi_z(\Delta)$ consisting of elements $g \in G$ such that z and $z \circ g^{-1}$ can be joined by a horizontal path λ, the projection $p(\lambda)$ of which into B is homotopic to zero.

Let $\psi_z(\Delta, U)$ denote the holonomy group of the connection Δ in the inverse image $p^{-1}(U)$ of some neighborhood U of the point $p(z) \in B$. The subgroup $\psi^*_z(\Delta) = \bigcap_{U \ni p(z)} \psi_z(\Delta, U) \subset G$ is said to be the *local holonomy group* of the connection Δ in E at $z \in E$. If the principal G-bundle E and connection Δ in it are analytic, then $\psi_z(\Delta) = \psi'_z(\Delta)$ at each point $z \in E$. In the nonanalytic case, if

$$\dim \psi'_z(\Delta) = \dim \psi^*_z(\Delta) \tag{3.18}$$

at all points of E, then $\psi'_z(\Delta) = \psi^*_z(\Delta) = \psi(\Delta)$. If (3.18) is satisfied at some point $z \in E$, then there exists a neighborhood $U \subset B$ of $p(z) \in B$ such that $\psi'_z(\Delta) = \psi_z(\Delta, U)$.

In the above cases the linear subspace $R^q_z \subset G'$, which is the Lie algebra of $\psi'_z(\Delta)$, uniquely determines the groups $\psi_z(\Delta)$ and $\psi_z(\Delta, U)$ respectively.

§4. Connections in homogeneous bundles

1. When studying connections Δ^* in G/H-bundles E^* we must rely on the correspondence that exists between them and the connections Δ in the corresponding principal G-bundles E. It is desirable to reduce the direct reversion to the connection Δ to a minimum in order to obtain formulas suitable for concrete applications. To achieve this aim we must solve a number of problems which are also of theoretical interest. In this section we establish (Theorem 3) necessary and sufficient conditions for a distribution Δ^* on E^*, complementary to the vertical distribution Θ^*, to be a horizontal distribution on E^*, or a connection in E^*.

2. Let a distribution Δ^*, complementary to the vertical distribution Θ^*, be defined on the homogeneous G/H-bundle E^*. We can define the plane

$\Delta^*_{z^*}$ of Δ^* at any point $z^* \in E^*$ by the system of covectors

$$'\omega^a_{z^*} = \omega^a_{z^*} - (\Gamma^a_i)_{z^*}\,\omega^i_{z^*}, \tag{4.1}$$

which are annihilated by this plane. The invariance of $\Delta^*_{z^*}$ at z^* is characterized by the condition

$$\delta'\omega^a_{z^*} = -\,'\omega^b_{z^*}\omega^a_b(\delta),$$

where δ is the fundamental vector field on E that was introduced in §2.2. Since, in view of (2.5) and (2.6), we find from (4.1) that

$$\delta'\omega^a_{z^*} = -\,'\omega^b_{z^*}\omega^a_b(\delta) - \{\delta(\Gamma^a_i)_{z^*} - (\Gamma^a_j)_{z^*}\omega^j_i(\delta) + (\Gamma^b_i)_{z^*}\omega^a_b(\delta) + \omega^a_i(\delta)\}\,\omega^i_{z^*},$$

it follows that this condition is equivalent to demanding that the coefficients in the Pfaffian equations

$$\omega^a - \Gamma^a_i\omega^i = 0, \tag{4.2}$$

(the system of these equations defines a distribution Δ^*) satisfy the conditions

$$d\Gamma^a_i - \Gamma^a_j\omega^j_i + \Gamma^b_i\omega^a_b + \omega^a_i = \Gamma^a_{bi}\omega^b + \Gamma^a_{ij}\omega^j, \tag{4.3}$$

where Γ^a_{bi} and Γ^a_{ij} are new quantities that are completely defined by the system of the Γ^a_i and their first partial derivatives. To derive a system of differential equations for the Γ^a_{bi} it is necessary to augment the structure equations (2.2), (2.8) and (2.9) by the equations

$$d\omega^i_j = \omega^k_j \wedge \omega^i_k + \omega^k \wedge \omega^i_{jk}$$

and

$$d\omega^a_i = \omega^j_i \wedge \omega^a_j + \omega^b_i \wedge \omega^a_b + \omega^b \wedge \tilde{\omega}^a_{bi} + \omega^j \wedge \omega^a_{ij}$$

which arise from (2.2) and (2.8) respectively. Then, after exterior differentiation, we obtain from (4.3) that

$$\left(\nabla\Gamma^a_{bi} + \frac{1}{2}C^a_{c\alpha}C^\alpha_{bd}\Gamma^c_i\omega^d\right) \wedge \omega^b + \nabla\Gamma^a_{ij} \wedge \omega^j = 0,$$

where, in particular,

$$\nabla\Gamma^a_{bi} = d\Gamma^a_{bi} - \Gamma^a_{bj}\omega^j_i - \Gamma^a_{ci}\omega^c_b + \Gamma^c_{bi}\omega^a_c + \Gamma^c_i\omega^a_{cb} + \omega^a_{bi}. \tag{4.4}$$

Hence

$$\nabla \Gamma^a_{bi} = \Gamma^a_{bc;i}\omega^c + \Gamma^a_{bi;j}\omega^j. \tag{4.5}$$

The forms ω^a_{cb} in (4.4) are defined by taking $u = 2$ in (2.12), and the forms ω^a_{bi} are given by $\omega^a_{bi} = - C^a_{b\rho}\omega^\rho_i$.

3. To carry out the further extensions, which turn out to be necessary in the case of general G/H-bundles, we must find the structure equations for the latter forms and, in general, for forms such as

$$\omega^a_{b_1 \ldots b_u i} = - C^a_{b_1 \ldots b_u \rho}\omega^\rho_i, \text{ where } C^a_{b_1 \ldots b_u \rho} = (-1)^{u-1} C^a_{b_1 \sigma_1} C^{\sigma_1}_{b_2 \sigma_2} \ldots C^{\sigma_{u-1}}_{b_u \rho}.$$

Lemma 4. *The following holds:*

$$C^a_{b_1 \ldots b_u \rho} C^\rho_{\sigma\alpha} = \sum_{\lambda(s,t,u)}^{0 \leqslant s < t \leqslant u} \left(C^c_{b_{s+1} b_{\lambda_{s+2}} \ldots b_{\lambda_t} \sigma} C^a_{c b_1 \ldots b_s b_{\lambda_{t+1}} \ldots b_{\lambda_u} \alpha} \right.$$
$$\left. - C^c_{b_{s+1} b_{\lambda_{s+2}} \ldots b_{\lambda_t} \alpha} C^a_{b_1 \ldots b_s c b_{\lambda_{t+1}} \ldots b_{\lambda_u} \sigma} \right) + (-1)^{u-1} C^a_{\sigma \beta_1} C^{\beta_1}_{b_1 \beta_2} \ldots C^{\beta_u}_{b_u \alpha}. \tag{4.6}$$

Proof. It follows from the Jacobi identity that

$$C^a_{b\rho} C^\rho_{\sigma\alpha} = C^c_{b\sigma} C^a_{c\alpha} - C^c_{b\alpha} C^a_{c\sigma} + C^a_{\sigma\beta} C^\beta_{b\alpha}; \tag{4.7}$$

hence (4.6) holds when $u = 1$.

Let us assume that (4.6) is valid for some value u. In view of the Jacobi identity and (4.6) we have that

$$C^a_{b_1 \ldots b_{u+1} \rho} C^\rho_{\sigma\alpha} = C^a_{b_1 \ldots b_u \tau} \left(C^\tau_{c\sigma} C^c_{b_{u+1}\alpha} + C^\tau_{\sigma\beta} C^\beta_{\alpha b_{u+1}} - C^\tau_{\rho\alpha} C^\rho_{b_{u+1}\sigma} \right)$$
$$= - C^c_{b_{u+1}\alpha} C^a_{b_1 \ldots b_u c \sigma} -$$
$$- \left\{ \sum_{\lambda(s,t,u)}^{0 \leqslant s < t \leqslant u} \left(C^c_{b_{s+1} b_{\lambda_{s+2}} \ldots b_{\lambda_t} \sigma} C^a_{c b_1 \ldots b_s b_{\lambda_{t+1}} \ldots b_{\lambda_u} \beta} \right. \right.$$
$$\left. \left. - C^c_{b_{s+1} b_{\lambda_{s+2}} \ldots b_{\lambda_t} \beta} C^a_{b_1 \ldots b_s c b_{\lambda_{t+1}} \ldots b_{\lambda_u} \sigma} \right) + (-1)^{u-1} C^a_{\sigma\beta_1} C^{\beta_1}_{b_1\beta_2} \ldots C^{\beta_u}_{b_u \beta} \right\} C^\beta_{b_{u+1}\alpha}$$
$$- \left\{ \sum_{\lambda(s,t,u)}^{0 \leqslant s < t \leqslant u} \left(C^c_{b_{s+1} b_{\lambda_{s+2}} \ldots b_{\lambda_t} \rho} C^a_{c b_1 \ldots b_s b_{\lambda_{t+1}} \ldots b_{\lambda_u} \alpha} - \right. \right.$$
$$\left. \left. - C^c_{b_{s+1} b_{\lambda_{s+2}} \ldots b_{\lambda_t} \alpha} C^a_{b_1 \ldots b_s c b_{\lambda_{t+1}} \ldots b_{\lambda_u} \rho} \right) + (-1)^{u-1} C^a_{\rho\beta_1} C^{\beta_1}_{b_1\beta_2} \ldots C^{\beta_u}_{b_u \alpha} \right\} C^\rho_{b_{u+1}\sigma},$$

and it is not difficult to see that the last expression is the same as that on the right-hand side of (4.6) for the value $u+1$. The lemma is thus proved.

The extension of the structure equations (1.10) gives

$$d\omega_i^\rho = \omega_i^j \wedge \omega_j^\rho - C_{\sigma\tau}^\rho \omega_i^\sigma \wedge \omega^\tau + \omega^j \wedge \omega_{ij}^\rho.$$

It is not not difficult, by using (4.6), to find the following structure equations:

$$d\omega_{b_1\ldots b_u i}^a = -C_{b_1\ldots b_u \rho}^a (\omega_i^j \wedge \omega_j^\rho - C_{\sigma c}^\rho \omega_i^\sigma \wedge \omega^c - C_{\sigma\alpha}^\rho \omega_i^\sigma \wedge \omega^\alpha + \omega^j \wedge \omega_{ij}^\rho),$$

that is,

$$\begin{aligned} d\omega_{b_1\ldots b_u i}^a &= \omega_i^j \wedge \omega_{b_1\ldots b_u j}^a \\ &+ \sum_{\lambda(s,t,u)}^{0\leqslant s<t\leqslant u} (\omega_{b_{s+1} b_{\lambda_{s+2}}\ldots b_{\lambda_t} i}^c \wedge \omega_{c b_1\ldots b_s b_{\lambda_{t+1}}\ldots b_{\lambda_u}}^a + \omega_{b_{s+1} b_{\lambda_{s+2}}\ldots b_{\lambda_t}}^c \\ &\wedge \omega_{b_1\ldots b_s c b_{\lambda_{t+1}}\ldots b_{\lambda_u} i}^a) + \omega^c \wedge \omega_{b_1\ldots b_u c i}^a + \omega_i^c \wedge \omega_{c b_1\ldots b_u}^a + \omega^j \wedge \omega_{b_1\ldots b_u i j}^a. \end{aligned} \tag{4.8}$$

4. We can now begin to carry out the successive extensions of the system (4.5). At the first stage we will only be interested in the quantities $\Gamma_{b_1\cdots b_u i}^a$ $(u = 1, 2, \cdots)$–the coefficients of the forms ω^c on the right-hand sides of the systems of differential equations analogous to (4.3) and (4.5).

We can represent the results of these extensions by the general formula

$$\begin{aligned} d\Gamma_{b_1\ldots b_u i}^a - \Gamma_{b_1\ldots b_u j}^a \omega_i^j - \sum_{\lambda(s,t,u)}^{0\leqslant s<t\leqslant u} \omega_{b_{s+1} b_{\lambda_{s+2}}\ldots b_{\lambda_t}}^c \Gamma_{b_1\ldots b_s c b_{\lambda_{t+1}}\ldots b_{\lambda_u} i}^a \\ + \sum_{\lambda(-1,t,u)}^{0\leqslant t\leqslant u} \Gamma_{b_{\lambda_1}\ldots b_{\lambda_t} i}^c \omega_{c b_{\lambda_{t+1}}\ldots b_{\lambda_u}}^a + \omega_{b_1\ldots b_u i}^a = \Gamma_{b_1\ldots b_u c i}^a \omega^c + \Gamma_{b_1\ldots b_u i j}^a \omega^j, \end{aligned} \tag{4.9}$$

whose validity can be verified by induction.

5. Let us assume that our distribution Δ^* on E^*, complementary to the vertical distribution Θ, is a horizontal distribution. Then there is a horizontal distribution Δ on the corresponding principal G-bundle E whose image under the differential π' of π: $E \to E^* = E/H$ is Δ^*. The distribution Δ annihilates some connection form $\omega' = {}'\omega^a \otimes \epsilon_a + {}'\omega^\alpha \otimes \epsilon_a$ on E, where ${}'\omega^a = \omega^a - \Gamma_i^a \omega^i$ and ${}'\omega^\alpha = \omega^\alpha - \Gamma_i^\alpha \omega^i$ $(a, b, \cdots = n+1, \cdots, n+m;$ $\alpha, \beta, \cdots = n+m+1, \cdots, n+r)$. Here we choose the basis $\{\epsilon_{n+1}, \cdots, \epsilon_{n+r}\}$ in the Lie algebra G' so that the last $r-m$ of its elements belong to H'. Then the plane $\Delta_{z^*}^*$ of the horizontal distribution Δ^* at any point $z^* \in E^*$

annihilates the system of covectors $'\omega^a_{z*}$, and Δ^* is consequently defined by the equations

$$\omega^a - \Gamma^a_i \omega^i = 0. \tag{4.10}$$

The coefficients Γ^a_i in these equations are part of the components of an object of a connection in E, and so satisfy the system of differential equations

$$d\Gamma^a_i - \Gamma^a_j \omega^j_i + \Gamma^b_i \omega^a_b + \omega^a_i = -(C^a_{be}\Gamma^e_i + C^a_{b\alpha}\Gamma^\alpha_i)\omega^b + \Gamma^a_{ij}\omega^j, \tag{4.11}$$

which arises from the system (3.9). Here the Γ^α_i are the remaining components of the object of the connection; they satisfy the system

$$d\Gamma^\alpha_i - \Gamma^\alpha_j \omega^j_i - \Gamma^\beta_i C^\alpha_{\beta\gamma}\omega^\gamma - \Gamma^a_i C^\alpha_{a\gamma}\omega^\gamma + \omega^\alpha_i = -(C^\alpha_{ba}\Gamma^a_i + C^\alpha_{b\beta}\Gamma^\beta_i)\omega^b + \Gamma^\alpha_{ij}\omega^j, \tag{4.12}$$

which also arises from (3.9).

We now have two systems of differential equations for the coefficients Γ^a_i in the equations (4.10) for Δ^*: on the one hand the system (4.3), and on the other hand the system (4.11). A comparison yields the following expressions for Γ^a_{bi}: $\Gamma^a_{bi} = -C^a_{b\rho}\Gamma^\rho_i$. Substituting them in (4.4) now gives, in view of (4.7) and (3.9), the equations $\Gamma^a_{bci} = C^a_{b\alpha}C^\alpha_{c\rho}\Gamma^\rho_i - C^a_{bd}\Gamma^d_{ci}$. We can substitute these expressions for Γ^a_{bci} in the differential equations (4.9) (with $u = 2$) and find expressions for Γ^a_{bcdi}. By proceeding in an analogous way, after the $(u-1)$th stage we obtain the following expressions:

$$\Gamma^a_{b_1\ldots b_u i} = C^a_{b_1\ldots b_{u-1}\alpha}C^\alpha_{b_u\rho}\Gamma^\rho_i + C^a_{b_1\ldots b_{u-2}\alpha}C^\alpha_{b_{u-1}c}\Gamma^c_{b_u i} + \ldots$$
$$\ldots + C^a_{b_1\alpha}C^\alpha_{b_2 c}\Gamma^c_{b_3\ldots b_u i} - C^a_{b_1 c}\Gamma^c_{b_2\ldots b_u i}, \quad u = 2, 3, \ldots .$$

Here we must assume that $\Gamma^a_{b_u i}, \ldots, \Gamma^a_{b_2\ldots b_u i}, \Gamma^a_{b_1\ldots b_u i}$ are defined when Δ^* is defined. We obtain a system of linear equations that must be satisfied by the components Γ^α_i of the object of the connection.

5. Theorem 3. *For a distribution Δ^*, on the G/H-bundle E^*, defined by the equations* (4.2):

$$\omega^a - \Gamma^a_i \omega^i = 0$$

(*and, hence, complementary to the vertical distribution Θ^* on E^**) *to be a horizontal distribution of some connection in E^*, it is necessary and* *sufficient that the system of linear equations*

$$
\begin{aligned}
C^a_{b\alpha}\Gamma^\alpha_i &= -C^a_{bc}\Gamma^c_i - \Gamma^a_{bi},\\
C^a_{b_1b_2\alpha}\Gamma^\alpha_i &= C^a_{b_1\alpha}C^\alpha_{b_2c}\Gamma^c_i - C^a_{b_1c}\Gamma^c_{b_2i} - \Gamma^a_{b_1b_2i},\\
&\dots\dots\dots\dots\dots\dots\\
C^a_{b_1b_2\dots b_{p+1}\alpha}\Gamma^\alpha_i &= C^a_{b_1\dots b_p\alpha}C^\alpha_{b_{p+1}c}\Gamma^c_i + C^a_{b_1\dots b_{p-1}\alpha}C^\alpha_{b_pc}\Gamma^c_{b_{p+1}i}\\
&\dots + C^a_{b_1\alpha}C^\alpha_{b_2c}\Gamma^c_{b_3\dots b_{p+1}i} - C^a_{b_1c}\Gamma^c_{b_2\dots b_{p+1}i} - \Gamma^a_{b_1\dots b_{p+1}i}
\end{aligned}
\tag{4.13}
$$

in the unknowns Γ^α_i *be compatible; here* p *is the order of the isotropy of the standard fiber* G/H *and* $\Gamma^a_{bi}, \Gamma^a_{b_1b_2i}, \dots, \Gamma^a_{b_1\dots b_{p+1}i}$ *are the objects arising from successive differential extensions of the objects* Γ^a_i, *starting from the system of differential equations* (4.3) *for the invariance of the distribution given by* (4.2).

Proof. Necessity was proved above. Let us note here that the successive formation of linear equations for the Γ^α_i, in accordance with the above scheme, breaks down when $u = p + 1$ for the same reason that the system indicated in Theorem 3 is the first (as follows from Remark 2, §2.4) which turns out to be incompatible.

We first prove the sufficiency of the condition for the simplest case when $p = 1$; that is, when the representation of the stationary subgroup H in the group of linear transformations of the tangent space $T_{y_0}(G/H)$ of the first order is faithful.

In this case the linear system given in the theorem reduces to

$$C^b_{b\alpha}\Gamma^\alpha_i = -C^a_{bc}\Gamma^c_i - \Gamma^a_{bi}, \tag{4.14}$$

$$C^a_{bc\alpha}\Gamma^\alpha_i = C^a_{b\alpha}C^\alpha_{cd}\Gamma^d_i - C^a_{bd}\Gamma^d_{ci} - \Gamma^a_{bci}, \tag{4.15}$$

where the Γ^α_i are defined uniquely from the first group of equations (4.14) (this follows from a remark in §2.4). We differentiate the equations (4.14) of this first group, use (4.3) and (4.5), substitute for Γ^a_{bi} and Γ^a_{bci} from (4.14) and (4.15) and take the Jacobi identity into account; after a number of transformations we obtain the following system:

$$C^a_{b\alpha}\Delta\Gamma^\alpha_i = \tilde{\Gamma}^a_{bij}\omega^j,$$

where

$$\Delta\Gamma^\alpha_i = d\Gamma^\alpha_i - \Gamma^\alpha_j\omega^j_i - \Gamma^\beta_i C^\alpha_{\beta\gamma}\omega^\gamma - \Gamma^a_i C^\alpha_{a\gamma}\omega^\gamma + \omega^\alpha_i + (C^\alpha_{ba}\Gamma^a_i + C^\alpha_{b\beta}\Gamma^\beta_i)\,\omega^b.$$

When $p = 1$ the total rank of the matrix $\|C^a_{ba}\|$ is $r - m$ (see the remark in §2.4); hence it follows from this system that the Γ^α_i, found from (4.14), satisfy (4.12) and hence the Γ^a_i and Γ^α_i together satisfy (3.9). Thus these quantities define a connection Δ in E which, under $\pi: E \to E^* = E/H$, is projected into a connection in E^* that obviously coincides with Δ^*.

The sufficiency of our condition in the general case is proved by an analogous scheme. If the system of linear equations (4.13) for the Γ^α_i is compatible it determines the Γ^α_i uniquely. By differentiating the equations of the first p groups we obtain, in the end, the following system:

$$C^a_{ba}\Delta\Gamma^\alpha_i = \widetilde{\Gamma}^a_{bij}\omega^j,$$

$$\cdots\cdots\cdots\cdots\cdots$$

$$C^a_{b_1\ldots b_p\alpha}\Delta\Gamma^\alpha_i = \widetilde{\Gamma}^a_{b_1\ldots b_p ij}\omega^j.$$

The matrix of this system in the $\Delta\Gamma^\alpha_i$ has the full rank (see Remark 2, §2.6); hence $\Delta\Gamma^\alpha_i = \Gamma^\alpha_{ij}\,\omega^j$; that is, the Γ^α_i satisfy the system (4.12), as we were required to prove. Thus the theorem is proved.

The object Γ^a_i (the totality of coefficients in the equations (4.2) for the distribution Δ^* on E^*) that satisfies the condition of Theorem 3 will be called the *object of the connection* Δ^* *in* E^*.

Remark 3. The quantities $\Gamma^a_{b_1\ldots b_u i}$ $(u = 1, 2, \cdots, p + 1)$ found on the right-hand sides of the equations of the system (4.13) were obtained by successively extending the system (4.2) for Δ^* on E^*. Hence they essentially depend only on the coefficients Γ^a_i in these equations and on their partial derivatives, up to order u, in the local coordinates in E^* that determine the position of $z^* \in E^*$ in the given fiber. Because we are using free nonholonomic systems of frames on E^*, our expressions for $\Gamma^a_{b_1\ldots b_u i}$ contain functions that determine the position of a frame relative to a natural system of frames connected with the chosen system of local coordinates, and the partial derivatives of these functions; however, the dependence on them is not essential and, when desired, can be removed by going over to this natural system of frames.

It follows from what we have said that the condition for the coefficients Γ^a_i in the equations (4.2) for Δ^* to form the object of a connection in E^*, is a system of partial differential equations, with derivatives up to order $p + 1$, for the functions Γ^a_i.

Example. Let us assume that the structure of the homogeneous standard fiber G/H of E^* is such that the basis for the Lie algebra G' of G can be selected so that, in addition to (2.1), the following equations hold:

$$C^{\alpha}_{a\beta} = 0. \tag{4.16}$$

If k denotes the linear subspace of G' spanned by the basis elements ϵ_a $(a = n+1, \cdots, n+m)$, then equations (4.16) are equivalent to the inclusion $[kH'] \subset k$; that is, that G/H, this structure, is a so-called *reductive homogeneous space* [22], [23].

In view of (4.16) it follows from (2.12) and (2.13) that

$$C^{a}_{b_1 \ldots b_u \alpha} = 0 \ (u \geqslant 2), \quad \omega^{a}_{b_1 \ldots b_u} = 0 \ (u \geqslant 2). \tag{4.17}$$

Hence the order of the isotropy of the reductive space G/H is $p = 1$. In fact, equations (4.17) imply that $H'_2 \equiv H'_1$. Since, in view of (2.18), $[H'_2 G'] \subset H'_1$, there is only one possibility when the group G acts effectively: $H_2 = H_1 = \{e\}$.

It now follows from Theorem 3 that the condition for the horizontal character of the distribution Δ^*, defined by $\omega^a - \Gamma^a_i \omega^i = 0$ on the G/H-bundle E^* with reductive G/H, reduces to the compatibility condition for the system

$$C^{a}_{b\alpha}\Gamma^{\alpha}_{i} = -C^{a}_{bc}\Gamma^{c}_{i} - \Gamma^{a}_{bi} \tag{4.18}$$

for the Γ^a_i and to the equations

$$\Gamma^{a}_{bci} = C^{a}_{b\alpha}C^{\alpha}_{cd}\Gamma^{d}_{i} - C^{a}_{bd}\Gamma^{d}_{ci}. \tag{4.19}$$

If we represent the relations of linear dependence between the forms $\omega^a_b = -C^a_{b\alpha}\omega^\alpha$, which exist in this case, as $\lambda^{ab}_{a}\omega^a_b = 0$, then the compatibility condition for the system (4.18) reduces to

$$\lambda^{ab}_{a}(C^{a}_{bc}\Gamma^{c}_{i} + \Gamma^{a}_{bi}) = 0, \tag{4.20}$$

and so the condition for Δ^* to be horizontal reduces to the system (4.19) and (4.20).

This system, which is a system of second-order partical differential equations in the functions Γ^a_i, can be simplified in the following special cases.

If G/H is a symmetric homogeneous space, then, in addition to (2.1) and (4.16), we have $C^a_{bc} = 0$. The system (4.19), (4.20) reduces to

$$\Gamma^a_{bci} = C^a_{b\alpha} C^\alpha_{cd} \Gamma^d_i, \tag{4.21}$$

$$\lambda^{ab}_a \Gamma^a_{bi} \doteq 0. \tag{4.22}$$

If $k \subset G'$ turns out to be a subalgebra it is necessarily an ideal, and so G is, at least locally, a semidirect product $k * H$. In this case the homogeneous space G/H is said to be *locally grouped*; also, in addition to (2.1) and (4.16), we have $C^\alpha_{cd} = 0$ and the equations (4.19) have the form

$$\Gamma^a_{bci} = - C^a_{bd} \Gamma^d_{ci}. \tag{4.23}$$

§5. Curvature and torsion

1. When investigating the deeper properties of a connection Δ^* in a G/H-bundle E^* we must take account of the particular structure that is determined in the Lie algebra G' by the homogeneous space G/H.

We can identify the linear factor-space G'/H' with the tangent space to G/H at the point with stationary subgroup H. The following sequence of ideals, embedded in one another, is defined in H':

$$H'_1 \supset H'_2 \supset \ldots \supset H'_p = \{0\};$$

these ideals are the Lie algebras of the kernels of the natural homomorphisms of H into the nonholonomy differential groups $\widetilde{D}_u(m, R)$ acting in higher tangent spaces to G/H of orders $u = 1, \cdots, p$. Here, in view of a theorem on homomorphisms, we can identify H'/H'_u with the Lie algebra of the image of H in $\widetilde{D}_u(m, R)$ under this homomorphism.

If a connection Δ^* is defined in the G/H-bundle E^*, then the Lie algebra R^q_z of the infinitesimal holonomy group $\psi'_z(\Delta)$ of the corresponding connection Δ in the principal G-bundle E, as a subalgebra of G' with this particular structure, determines a number of images in Δ^* which are invariantly associated to points $z^* \in E^*$. An invariant linear subspace of the tangent space $\Theta^*_{z^*}$ to the fiber is defined at each point $z^* = \pi(z)$; this subspace is tangential to the orbit of z^* under the group $\psi'_z(\Delta)$ acting in $p^{*-1}(x) \ni z^*$. When z^* is a fixed point of $\psi'_z(\Delta)$, to it are associated, in an invariant way, the subalgebras of the Lie algebras H'/H'_u of all the groups that are the images of the stationary subgroup H under the natural homomorphisms into the groups $\widetilde{D}_u(m, R)$.

Analogous constructions can be carried out on the base of each of the linear subspaces $R^1_z, \dots, R^{q-1}_z$ of R^q_z that were found in §3.4. There is a whole series of invariant subspaces, subalgebras, and objects defined by them associated with the connection Δ^*; we study these in the present section.

2. We are going to consider the set of components R^a_{ij} of the curvature object R^ρ_{ij} of Δ in E. In view of (2.1) it follows from (3.6) that

$$dR^a_{ij} = R^a_{kj}\omega^k_i + R^a_{ik}\omega^k_j - R^b_{ij}\omega^a_b + R^\rho_{ij}C^a_{\rho b}\omega^b + \widetilde{R}^a_{ijk}\omega^k. \tag{5.1}$$

Hence, at each fixed point $z^* \in E^*$, the components R^a_{ij} form an independent object, called the *torsion object* of the connection Δ^* at z^*.

It is not difficult to derive an expression for them directly in terms of the components Γ^a_i of the object of the connection and of the extended objects of this object; namely, it follows from (3.10) and (4.13) that

$$R^a_{ij} = -2(\Gamma^a_{[ij]} + \Gamma^a_{b[i}\Gamma^b_{j]}) + C^a_{bc}\Gamma^b_i\Gamma^c_j. \tag{5.2}$$

We can use the torsion object to define a linear subspace $R^{*0}_{z^*}$ in the tangent space $\Theta^*_{z^*}$ to a fiber; $R^{*0}_{z^*}$ is spanned by the vectors $e_{ij} = R^a_{ij}e_a$. It follows from (5.1) and (2.6′) that $\delta^* e_{ij} = e_{ki}\omega^k_i(\delta^*) + e_{ik}\omega^k_j(\delta^*)$; hence $R^{*0}_{z^*}$ is invariantly associated to z^*. We can regard it as the domain of values of the Θ_{z^*}-valued 2-form $\Sigma = \Sigma^a e_a$, where $\Sigma^a = -\frac{1}{2} R^a_{ij}\omega^i \wedge \omega^j$. By using (5.1) and (2.6) it is easy to check that $\delta^*\Sigma^a = -\Sigma^b\omega^a_b(\delta^*)$, and hence, in view of (2.6′), that $\delta^*\Sigma = 0$.

The forms Σ^a are said to be the *torsion forms* of the connection Δ^* in E^*. They have been considered earlier for certain concrete (projective, conformal, etc.) homogeneous standard fibers G/H. The concept of the torsion of a connection Δ^* for more general classes of G/H-bundles E^* has been introduced only under the assumption that $\dim G/H = \dim B$: in [24] and [25] when the spaces G/H are reductive; in [26] for general G/H; the form Σ has been considered only at points of a given intersecting surface pasted to the bundle E^*.

The association of the forms Σ^a to all the points $z^* \in E^*$ in the manner indicated above allows us to prove

Theorem 4. *The torsion forms Σ^a of the connection Δ^* in the G/H-bundle E^* vanish at all the points of the manifold E^* if and only if the horizontal distribution Δ^* is involutory. For this, in turn, it is necessary and sufficient that the distribution Δ on the principal G-bundle E, whose image under the differential π' of the projection $\pi: E \longrightarrow E^* = E/H$ is Δ^*, be involutory.*

Proof. The distribution Δ^* on E^* is defined by the Pfaffian system (4.2), and Δ^* is involutory if and only if this system is completely integrable. By using (2.2), (2.8), (4.3) and (5.2) it is not difficult to derive the corresponding condition: $R^a_{ij} = 0$.

If Δ is involutory then in (1.1) we must have $R^\rho_{ij} = 0$, and it is obvious that the condition $R^a_{ij} = 0$ for Δ^* on E^* to be involutory is satisfied.

Conversely, let Δ^* be involutory; that is, let $R^a_{ij} = 0$. Then we find from (5.1) that $C^a_{b\alpha} R^\alpha_{ij} = 0$. From here, after differentiation in accordance with the formula

$$dR^\alpha_{ij} = R^\alpha_{kj}\omega^k_i + R^\alpha_{ik}\omega^k_j + R^\sigma_{ij}C^\alpha_{\sigma\gamma}\omega^\gamma + R^\sigma_{ij}C^\alpha_{\sigma c}\omega^c + \widetilde{R}^\alpha_{ijk}\omega^k, \tag{5.3}$$

which follows from (3.6), in view of (2.14) (with $u = 1$) and (2.13), we find that $C^a_{b_1 b_2 \alpha} R^\alpha_{ij} = 0$ when $R^a_{ij} = 0$. Similarly, after the $(p-1)$th step we obtain that $C^a_{b_1 \dots b_p \alpha} R^\alpha_{ij} = 0$.

It follows from here that $R^\alpha_{ij} = 0$ (see Remark 2, §2.6). As a result we have that $R^\rho_{ij} = 0$, and consequently the distribution Δ is involutory and our theorem is proved.

3. By analogy with the way in which we introduced the torsion object, at each point $z^* \in E^*$ we can define a series of objects which we naturally call the *torsion objects of higher order.*

Namely, in view of (2.1) we find from (3.14) that

$$\begin{aligned} dR^a_{ijk_1\dots k_u} = &\sum_{\lambda_{(-1,t,u)}}^{0\leqslant t\leqslant u} \left(R^a_{ilk_{\lambda_1}\dots k_{\lambda_t}}\omega^l_{jk_{\lambda_{t+1}}\dots k_{\lambda_u}} + R^a_{ljk_{\lambda_1}\dots k_{\lambda_t}}\omega^l_{ik_{\lambda_{t+1}}\dots k_{\lambda_u}}\right) \\ &+ \sum_{\lambda_{(s,t,u)}}^{0\leqslant s<t\leqslant u} R^a_{ijk_1\dots k_s lk_{\lambda_{s+2}}\dots k_{\lambda_t}}\omega^l_{k_{s+1}k_{\lambda_{t+1}}\dots k_{\lambda_u}} \\ &- R^b_{ijk_1\dots k_u}\omega^a_b + R^\rho_{ijk_1\dots k_u}C^a_{\rho b}\omega^b + \widetilde{R}^a_{ijk_1\dots k_u k}\omega^k. \end{aligned} \tag{5.4}$$

It follows from here that, at each point $z^* \in E^*$, the set of quantities R^a_{ij},

$R^a_{ijk_1}, \cdots, R^a_{ijk_1\dots k_u}$ forms an independent object, which we call the *torsion object of order* u of the connection Δ^* in E^*.

In the space $\Theta^*_{z^*}$ tangential to a fiber we can define the linear subspace $R^{*s}_{z^*}$ spanned by the vectors

$$e_{ij} = R^a_{ij}e_a, e_{ijk_1} = R^a_{ijk_1}e_a, \dots, e_{ijk_1\dots k_s} = R^a_{ijk_1\dots k_s}e_a.$$

It follows from (5.4) and (2.6′) that the $\delta^* e_{ijk_1\dots k_u}$ can be expressed in terms of the vectors $e_{i'j'}, e_{i'j'k_1}, \cdots, e_{i'j'k'_1\dots k'_u}$ for each $u = 1, \cdots, s$. Hence $R^{*s}_{z^*}$ is invariantly associated to $z^* \in E^*$.

Let $\overline{R}^s_z$ denote the linear subspace of the tangent space Θ_z to a fiber that is spanned by the vectors at the point z of those fundamental vector fields on E that correspond to elements from $R^s_z \subset G'$. If $z^* = \pi(z)$ under the projection π: $E \longrightarrow E^*$, then it is obvious that $R^{*s}_{z^*}$ is the image of $\overline{R}^s_z$: $R^{*s}_{z^*} = \pi'(\overline{R}^s_z)$.

The Lie algebra of the infinitesimal holonomy group $\psi'_z(\Delta)$ of the connection Δ at the point $z \in E$ coincides with some subspace $R^q_z \subset G'$ (see §3.5). The corresponding subspace $\overline{R}^q_z \subset \Theta_z$ is consequently the tangent space at $z \in E$ to the orbit of this point under the action of the group $\psi'_z(\Delta)$ in $p^{-1}(x) \ni z$. Hence its image $R^{*q}_{z^*}$ is the tangent space at $z^* = \pi(z)$ to the orbit of this point under the action of $\psi'_z(\Delta)$ in $p^{*-1}(x) \ni z^*$.

For example, in the case examined in Theorem 4, the space $R^{*q}_{z^*}$ is zero-dimensional at each point $z^* \in E^*$; hence any point z^* is fixed relative to the group $\psi'_z(\Delta)$; that is, $\psi'_z(\Delta)$ consists of only the identity element. In this case the full holonomy group $\Psi_z(\Delta)$ is isomorphic to the fundamental group of the base B of the bundles E and E^* and, when B is simply connected, also consists of only the identity element.

In contrast to principal G-bundles, on G/H-bundles E^* there can exist horizontal distributions Δ^* that admit singular isolated intersecting surfaces of E^*consisting largely of the fixed points of the holonomy groups $\Psi_z(\Delta)$. For example, any horizontal surface (that is, an intersecting surface in E^* whose tangent plane at each point z^* coincides with $\Delta^*_{z^*}$), if it exists, is such a surface. In fact, a horizontal curve over any closed curve in B, starting at a point z of a horizontal surface, obviously completely belongs to this surface, and hence z is fixed relative to $\Psi_z(\Delta)$.

If Δ^* on E^* admits a horizontal intersecting surface only over a domain $U \subset B$, then any point $z^* = \pi(z)$ of this surface is fixed relative to the holonomy group $\Psi_z(\Delta, U)$; in particular, relative to the local and infinitesimal holonomy groups $\psi^*_z(\Delta)$ and $\psi'_z(\Delta)$. It is not difficult to verify this last statement analytically. In fact, it is obvious that at the points of such a surface we have $R^a_{ij} = 0$ and $\omega^a = \Gamma^a_i \omega^i$, where the 1-forms ω^i are linearly independent. Hence, in view of (5.1), (5.3) and (3.15), along the hroizontal surface we have

$$R^a_{ijk_1\ldots k_u} = 0 \quad (u = 0, 1, \ldots, q), \tag{5.5}$$

and so its points z^* are fixed relative to $\psi'_z(\Delta)$.

Conversely, if the points z^* of some intersecting surface of the bundle E^* are fixed relative to $\psi'_z(\Delta)$, then in general this surface is horizontal. In fact, at the points of such a surface equations (5.5) hold and $\omega^a = \gamma^a_i \omega^i$, where the ω^i $(i = 1, \cdots, n)$ are linearly independent. It follows from (3.15), (5.3) and (5.5) that along the surface we have

$$R^\alpha_{ijk_1\ldots k_u} C^a_{\alpha b} (\gamma^b_k - \Gamma^b_k) = 0 \quad (u = 0, 1, \ldots .$$

If the rank of this linear homogeneous system in $\gamma^b_k - \Gamma^b_k$ has the maximum possible value m (in general this situation is possible), then at the points of the intersecting surface being considered we have $\gamma^b_k = \Gamma^b_k$, and so the surface is horizontal.

Depending on the properties of the G/H-bundle E^* and the connection Δ^* in E^*, there can be a discrete (in particular, empty) set of horizontal intersecting surfaces or a set of the power of the continuum. In other words, in the somewhat old-fashioned terminology of S. Lie, the action of $\Psi_z(\Delta)$ on a fiber can be asystatic or systatic [27]. In the first case the connection Δ^* in E^* is said to be *asystatic*, and in the second case, *systatic*.

Let us note that the curvature of a systatic connection Δ in a principal G-bundle E is necessarily zero, and that a horizontal intersecting surface of the connection passes through each point $z \in E$. Moreover, the presence of at least one horizontal intersecting surface in E reduces to this case. However there is a greater variety of connections Δ^* in the G/H-bundles E^*.

4. As we explained in §3.4, the Lie algebra R^q_z of the infinitesimal holonomy group $\psi'_z(\Delta)$ is spanned by the elements

$$\varepsilon_{ij} = R^\rho_{ij}\varepsilon_\rho, \; \varepsilon_{ijk} = R^\rho_{ijk}\varepsilon_\rho, \; \ldots, \; \varepsilon_{ijk_1\ldots k_q} = R^\rho_{ijk_1\ldots k_q}\varepsilon_\rho.$$

It intersects the Lie algebra H' in some subalgebra ${}^H R^q_z = R^q_z \cap H' \subset G'$. We have given above an interpretation of the factor-space $R^q_z / {}^H R^q_z$, which turned out to be isomorphic to the tangent space to the orbit of $z^* = \pi(z)$ under $\psi'_z(\Delta)$. By using the invariant subspaces of this tangent space analogous interpretations can be given to the linear subspaces R^u_z of R^q_z.

It remains for us to explain the significance of the subalgebra ${}^H R^q_z$. To do this we use the natural representations, defined by (2.12), of the stationary subgroup H in the nonholonomy differential groups $\widetilde{D}_u(m, R)$.

The component of an arbitrary element $\epsilon_{ijk_1 \dots k_u}$ of R^q_z in H' has the coordinates $R^{\alpha}_{ijk_1 \dots k_u}$. Hence the following element of the Lie algebra of $\widetilde{D}_u(m, R)$ corresponds to an arbitrary element of ${}^H R^q_z$ under the homomorphism (2.12):

$$R^a_{b_1 \dots b_{u'} ijk_1 \dots k_u} = -C^a_{b_1 \dots b_{u'} \alpha} R^{\alpha}_{ijk_1 \dots k_u}. \tag{5.6}$$

By using the expression for $dR^{\alpha}_{ijk_1 \dots k_u}$ which follows from (3.14), and also (2.14), we arrive at the following equations:

$$\begin{aligned} dR^a_{b_1 \dots b_{u'} ijk_1 \dots k_u} = &\sum_{\lambda_{(s,t,u)}}^{0 \leqslant s < t \leqslant u} R^a_{b_1 \dots b_{u'} ijk_1 \dots k_s l k_{\lambda_{s+2}} \dots k_{\lambda_t}} \omega^l_{k_{s+1} k_{\lambda_{t+1}} \dots k_{\lambda_u}} \\ &+ \sum_{\lambda_{(-1,t,u)}}^{0 \leqslant t \leqslant u} (R^a_{b_1 \dots b_{u'} ilk_{\lambda_1} \dots k_{\lambda_t}} \omega^l_{jk_{\lambda_{t+1}} \dots k_{\lambda_u}} + R^a_{b_1 \dots b_{u'} ljk_{\lambda_1} \dots k_{\lambda_t}} \omega^l_{ik_{\lambda_{t+1}} \dots k_{\lambda_u}}) \\ &- \sum_{\lambda_{(s,t,u')}}^{0 \leqslant s < t \leqslant u'} (R^c_{b_{s+1} b_{\lambda_{s+2}} \dots b_{\lambda_t} ijk_1 \dots k_u} \omega^a_{b_1 \dots b_s c b_{\lambda_{t+1}} \dots b_{\lambda_{u'}}} \\ &- \omega^c_{b_{s+1} b_{\lambda_{s+2}} \dots b_{\lambda_t} ijk_1 \dots k_u} R^a_{b_1 \dots b_s c b_{\lambda_{t+1}} \dots b_{\lambda_{u'}}}) \\ &- R^c_{ijk_1 \dots k_u} \omega^a_{b_1 \dots b_{u'} c} - R^{\sigma}_{ijk_1 \dots k_u} C^a_{b_1 \dots b_{u'} \alpha} C^{\alpha}_{\sigma c} \omega^c + R^a_{b_1 \dots b_{u'} ijk_1 \dots k_u k} \omega^k. \end{aligned} \tag{5.7}$$

They show that the elements (5.6), together with similar elements corresponding to all smaller nonnegative integral values of u and u', form an independent object, which is called the *curvature object* of Δ^* of order (u', u). The components of the curvature object of order (u', q) define a natural representation of the infinitesimal holonomy group $\psi'_z(\Delta)$ in the nonholonomy differential group $\widetilde{D}_u(m, R)$ acting in the higher tangent space of order u' to $p^{*-1}(x) \ni z^*$ at $z^* = \pi(z)$. Let us note that the curvature object of order $(0, u)$ is the same as the torsion object of order u'.

The simplest of the remaining curvature objects is that of order (1, 0); this is defined by the components R^a_{ij} and R^a_{bij}. For the latter we can derive expressions, analogous to (5.2), which give them directly in terms of the components Γ^a_i of the object of the connection and of its extended objects. Indeed from (3.10) and (2.14) (with $u' = 1$) we obtain that

$$R^a_{bij} = 2C^a_{b\alpha}\Gamma^\alpha_{[ij]} + 2C^a_{bc\alpha}\Gamma^\alpha_{[i}\Gamma^c_{j]} - 2\,(C^a_{c\beta}\Gamma^\beta_{[i})\,(C^c_{|b\gamma|}\Gamma^\gamma_{j]}) + C^a_{b\alpha}C^\alpha_{cd}\Gamma^c_i\Gamma^d_j, \quad (5.8)$$

and it remains to substitute from (4.13) into these equations. In general we can express, in principle, all the quantities (5.6) in terms of the components Γ^a_i of the object and of its extended objects. The same assertion can be made for the components of the uth order torsion object $R^a_{ijk_1\ldots k_u}$. For example, in view of (5.6) (with $u' = 0$) and (4.14), we find from (3.12) that $R^a_{ijk} = \widetilde{R}^a_{ijk} - R^b_{ij}\Gamma^a_{bk} - R^a_{bij}\Gamma^b_k$.

In other words, there are formulas that allow us, with the help of the object Γ^a_i of Δ^* in E^*, and by using differential extensions and rational integral operations, to calculate all the objects which completely characterize the action of $\psi'_z(\Delta)$ at $z^* = \pi(z)$.

When $u' = 1$ and $u = 0$, equations (5.7) give

$$dR^a_{bij} = R^a_{bkj}\omega^k_i + R^a_{bik}\omega^k_j$$
$$+ R^a_{cij}\omega^c_b - R^c_{bij}\omega^a_c - R^c_{ij}\omega^a_{bc} - R^\sigma_{ij}C^a_{b\alpha}C^\alpha_{\sigma c}\omega^c + R^a_{bijk}\omega^k. \quad (5.9)$$

If next we form the 2-forms

$$\Omega^a_b = -\frac{1}{2}R^a_{bij}\,\omega^i \wedge \omega^j, \quad (5.10)$$

then from (5.9) and (2.6) we obtain that

$$\delta^*\Omega^a_b = \Omega^a_c\omega^c_b(\delta^*) - \Omega^c_b\omega^a_c(\delta^*) - \Sigma^c\omega^a_{bc}(\delta^*). \quad (5.11)$$

The forms Ω^a_b are called the *curvature forms of the first order.*

More generally the 2-forms

$$\Omega^a_{b_1\ldots b_u} = -\frac{1}{2}R^a_{b_1\ldots b_u ij}\omega^i \wedge \omega^j \quad (5.12)$$

are called the *curvature forms of order* u. It follows from (5.7) (with $u = 0$)

and (2.6) that

$$\delta^*\Omega^a_{b_1\dots b_u} = \sum_{\lambda\,(s,t,u)}^{0\leqslant s<t\leqslant u} (\omega^c_{b_{s+1}b_{\lambda_{s+2}}\dots b_{\lambda_t}}(\delta^*)\,\Omega^a_{b_1\dots b_s c b_{\lambda_{t+1}}\dots b_{\lambda_u}}$$

$$- \Omega^c_{b_{s+1}b_{\lambda_{s+2}}\dots b_{\lambda_t}}\omega^a_{b_1\dots b_s c b_{\lambda_{t+1}}\dots b_{\lambda_u}}(\delta^*)) - \Sigma^c\omega^a_{b_1\dots b_u c}(\delta^*). \tag{5.13}$$

It is clear that, in general, they form an independent object at the point z^* only together with the preceding curvature forms and components of the torsion form. The vanishing of the curvature forms of order u has an invariant sense only in the following cases: either 1) at a given point $z^* \in E^*$ at which $\Sigma^c = 0$, or 2) for a G/H-bundle on which the forms $\omega^{ab}_{b_1 b_2}$ can vanish (then the forms $\omega^a_{b_1 b_2 b_3}, \dots, \omega^a_{b_1\dots b_u}$ are also zero).

5. In the second case, in view of (2.10), there must be a basis in G' in which

$$C^a_{b\beta}C^b_{ca} = 0. \tag{5.14}$$

It turns out that the corresponding class of homogeneous spaces G/H coincides with the class of reductive spaces G/H (see the example in §4.5). In fact the subalgebra H'_2, which, in general, in view of (2.19), is defined by the system $C^a_{b\beta}C^\beta_{ca}\xi^a = 0$, in this case coincides with H'_1; further, because $[H'_2 G']$ (in view of (2.18)), when G is effective there is only one possibility: $H'_2 = H'_1 = \{0\}$; in other words the order of the isotropy of G/H is $p = 1$. Hence when $u = 1$ the system (2.21) has only the trivial solution, and so it follows from (5.14) that

$$C^\beta_{ca} = 0. \tag{5.15}$$

The space G/H is reductive. Hence we have proved

Theorem 5. *The curvature forms Ω^a_b of the connection Δ^* in the G/H-bundle E^* form an independent object at each point $z^* \in E^*$ if and only if G/H is a reductive homogeneous space.*

In other words, it is only when G/H is reductive that there are connections Δ^* in E^* that have, at some point, zero curvature and nonzero torsion.

Let us note that in this case, since the order of the isotropy of G/H is $p = 1$, the torsion form Σ and the curvature forms Ω^a_b allow us to completely recover the curvature object R^ρ_{ij} of the connection Δ.

An object defined at each point of a G/H-bundle E^* is said to be *semibasic* if its components are transformed amongst themselves when a fiber is fixed (that is, when $x \in B$ is fixed) and not only when the point z^* in a fiber is fixed. If an object is given by a system of differential equations that determine the transformation of its components under an infinitesimal displacement of the frame, then the object is semibasic if and only if the differential of any one of its components is a linear combination of only the components of the object when the 1-forms ω^i are zero.

Theorem 6. *The curvature forms* Ω^a_b *of the connection* Δ^* *in the* G/H-*bundle* E^* *form a semibasic object on* E^* *if and only if* G/H *is locally grouped.*

In other words, the identical vanishing of all the curvature forms when the torsion forms are nonzero can only occur when G/H is locally grouped.

Proof. It follows from (5.9) that the R^a_{bij} form a semibasic object only when the forms $\omega^a_{bc} = C^a_{b\beta} C^\beta_{c\alpha}\omega^\alpha$ and the quantities $C^a_{b\alpha} C^\alpha_{cd}$ are zero; that is, the space G/H must be, firstly, reductive and, secondly, such that $C^\alpha_{cd} = 0$. These conditions are also sufficient. The theorem is proved.

§6. Examples

1. First we point out two ways of interpreting a linear connection on a differentiable manifold V_n as connections Δ^* in certain concrete G/H-bundles E^*.

A. We can treat a linear connection on V_n as a horizontal distribution Δ^* on the tangent vector bundle $T(V_n)$ of V_n. The latter is a G/H-bundle E^* with base $B = V_n$, for which $G = GL(n, R)$ and H is the subgroup of matrices of the form $\left\|\begin{smallmatrix}1 & a\\ 0 & A\end{smallmatrix}\right\|$. A linear connection on V_n has been considered from this point of view, for example, in [28] and [29].

Let us note that we can similarly treat a higher connection (*surconnexion*) of order p on V_n as a horizontal distribution on the tangent hypervector bundle $T_p(V_n)$ of order p (cf. [17], [30]). Here $T_p(V_n)$ is the G/H-bundle in which G/H is the same as the homogeneous space considered in the example in §2.6.

B. We can also interpret a linear connection on V_n with the help of a connection in an affine bundle in which V_n plays the role of an intersecting surface.

A G/H-bundle is called an *affine bundle* if G/H is an affine space; that is, if G is the semidirect product $GA(n, R) = T(n) * GL(n, R)$ of the additive group of vectors $T(n)$ by $GL(n, R)$ and H coincides with $GL(n, R)$.

The structure equations of an affine bundle have the form

$$d\omega^i = \omega^j \wedge \omega^i_j, \tag{6.1}$$

$$d\omega^a = \omega^b \wedge \omega^a_b + \omega^i \wedge \omega^a_i, \tag{6.2}$$

$$d\omega^a_b = \omega^c_b \wedge \omega^a_c + \omega^i \wedge \omega^a_{bi}; \tag{6.3}$$

by comparing these with (2.7), (2.8) and (2.9) we see that, in this case, G/H is both a symmetric and a locally grouped reductive space (see the example in §4.5).

Hence the condition for the distribution Δ^*, defined by the equations $\omega^a - \Gamma^a_i\,\omega^i = 0$, on an affine bundle to be horizontal reduces to the equations $\Gamma^a_{bci} = 0$.

In view of (5.2) the torsion object is given by

$$R^a_{ij} = -2\,(\Gamma^a_{[ij]} + \Gamma^a_{b[i}\Gamma^b_{j]}), \tag{6.4}$$

and, in view of (5.8) and (4.13), the curvature object is expressed by

$$R^a_{bij} = -2\,(\Gamma^a_{b[ij]} + \Gamma^a_{c[i}\Gamma^c_{|b|j]}). \tag{6.5}$$

Let us note that an affine space is an example of a grouped homogeneous space, and hence, in view of Theorem 6, the curvature object (6.5) of a connection Δ^* in an affine bundle is semibasic.

From here we obtain a vector bundle and a linear connection in it if we fix some intersecting surface in the affine bundle (assuming this surface exists globally). At points of this surface we have

$$\omega^a = \gamma^a_i\,\omega^i, \tag{6.6}$$

where, in view of (6.2), $d\gamma^a_i = \gamma^a_j\,\omega^j_i - \gamma^b_i\,\omega^a_b - \omega^a_i + \gamma^a_{ij}\,\omega^j$.

There remain the structure equations (6.1), (6.3) that determine the structure of a principal $GL(m, R)$-bundle.

We can choose a frame in the tangent spaces to an affine bundle at points z^* of an intersecting surface, so that the vectors e_i $(i = 1, \cdots, n)$ belong to the horizontal distribution Δ^*. Then the planes $\Delta^*_{z^*}$ are determined by the equations $\omega^a = 0$, and consequently in (6.4) we have

$$\Gamma^b_j = 0.$$

We obtain the classical case from this result if we assume that

1) $\dim B = \dim G/H$ (that is, $n = m$) and 2) the tangent plane to the intersecting surface at the point z^* intersects the horizontal plane $\Delta^*_{z^*}$ only at z^*. Then the matrix $\|\gamma^a_i\|$ is nonsingular, and, by a suitable transformation of the vectors $e_a (a = n + 1, \cdots, 2n)$, we can reduce equations (6.6) to $\omega^{n+i} = \omega^i$.

Hence at points of the intersecting surface we can simply identify the indices $n+i$ and i, the forms ω^{n+i} and ω^i, the vectors e_{n+1} and e_i, etc. From (6.4) and (6.5) we obtain the classical equations

$$R^k_{ij} = -2\Gamma^k_{[ij]}, \qquad R^l_{kij} = -2(\Gamma^l_{k[ij]} + \Gamma^l_{m[i}\Gamma^m_{|k|j]})$$

for the torsion and curvature tensors of the linear connection defined by the object Γ^k_{ij}.

3. Let E^* be a projective bundle; that is, the standard fiber G/H is a real projective space P_n in which the group $GP(m, R)$ of projective transformations acts; as is well known, this group is isomorphic to the factor-group $GL(m+1, R)/Z$ of the full linear group $GL(m+1, R)$ with respect to its center Z of scalar matrices $\|\lambda\delta^b_a\|$. We can reduce the Maurer-Cartan structure equations of $GP(m, R)$ to the form

$$d\omega^a = \omega^b \wedge \omega^a_b, \tag{6.7}$$

$$d\omega^a_b = \omega^c_b \wedge \omega^a_c + (\delta^a_b\omega_c + \delta^a_c\omega_b) \wedge \omega^c, \tag{6.8}$$

$$d\omega_b = \omega^a_b \wedge \omega_a \tag{6.9}$$

$$(a, b, \ldots = 1, \ldots, m).$$

The stationary subgroup H of a point of P_n is distinguished by the equations $\omega^a = 0$. Hence the set of values of the index α splits into a collection of m^2 values of the pair of indices $\alpha' \equiv (a, b)$ and of m values of the lower index $\alpha'' \equiv c$ in the forms ω_c.

It follows from the structure equations (6.8) and (6.9) that

$$\omega^a_{bc} = \delta^a_b\omega_c + \delta^a_c\omega_b, \qquad \omega^a_{bcd} = 0.$$

Hence $p = 2$. In this case the condition for a distribution Δ^* on E^* complementary to the vertical distribution Θ^* to be horizontal is the compatibility condition for the system

$$C^a_{b\alpha}\Gamma^\alpha_i = -C^a_{bc}\Gamma^c_i - \Gamma^a_{bi},$$
$$C^a_{bc\alpha}\Gamma^\alpha_i = C^a_{b\alpha}C^\alpha_{cd}\Gamma^d_i - C^a_{bd}\Gamma^d_{ci} - \Gamma^a_{bci},$$
$$C^a_{bcd\alpha}\Gamma^\alpha_i = C^a_{bc\beta}C^\beta_{de}\Gamma^e_i + C^a_{b\alpha}C^\alpha_{ce}\Gamma^e_{di} - C^a_{be}\Gamma^e_{cdi} - \Gamma^a_{bcdi}.$$

In the case being considered, since

$$C^a_{bc} = 0, \quad C^\alpha_{bc} = 0, \quad C^a_{bc\alpha} = \delta^a_b\delta_{c\alpha} + \delta^a_e\delta_{b\alpha}, \quad C^a_{bcd\alpha} = 0,$$

this system reduces to the equations $\Gamma^a_{bci} = -(\delta^a_b\Gamma_{ci} + \delta^a_c\Gamma_{bi})$, $\Gamma^a_{bcdi} = 0$, where we have used the notation $\Gamma_{ci} = -(1/(m+1))\Gamma^a_{aci}$.

In view of (5.2) the torsion object of the connection Δ^* is given by

$$R^a_{ij} = -2\,(\Gamma^a_{[ij]} + \Gamma^a_{b[i}\Gamma^b_{j]}). \tag{6.10}$$

The curvature object of order (1, 0) is formed by the quantities (6.10) and

$$R^a_{bij} = -2\,(\Gamma^a_{b[ij]} + \Gamma^a_{c[i}\Gamma^c_{|b|j]} + \delta^a_b\Gamma_{c[i}\Gamma^c_{j]} + \Gamma_{b[i}\Gamma^a_{j]}), \tag{6.11}$$

which are derived from (5.8) and (4.13), and the curvature object of order (2, 0) is formed by the quantities (6.10), (6.11) and $R^a_{bcij} = \delta^a_b R_{cij} + \delta^a_c R_{bij}$, where

$$R_{aij} = -2\,(\Gamma_{a[ij]} + \Gamma_{b[i}\Gamma^b_{|a|j]}), \tag{6.12}$$

as is shown by the appropriate calculations.

The classical case of the space of a projective connection is obtained from here as well as in the preceding example, when $\dim B = \dim G/H$, by taking a fixed intersecting surface in the projective bundle E^*.

4. Let E^* be a conformal bundle; that is, the standard fiber is the real conformal space C_n in which the group of conformal transformations acts.

For simplicity we restrict ourselves to the case of a positive definite angular metric; our results can be carried over to the general case without difficulty. If we replace the indices $a, b, \cdots = 1, \cdots, m$ in (6.7), (6.8) and (6.9) by $A, B, \cdots = 1, \cdots, m+1$, then the following completely integrable Pfaffian system distinguishes a subgroup of conformal transformations:

$$\omega^{m+1} = \omega_{m+1} = 0,$$

$$\omega_a^{m+1} = -\omega^a, \quad \omega_{m+1}^a = -\omega_a,$$

$$\omega_a^b + \omega_b^a = \omega_{m+1}^{m+1}\delta_{ab},$$

and hence the subgroup has the structure equations

$$d\omega^a = \omega^b \wedge \omega_b^a,$$

$$d\omega_b^a = \omega_b^c \wedge \omega_c^a + \omega^c \wedge (\delta_{bc}\omega_a - \delta_b^a\omega_c - \delta_c^a\omega_b),$$

$$d\omega_b = \omega_b^a \wedge \omega_a$$

$$(a, b, \ldots = 1, \ldots, m).$$

The stationary subgroup H of a point of C_n is distinguished by the equations $\omega^a = 0$. It is clear that in this case we also have $p = 2$.

The conditions for the horizontal character of the distribution Δ^* defined by

$$\omega^a - \Gamma_i^a \omega^i = 0$$

on a conformal bundle E^* take the form

$$\Gamma_{ai}^b + \Gamma_{bi}^a = \delta_{ab}\Gamma_i,$$

$$\Gamma_{bci}^a = -(\delta_b^a\Gamma_{ci} + \delta_c^a\Gamma_{bi} - \delta_{bc}\Gamma_{ai}),$$

$$\Gamma_{bcdi}^a = 0, \tag{6.13}$$

where we have used the notation

$$\Gamma_{ai} = -\frac{1}{m}\Gamma_{bai}^b, \quad \Gamma_i = \frac{2}{m}\Gamma_{ai}^a.$$

In addition it is clear that the conditions obtained from (6.13) by a differential prolongation are satisfied.

The torsion object is expressed, as before, by (6.10). To derive the curvature object of order (1, 0) we must add the equation

$$R_{bij}^a = -2\,(\Gamma_{b[ij]}^a + \Gamma_{c[i}^a\Gamma_{|b|j]}^c + \delta_b^a\Gamma_{c[i}\Gamma_{j]}^c + \Gamma_{b[i}\Gamma_{j]}^a - \Gamma_{a[i}\Gamma_{j]}^b);$$

here $R_{bij}^a + R_{aij}^b = \delta_{ab}R_{cij}^c$. To obtain the curvature object of order (2, 0) we must add the equation

$$R_{bcij}^a = \delta_b^a R_{cij} + \delta_c^a R_{bij} - \delta_{bc}R_{aij},$$

where the R_{aij} are given by (6.12).

As above the classical case is obtained by selecting an intersecting surface of the bundle when $\dim B = \dim G/H$.

BIBLIOGRAPHY

[1] V. V. Nemyckiĭ and V. V. Stepanov, *Qualitative theory of differential equations,* 2nd ed., GITTL, Moscow, 1949; English transl., Princeton Math. Series, no. 22, Princeton Univ. Press, Princeton, N. J., 1960. MR 10, 612; 22 #12258.

[2] K. Nomizu, *Lie groups and differential geometry,* The Mathematical Society of Japan, Tokyo, 1956; Russian transl., IL, Moscow, 1960. MR 18, 821; 22 #8456.

[3] A. Lichnerowicz, *Théorie globale des connexions et des groupes d'holonomie,* Edizioni Cremonese, Rome, 1955; Russian transl., IL, Moscow, 1960. MR 19, 453; 23 #A1320.

[4] G. F. Laptev, *Differential geometry of imbedded manifolds. Group theoretical method of differential geometric investigations,* Trudy Moskov. Mat. Obšč. 2 (1953), 275–382. (Russian) MR 15, 254.

[5] Ju. G. Lumiste, *On the foundations of the global theory of connections,* Tartu Riikl. Ül. Toimetised No. 150 (1964), 69–108. (Russian) MR 33 #4842.

[6] Ch. Ehresmann, *Les connexions infinitésimales dans un espace fibré différentiable,* Colloq. Topologie (espaces fibrés, Brussels, 1950), George Thone, Liège et Masson, Paris, 1951, pp. 29–55. MR 13, 159.

[7] S. P. Finikov, *Cartan's method of exterior forms in differential geometry,* OGIZ, Moscow, 1948. (Russian) MR 11, 597.

[8] A. M. Vasil'ev, *General invariant methods in differential geometry,* Dokl. Akad. Nauk SSSR 79 (1951), 5–7. (Russian) MR 13, 159.

[9] Ju. G. Lumiste, *Induced connections in imbedded projective and affine bundles,* Tartu Riikl. Ül. Toimetised. No. 177 (1965), 6–42. (Russian) MR 35 #2238.

[10] J. A. Schouten, *Ricci-calculus. An introduction to tensor analysis and its geometrical applications,* Die Grundlehren der math. Wissenschaften, Band 10, Springer-Verlag, Berlin, 1954. MR 16, 521.

[11] R. König, *Beiträge zu einer allgemeinen linearen Mannigfaltigkeitslehre,* Jber. Deutsch Math. Verein. 28 (1919), 213–228.

[12] É. Cartan, *Sur les variétés à connexion affine et la théorie de la relativité généralisée.* I, II, Ann. Sci. École Norm. Sup. (3) 40 (1923), 325–412; 41 (1924), 1–25; 42 (1925), 17–88; *Sur les variétés à connexion projective,* Bull. Soc. Math. France 52 (1924), 205–241; *Les espaces à connexion conforme,* Ann. Soc. Polon Math. 2 (1923), 171–221; Russian transl., *Spaces of affine, projective and conformal connectivity,* Studies in Non-Euclidean Geometry, no. 3, Izdat. Kazan, Univ., Kazan, 1962. MR 36 #2080.

[13] L. Schlesinger, *Parallelverschiebung und Krummungstensor*, Math. Ann. 99 (1928), 413–434.

[14] I. Cattaneo Gasparini, *Sulle connessioni infinitesimali nello spazio fibrato dei riferimenti affini di una* V_n, Rend. Mat. e Appl. (5) 17 (1958), 327–404. MR 24 #A512.

[15] A. Švec, *L'application des variétés à connexion à certains problèmes de la géométrie différentielle*, Czechoslovak Math. J. 10 (85) (1960), 523–550. MR 26 #2974.

[16] B. Cenkl, *Les variétés de König généralisées*, Czechoslovak Math. J. 14 (89) (1964), 1–21. MR 29 #5197.

[17] G. F. Laptev, *Manifolds imbedded in generalized spaces*, Proc. Fourth All-Union Math. Congress (Leningrad, 1961) vol. 2, "Nauka", Leningrad, 1964, pp. 226–233. (Russian) MR 36 #4473.

[18] P. K. Raševskiĭ, *On linear representations of differential groups and Lie groups with nilpotent radical*, Trudy Moskov. Mat. Obšč. 6 (1957), 337–370. (Russian) MR 20 #929.

[19] S.-S. Chern, *Complex manifolds*, Textos de Matemática, no. 5, Instituto de Fisca e Matemática, Universidade do Recife, 1959; Russian transl., IL, Moscow, 1961. MR 22 #1920.

[20] H. Ozeki, *Infinitesimal holonomy groups of bundle connections*, Nagoya Math. J. 10 (1956), 105–123. MR 18, 232.

[21] A. Nijenhuis, a) *On the holonomy groups of linear connections.* I, Nederl. Akad. Wetensch. Proc. Ser. A. 56 = Indag. Math. 15 (1953), 233–249. MR 16, 171.
b) II, Nederl. Akad. Wetensch. Proc. Ser. A. 57 = Indag. Math. 16 (1954), 17–25. MR 16, 172.

[22] P. K. Raševskiĭ, *On the geometry of homogeneous spaces*, Trudy Sem. Vektor. Tenzor Anal. 9 (1952), 49–74. (Russian) MR 14, 795.

[23] K. Nomizu, *Invariant affine connections on homogeneous spaces*, Amer. J. Math. 76 (1954), 33–65. MR 15, 468.

[24] S. Kobayashi, *On connections of Cartan*, Canad. J. Math. 8 (1956), 145–156. MR 17, 1126.

[25] ———, *Theory of connections*, Ann. Mat. Pura Appl. (4) 43 (1957), 119–194. MR 20 #2760.

[26] A. I. Sirota, *On torsion spaces with an infinitesimal connection*, Akad. Nauk Gruzin. SSR Trudy Tbiliss. Mat. Inst. Razmadze 27 (1960), 3–9. (Russian)

[27] N. G. Čebotarev, *The theory of Lie groups*, GITTL, Moscow, 1940. (Russian)

[28] T. Ōtsuki, *On the tangent bundles of order* 2 *and affine connections*, Proc. Japan Acad. 34 (1958), 325–330. MR 21 #3018.

[29] M. Rahula, *On differential geometry of higher order,* Tartu Riikl. Ül. Toimetised No. 150 (1964), 122–131. (Russian) MR 33 #4834.

[30] P. Libermann, *Surconnexions. Propriétés générales,* C. R. Acad. Sci. Paris 258 (1964), 6327–6330. MR 29 #3997.

Translated by:
L. Longdon.

Amer. Math. Soc. Transl.
(2) Vol. 92, 1970

THE THEORY OF THE CURVATURE OF X_2 IN E_4*

UDC 513.82

K. Š. RAMAZANOVA

Individual problems on the curvature theory of two-dimensional surfaces in a four-dimensional space have been considered in [3], [4], [6] and [7]: a generalization of Meusnier's Theorem is considered in [6] and [4]; the normal curvature vector is examined in [3], [4], and [6]; Cartan [3] has derived the parametric equations for the ellipse of normal curvature for X_2 in R_4 and, for these surfaces, has introduced the concepts of lines of curvature and of asymptotic lines. By constructing the curvature theory for a dual normalized X_2 in E_4, Čakmazjan [4] has obtained an analog of Euler's formula for these surfaces.

In the present article we use a tensor method to obtain the equation of the ellipse of normal curvature, we introduce the concept of principal curvatures and principal directions of a surface, and we give the equations of the lines of curvature and obtain an analog of Euler's formula for X_2 in E_4. We also consider some special classes of surfaces.

§1. The ellipse of normal curvature

We consider the two-dimensional surface in four-dimensional euclidean space defined by

$$\boldsymbol{r}=\boldsymbol{r}(u^1,\ u^2)=\boldsymbol{r}(u^i). \tag{1.1}$$

Its intersection with some hyperplane gives a curve on the surface; we call the curve which is the section of the surface by a normal hyperplane a normal section. Since the curvature of any section can easily be expressed in terms of the curvature of the normal section with the same direction [6], [4], in what follows we consider only normal sections of a surface. It is obvious that a normal section is completely determined by giving its direction.

Let the normal section have the equation

$$u^i=u^i(s), \tag{1.2}$$

---

*Translation of Izv. Vysš. Učebn. Zaved. Matematika 1966, no. 6 (55), 137–143.

and let us denote its unit tangent vector by v. Then $v = r' = r_i v^i$, where $v^i = du^i/ds$ are the coordinates of the tangent vector in the system (r_1, r_2).

We consider the projection of the curvature vector r'' of the curve (1.2) onto the normal plane; following Cartan [3], the vector we obtain is said to be the *normal curvature vector of the surface* corresponding to the given direction v, and its length is called the *normal curvature of the surface*. We can write the normal curvature vector k_n as ([4], §12)

$$k_n = b_{ij} v^i v^j n + c_{ij} v^i v^j m, \tag{1.3}$$

where b_{ij} and c_{ij} are the fundamental tensors of the surface[1)] and n and m are mutually perpendicular unit vectors lying in the normal plane of the surface.

When the direction v varies the end point of k_n describes some curve in the normal plane; following Cartan we call this curve the ellipse of normal curvature. We are going to derive the equation of this curve.

Let x, y be the coordinates of the vector k_n localized at a point on the surface and lying in the plane (n, m). Then

$$x = b_{ij} v^i v^j, \quad y = c_{ij} v^i v^j. \tag{1.4}$$

We can write the coordinates v^i of the tangent vector of a curve lying on the surface as follows ([1], §20):

$$v^1 = \frac{g \cos \psi - g_{12} \sqrt{g} \sin \psi}{g \sqrt{g_{11}}}; \quad v^2 = \sqrt{\frac{g_{11}}{g}} \sin \psi, \tag{1.5}$$

where ψ is the angle between the tangent vector v and the coordinate axis vector r_1; $g_{ij} = r_i r_j$ is the metric tensor of the surface and g is the discriminant of this tensor. By substituting (1.5) into (1.4) we obtain the parametric equations of the ellipse of normal curvature:

$$x = \frac{b_i^i}{2} + \frac{b_{11} - g b^{22}}{2 g_{11}} \cos 2\psi + \frac{\sqrt{g}\, b_1^2}{g_{11}} \sin 2\psi, \tag{1.6}$$
$$y = \frac{c_i^i}{2} + \frac{c_{11} - g c^{22}}{2 g_{11}} \cos 2\psi + \frac{\sqrt{g}\, c_1^2}{g_{11}} \sin 2\psi,$$

---

1) In [5] these are called asymptotic tensors.

where, here and in the future, the indices are raised by using g_{ij}.

We are going to eliminate the parameter ψ from the last equations; by writing (1.6) as $x = \bar{\alpha} + a\cos 2\psi + b\sin 2\psi$, $y = \bar{\beta} + c\cos 2\psi + d\sin 2\psi$ and eliminating ψ we obtain the equation

$$\begin{aligned}&(c^2+d^2)x^2 - 2(ac+bd)xy + (a^2+b^2)y^2\\ &-2[\bar{\alpha}(c^2+d^2) - \bar{\beta}(ac+bd)]x - 2[\bar{\beta}(a^2+b^2) - \bar{\alpha}(ac+bd)]y\\ &+\bar{\alpha}^2(c^2+d^2) + \bar{\beta}^2(a^2+b^2) - 2\bar{\alpha}\bar{\beta}(ac+bd) - (ad-bc)^2 = 0.\end{aligned}$$

By taking (1.6) into account it is not difficult to calculate the coefficients in this equation:

$$\begin{aligned} a^2+b^2 &= \left(\frac{b_{11}-gb^{22}}{2g_{11}}\right)^2 + \left(\frac{\sqrt{g}\,b_1^2}{g_{11}}\right)^2 = \frac{1}{4}[(b_i^i)^2 - 4N_b],\\ c^2+d^2 &= \left(\frac{c_{11}-gc^{22}}{2g_{11}}\right)^2 + \left(\frac{\sqrt{g}\,c_1^2}{g_{11}}\right)^2 = \frac{1}{4}[(c_i^i)^2 - 4N_c],\\ ac+bd &= \frac{1}{4}(b_i^i c_j^j - 2e^{mi}a^{nj}b_{mn}c_{ij}),\\ bc-ad &= -\frac{1}{4}e^{mn}s_{mn} = -\frac{1}{4}\alpha, \end{aligned} \tag{1.7}$$

where e_{ij} is the discriminant bivector, b_i^i and N_b (c_i^i and N_c) are the trace and norm of the tensor b_{ij} (c_{ij}), and $s_{mn} = b_m^k c_{nk}$.

By substituting the expressions we have found for the coefficients in the previous equation, we obtain

$$\begin{aligned}&[(c_i^i)^2 - 4N_c]x^2 - 2(b_i^i c_j^j - 2e^{mi}e^{nj}b_{mn}c_{ij})xy + [(b_i^i)^2 - 4N_b]y^2\\ &+2(2N_c b_i^i - c_i^i e^{mp}e^{nq}b_{mn}c_{pq})x + 2(2N_b c_i^i - b_i^i e^{mp}e^{nq}b_{mn}c_{pq})y\\ &+b_i^i c_j^j e^{mp}e^{nq}b_{mn}c_{pq} - N_c(b_i^i)^2 - N_b(c_i^i)^2 - (e^{mn}s_{mn})^2 = 0.\end{aligned} \tag{1.8}$$

We take the normals n and m to be parallel to the principal axes of the ellipse (1.8); that is, so that

$$e^{mp}e^{nq}b_{mn}c_{pq} = \frac{1}{2}b_i^i c_j^j. \tag{1.9}$$

By using the equation $e^{mn}e^{pq} = g^{mp}g^{nq} - g^{mq}g^{np}$ [1], we find that

$$\alpha^2 = e^{mn}e^{pq}s_{mn}s_{pq} = b^{mp}b_{pq}c^{qn}c_{nm} - b^{mp}c_{pq}b^{qn}c_{nm}.$$

On the other hand, we have

$$e^{mn}e^{ij}s_{mi}s_{nj} = 2N_bN_c = (g^{mi}g^{nj} - g^{mj}g^{ni})s_{mi}s_{nj} = (b^{ij}c_{ij})^2 - b^{im}c_{mn}b^{nk}c_{ki},$$

from which, in view of (1.9) and the equation $b^{ij} = b^k_k g^{ij} - N_b \overset{\vee}{b}{}^{ij}$, we have

$$b^{im} c_{mn} b^{nk} c_{ki} = (b^{ij} c_{ij})^2 - 2N_b N_c$$

$$= (b^i_i c^j_j - e^{mp} e^{nq} b_{mn} c_{pq})^2 - 2N_b N_c = \frac{(b^i_i c^j_j)^2}{4} - 2N_b N_c.$$

Finally, by noting that the conjugate tensor $\overset{\vee}{b}{}^{ij}$ satisfies $\overset{\vee}{b}{}^{ik} b_{kj} = \delta^i_j$ (δ^i_j is the Kronecker symbol) [1], we find that

$$\alpha^2 = (b^n_n g^{mi} - N_b \overset{\vee}{b}{}^{mi})(c^j_j g^{sk} - N_c \overset{\vee}{c}{}^{sk}) b_{ms} c_{ki}$$

$$+ 2N_b N_c - \frac{(b^i_i c^j_j)^2}{4} = 4N_b N_c - N_b (c^i_i)^2 - N_c (b^i_i)^2 + \frac{(b^i_i c^j_j)^2}{4}.$$

By introducing the principal value $\rho_i (\sigma_i)$ of the tensor b_{ij} (c_{ij}) and bearing in mind that $N_b = \rho_1 \rho_2$, $b^i_i = \rho_1 + \rho_2$, $N_c = \sigma_1 \sigma_2$ and $c^i_i = \sigma_1 + \sigma_2$ [1], we finally obtain

$$\alpha^2 = \frac{1}{4} (\rho_1 - \rho_2)^2 (\sigma_1 - \sigma_2)^2, \tag{1.10}$$

and the equation (1.8) of the ellipse of normal curvature can be written as

$$(\sigma_1 - \sigma_2) x^2 + (\rho_1 - \rho_2)^2 y^2 - (\rho_1 + \rho_2)(\sigma_1 - \sigma_2)^2 x$$

$$- (\sigma_1 + \sigma_2)(\rho_1 - \rho_2)^2 y + \frac{1}{4} (\rho_1 + \rho_2)^2 (\sigma_1 + \sigma_2)^2 - 4\rho_1 \rho_2 \sigma_1 \sigma_2 = 0. \tag{I}$$

Let us note, that, under (1.9), the equation of the normal plane indicatrix [5] is of the form

$$N_b x^2 + \frac{1}{2} b^i_i c^j_j xy + N_c y^2 - b^i_i x - c^i_i y + 1 = 0. \tag{II}$$

It is not difficult to see that the invariants of the curves (I) and (II) are equal:

$$\underset{\text{I}}{\delta} = 4\alpha^2, \quad \underset{\text{I}}{\Delta} = -4\alpha^4,$$

$$\underset{\text{II}}{\delta} = \frac{1}{16} [16 N_b N_c - (b^i_i c^j_j)^2], \quad \underset{\text{II}}{\Delta} = \frac{1}{4} \alpha^2, \tag{1.11}$$

where δ (Δ) denotes the second (third) invariant of a curve.

It is clear from the expressions for Δ_{I} and Δ_{II} that both curves are simultaneously degenerate. It was shown in [5] that a dual normalized X_2 in E_4 is characterized by the degeneracy of the normal plane indicatrix (II); hence, the ellipse of normal curvature of a dual normalized surface is degenerate.

§2. Principal curvatures and principal directions of X_2 in E_4. Lines of curvature

We write the equation of the ellipse of normal curvature in canonical form:

$$\frac{\left(x-\frac{b_i^i}{2}\right)^2}{a^2}+\frac{\left(y-\frac{c_i^i}{2}\right)^2}{b^2}=1, \tag{I$'$}$$

where

$$a=\frac{\rho_1-\rho_2}{2}, \qquad b=\frac{\sigma_1-\sigma_2}{2}. \tag{2.1}$$

The normal curvatures corresponding to the end points of the principal axes of the ellipse of normal curvature are said to be the *principal curvatures* of the surface at a given point, and the corresponding direction are called the *principal directions*.

By noting that the end points have the coordinates

$$\left(\frac{b_i^i}{2};\ \frac{c_i^i}{2}\pm\frac{\sigma_1-\sigma_2}{2}\right), \tag{2.2}$$

$$\left(\frac{b_i^i}{2}\pm\frac{\rho_1-\rho_2}{2};\ \frac{c_i^i}{2}\right), \tag{2.3}$$

we can find the directions corresponding to these end points. To do this we consider the principal nets α_{ij} and β_{ij} of the fundamental nets b_{ij} and c_{ij} of the surface ([1], §21):

$$\alpha_{ij}=e^k_{\cdot(i}b_{j)k}, \quad \beta_{ij}=e^k_{\cdot(i}c_{j)k}. \tag{2.4}$$

In view of (1.9) the nets α_{ij} and β_{ij} are apolar, and, because they are orthogonal, each bisects the other.

Hence, if a, $\tilde{a}$ and b, $\tilde{b}$ are the principal directions of b_{ij} and c_{ij}, then $(\widehat{ab})=\pi/4$.

We express the vectors of the principal directions of b_{ij} in terms of b and $\tilde{b}$:

$$a^i=\frac{\sqrt{2}}{2}(b^i-\tilde{b}^i), \qquad \tilde{a}^i=\frac{\sqrt{2}}{2}(b^i+\tilde{b}^i).$$

Bearing in mind that the principal directions of a net are conjugate with respect to the tensor of this net, we find that the normal curvature vectors corresponding to the directions a^i and $\widetilde{a}^i$ are

$$\boldsymbol{k}_1 = b_{ij}a^ia^j\boldsymbol{n} + c_{ij}a^ia^j\boldsymbol{m} = \rho_1\boldsymbol{n} + \frac{1}{2}c_{ij}(b^i - \widetilde{b}^i)(b^j - \widetilde{b}^j)\boldsymbol{m}$$

$$= \rho_1\boldsymbol{n} + \frac{1}{2}(\sigma_1 + \sigma_2)\boldsymbol{m} = \rho_1\boldsymbol{n} + \frac{c_i^i}{2}\boldsymbol{m};$$

$$\boldsymbol{k}_2 = b_{ij}\widetilde{a}^i\widetilde{a}^j\boldsymbol{n} + c_{ij}\widetilde{a}^i\widetilde{a}^j\boldsymbol{m} = \rho_2\boldsymbol{n} + \frac{c_i^i}{2}\boldsymbol{n}.$$

Analogously, the normal curvature vectors corresponding to b and $\widetilde{b}$ are

$$\boldsymbol{k}_3 = b_{ij}b^ib^j\boldsymbol{n} + c_{ij}b^ib^j\boldsymbol{m} = \frac{b_i^i}{2}\boldsymbol{n} + \sigma_1\boldsymbol{m},$$

$$\boldsymbol{k}_4 = b_{ij}\widetilde{b}^i\widetilde{b}^j\boldsymbol{n} + c_{ij}\widetilde{b}^i\widetilde{b}^j\boldsymbol{m} = \frac{b_i^i}{2}\boldsymbol{n} + \sigma_2\boldsymbol{m},$$

that is, the vectors we have found correspond to the end points (2.2) and (2.3) of the ellipse of normal curvature. Thus, *the principal directions of a surface coincide with the principal directions of the fundamental nets* b_{ij} *and* c_{ij} *of the surface.*

Let us note the expressions for the principal curvatures of the surface:

$$k_1^2 = \left(\frac{c_i^i}{2}\right)^2 + \rho_1^2, \quad k_2^2 = \left(\frac{c_i^i}{2}\right)^2 + \rho_2^2, \tag{2.5}$$

$$k_3^2 = \left(\frac{b_i^i}{2}\right)^2 + \sigma_1^2, \quad k_4^2 = \left(\frac{b_i^i}{2}\right)^2 + \sigma_2^2. \tag{2.6}$$

A line which at each of its points touches a principal direction of the surface corresponding to this point is said to be a *line of curvature*. It follows from what has been said that the lines of curvature are the integral curves of the equations

$$\alpha_{ij}du^idu^j = 0, \tag{2.7}$$

$$\beta_{ij}du^idu^j = 0. \tag{2.8}$$

In general there are four families of lines of curvature on a surface; if a or b is zero, then there will be two families. Suppose, for example, that $2b = \sigma_1 - \sigma_2 = 0$, then the ellipse of normal curvature degenerates into a pair of points (ρ_1, σ), (ρ_2, σ) to which correspond the principal directions $\boldsymbol{a}$, $\widetilde{\boldsymbol{a}}$ of the net b_{ij}, and the principal curvatures

$$k_1^2 = \rho_1^2 + \sigma^2, \quad k_2^2 = \rho_2^2 + \sigma^2 \quad (\sigma = \sigma_1 = \sigma_2). \tag{2.9}$$

Equations (2.7) are the equations of the lines of curvature and the surface is a dual normalized surface. Here $\sigma \neq 0$, otherwise $X_2 \subset E_4$.

Next let us note a condition for the existence of asymptotic lines on X_2: that is, of lines of zero normal curvature. By observing that the normal curvature is zero when the ellipse passes through a point of the surface, we deduce from (I) that, for a family of asymptotic lines to exist on a surface X_2 in E_4, it is necessary and sufficient that

$$b_i^i c_j^j - 16N_b N_c = 0, \tag{2.10}$$

or, by comparing with δ_{I} from (1.11): for a family of asymptotic lines to exist on X_2 in E_4, it is necessary and sufficient that the normal plane indicatrix (II) should be a parabolic type curve.

It is obvious that not more than one family of asymptotic lines can exist on the surface.

§3. An analog of Euler's formula for X_2 in E_4

We are going to show how the normal curvature depends on the direction at a point on the surface. In the usual theory of surfaces this dependence is expressed by Euler's Theorem.

Let v be a direction at a point on the surface. Let ϕ denote the angle this direction makes with one of the principal directions on the surface, for example, a; we express v in terms of the vectors of the principal directions: $v^i = a^i \cos\phi + \tilde{a}^i \sin\phi$ or $v^i = b^i \cos(\pi/4 - \phi) - \tilde{b}^i \sin(\pi/4 - \phi)$.

Then, for the normal curvature vector corresponding to the direction v, we have

$$\begin{aligned}
\boldsymbol{k}_n &= b_{ij}v^i v^j \boldsymbol{n} + c_{ij}v^i v^j \boldsymbol{m} = b_{ij}(a^i \cos\varphi + \tilde{a}^i \sin\varphi)(a^j \cos\varphi + \tilde{a}^j \sin\varphi)\boldsymbol{n} \\
&\quad + c_{ij}\left[b^i \cos\left(\frac{\pi}{4} - \varphi\right) - \tilde{b}^i \sin\left(\frac{\pi}{4} - \varphi\right)\right]\left[b^j \cos\left(\frac{\pi}{4} - \varphi\right)\right. \\
&\quad \left. - \tilde{b}^j \sin\left(\frac{\pi}{4} - \varphi\right)\right]\boldsymbol{m} = (b_{ij}a^i a^j \cos^2\varphi + b_{ij}\tilde{a}^i \tilde{a}^j \sin^2\varphi)\boldsymbol{n} \\
&\quad + \left[c_{ij}b^i b^j \cos^2\left(\frac{\pi}{4} - \varphi\right) + c_{ij}\tilde{b}^i \tilde{b}^j \sin^2\left(\frac{\pi}{4} - \varphi\right)\right]\boldsymbol{m} \\
&= (\rho_1 \cos^2\varphi + \rho_2 \sin^2\varphi)\boldsymbol{n} + \frac{1}{2}[\sigma_1(\cos\varphi + \sin\varphi)^2 + \sigma_2(\cos\varphi - \sin\varphi)^2]\boldsymbol{m},
\end{aligned}$$

or

$$\boldsymbol{k}_n = \left(\frac{b_i^i}{2} + a\cos 2\varphi\right)\boldsymbol{n} + \left(\frac{c_i^i}{2} + b\sin 2\varphi\right)\boldsymbol{m}. \tag{3.1}$$

For the normal curvature we obtain

$$k_n^2 = \left(\frac{b_i^i}{2} + a\cos 2\varphi\right)^2 + \left(\frac{c_i^i}{2} + b\sin 2\varphi\right)^2. \tag{3.2}$$

We can express the last equation in terms of the principal curvatures (2.5) and (2.6):

$$k_n^2 = k_1^2\cos^4\varphi + (k_3^2 - k_4^2)\sin\varphi\cos\varphi + 2(3b^2 + K)\sin^2\varphi\cos^2\varphi + k_2^2\sin^4\varphi, \tag{3.3}$$

where K is the Gaussian curvature of the surface. This is the analog of Euler's formula for X_2 in E_4.

In the particular case of a dual normalized surface, for example when $\sigma_1 - \sigma_2 = 0$, as we have already noted the ellipse degenerates into a pair of points, and because $k_3 = k_4$ and $b = 0$, (3.3) becomes

$$k_n^2 = k_1^2\cos^4\varphi + 2K\sin^2\varphi\cos^2\varphi + k_2^2\sin^4\varphi.^{1)} \tag{3.4}$$

If $a = 0$, we must replace k_1 and k_2 in this formula by k_3 and k_4; here $\phi = (\widehat{vb})$.

When one of the fundamental tensors of $X_2 \subset E_4$ degenerates identically (for example, when $\sigma = \sigma_1 = \sigma_2 = 0$), (3.3) becomes Euler's formula in the usual theory of surfaces: $k_n = \rho_1\cos^2\phi + \rho_2\sin^2\phi$.

When one of the fundamental nets is orthogonal, for example b_{ij} $(b_i^i = 0)$, the equation of the ellipse of normal curvature is

$$\frac{x^2}{\rho^2} + \frac{\left(y - \frac{c_i^i}{2}\right)^2}{b^2} = 1 \quad (\rho = \rho_1 = -\rho_2); \tag{3.5}$$

that is, the center of the ellipse lies on one of the coordinate axes. For the normal curvature we have

$$k_n^2 = \rho^2\cos^2 2\varphi + \left(\frac{c_i^i}{2} + b\sin 2\varphi\right)^2. \tag{3.6}$$

When $b_i^i = c_i^i = 0$, the surface is a minimal surface and the equation (I)$'$ of the ellipse becomes

---

1) This formula was first obtained in [4].

$$\frac{x^2}{\rho^2}+\frac{y^2}{\sigma^2}=1 \quad (\sigma=\sigma_1=-\sigma_2); \tag{3.7}$$

that is, the center of the ellipse is a point of the surface; further, from (3.2) we find that

$$k_n^2=\rho^2\cos^2 2\varphi+\sigma^2\sin^2 2\varphi. \tag{3.8}$$

For a singular minimal surface $(\rho=\sigma)$ the ellipse of normal curvature is a circle whose center lies on the surface:

$$x^2+y^2=\rho^2, \tag{3.9}$$

and the normal curvature is independent of direction and is given by $k_n^2=\rho^2$. Bearing in mind that in this case the Gaussian curvature of the surface satisfies $K=N_b+N_c=-2\rho^2$, we obtain[1)]

$$k_n=\sqrt{-\frac{K}{2}}. \tag{3.10}$$

BIBLIOGRAPHY

[1] A. P. Norden, *Theory of surfaces*, GITTL, Moscow, 1956. (Russian) MR 18, 504.

[2] V. I. Šulikovskiĭ, *Classical differential geometry in tensor form*, Fizmatgiz, Moscow, 1963. (Russian) MR 29 #6396.

[3] É. Cartan, *Riemannian geometry in an orthogonal coordinate system*, Lectures (Sorbonne, 1926/27); Russian transl., Izdat. Moskov. Univ., Moscow, 1960. MR 23 #A1316.

[4] A. V. Čakmazjan, *On dual normalization*, Candidate's Dissertation, Erevan, 1962.

[5] K. Š. Ramazanova, *On the theory of two-dimensional surfaces in* E_4, Volž. Mat. Sb. No. 3 (1965), 296–311. (Russian) MR 34 #693.

[6] K. Kommerell, *Riemannsche Flächen im ebenen Raum von vier Dimensionen*, Math. Ann. 60 (1905), 548–596.

1) This result is obtained in [6]; the formula was first noted by Fässler [7].

[7] W. Fässler, *Über die Normaltorsion von Flächen im vierdimensionalen euklidischen Raum*, Comment. Math. Helv. 33 (1959), 89–108. MR 21 #3869.

Translated by:
L. W. Longdon